OUR ENERGY FUTURE

OUR ENERGY FUTURE

RESOURCES, ALTERNATIVES, AND THE ENVIRONMENT

Christian Ngô
Joseph B. Natowitz

WILEY

A JOHN WILEY & SONS, INC., PUBLICATION

Published by John Wiley & Sons, Inc., Hoboken, New Jersey
Published simultaneously in Canada

For general information on our other products and services or for technical support, please contact our Customer Care Department within the United States at 877-762-2974, outside the United States at 317-572-3993 or fax 317-572-4002.

Wiley also publishes its books in a variety of electronic formats. Some content that appears in print may not be available in electronic formats. For more information about Wiley products, visit our web site at www.wiley.com.

Library of Congress Cataloging-in-Publication Data:

Ngô, Christian.
   Our energy future : resources, alternatives, and the environment / Christian Ngô, Joseph B. Natowitz.
      p.   cm.—(Wiley survival guides in engineering and science)
   Includes bibliographical references and index.
   ISBN 978-0-470-11660-9 (cloth)
   1. Power resources–Forecasting.   2. Renewable energy sources.   3. Energy development–Environmental aspects.   I. Natowitz, J. B.   II. Title.
   TJ163.2.N49 2009
   333.79–dc22

                                                                              2008049894

Printed in the United States of America

10 9 8 7 6 5 4 3 2 1

�@@@ CONTENTS

Energy availability is a real concern for everyone. Without energy or with access to much less energy than we currently use, we could not live in the same way, and life would not be easy. For example, before the French Revolution in 1789, the average life expectancy in France was below 30 years and in the United States it was 34 years. Now it is 80 years in France and 78 years in the United States. This is due in a significant measure to a ready access to energy which spurred the development of the agricultural, industrial, and medical resources that played a key role in increasing this life expectancy. Unfortunately, energy resources are not evenly distributed throughout the world and a large part of the world's population has a very low standard of living and a short life span. The poorest among them have life expectancies just slightly above that of an inhabitant of France in 1789.

Since 1789, the world population has increased dramatically, from a bit less than a billion inhabitants to above 6.5 billion. The average energy needs of these inhabitants are much greater than those of two centuries ago. In addition, after a long period in which energy was relatively cheap, its price is now increasing, and this is very likely just the beginning of a long trend. As a consequence humankind is no longer a small perturbation on planet Earth, and every day we face the possibility of increasingly negative consequences of human activities for the environment. It is time to take care of our planet and to make use of its wealth more carefully than before.

In this new paradigm, energy plays a central role. Building an energy future which assures ample supplies of energy to meet our needs should be a major priority and of concern to all. But in order to do that rationally we need to be adequately informed. Energy supply is a complex subject and many considerations come into play: science, technology, the economy, politics, the environment, energetic independence, national security, and so on. Reflecting this, there already exist, in papers, reports, newspapers, and books and on the internet several millions of pages devoted to the subject. Some of these sources are general but most are devoted to a particular aspect of energy technology or energy policy. Of these, some are written to advocate particular agendas and present only the positive features of their subject matter. They avoid presenting information about some of the drawbacks. This book is devoted to energy. As part of the Wiley Survival Guides series, this book aims to provide the reader with a fundamental working knowledge of this subject matter. It is not encyclopedic. In writing this volume the authors felt that it was important to adopt a broad approach to discussing the problem of assuring an adequate future supply of energy. The reason is that there is no single solution to the problem but a choice of solutions that depend on many different parameters: the availability of energy reserves or resources and their location, existing or promising future technologies and their cost, the needs of individual consumers, the needs of the country or region, and the externalities which are not normally accounted for in the price. For example, when health considerations are taken into account, what is the real economic impact of a coal-powered electricity-generating plant?

We have tried not to be overly technical in our approach, but we have been determined to provide sufficient quantitative information and tools to allow the reader to make realistic comparisons of the different technologies. Being able to make reasonable first-order estimates to evaluate the suitability of a particular technology for the application under consideration is of primary importance in judging whether or not a given energy solution applies. The economic aspects of the problem are also of great importance. Except for a very small and very committed minority, people want access to energy at the lowest possible price. Finally, the impacts of greenhouse gas emissions and other pollutants are important issues in energy generation and they can be expected to take on increased importance in the future. The environmental and health impacts of the different energy technologies are dealt with throughout the book.

Harnessing energy resources and exploiting them to improve our living conditions are natural endeavors. Wasting energy resources or adopting energy supply solutions which have a large negative impact on health and on the environment is, given options, both foolish and unethical. We do believe that there exist sustainable energy supply solutions for each situation. The goal of this volume is not to try to promote any specific technology but rather to provide adequate background to prepare the readers to participate in choosing energy supply solutions appropriate to their own future needs and to those of the society they live in.

Change occurs slowly in the energy domain. It takes time to build a new power plant, to exploit oil from a newly found resource, to build or extend the electrical or natural gas grid, and so on. If we want to have the right energy at the right time and the right place, we have to anticipate our needs. It can take decades of research and years of development before significant technological changes are implemented. If we do not anticipate our future needs, we may be obliged to accept poor solutions to meet our energy requirements.

We start, in Chapter 1, by presenting basic energy concepts and discuss the evolution of the energy demand through the ages. Our standard of living and life expectancy have increased as our energy consumption increased. Many energy sources are available to us, but today's world is extremely dependent upon fossil fuels (oil, gas, and coal), which exist in finite quantities in the earth.

Issues of environmental impact, energy independence, and national security which are associated with our energy use practices are introduced in this chapter.

Fossil fuels (oil, natural gas, and coal) have allowed a vigorous development of our civilization. They currently satisfy most of our energy needs. Any change in price or decrease in the production of these fuels has significant consequences for the world economy. Chapters 2 and 3 describe the properties, production of, reserves of, transportation of, and utilization of fossil fuels. The problem of an impending peak in oil production is discussed. We also treat unconventional fuels, sources such as extra heavy oil, tar sands, oil shale, and so on. Impacts on the environment are also discussed. Climate change due to an increase of greenhouse gas emissions coming from human activities is a major concern. Since today we cannot avoid using fossil fuel to satisfy our energy demand, the issue of capturing, transporting, and sequestering CO_2 is presented. The greenhouse effect and its consequences are discussed in Chapter 4.

For a very long time renewable energies were the only energy sources that humans used to produce work or heat. Two such sources remain extensively used in current times: hydro power to produce electricity and biomass to provide heat. These sources produce a nonnegligible part of the world's total primary energy. They are examined in Chapters 5 and 6. Chapter 5, devoted to energy harnessed from water, deals with hydropower and the energy derived from the sea. Chapter 6 deals with biomass, which is extensively used today in many energy applications, for example, power generation, heating, and biofuels. The promise of new biofuels, in which there is presently a great interest, is considered in detail.

The renewable energies—solar energy, geothermal energy, and wind energy—are examined in Chapters 7, 8, and 9, respectively. Solar energy seems to promise a bright future. Geothermal energy is not renewable in the exact sense but is rather inexhaustible at the human level since 99% of the mass of the earth is at a temperature greater than 200 °C. Wind energy is currently seeing a very strong development. Renewable energies will take on more and more importance in the future and people must be prepared to use energy in a different way. They should also be ready to spend significant amounts of money to install such systems at home before they can get low-priced electricity or heat during operation of those systems. Unfortunately, some renewable energy sources are often available only intermittently and are currently expensive compared to fossil fuels. This is the case for wind energy and solar energy. Both of these also have relatively low energy densities and delivered power is sometimes not sufficient to satisfy modern-day energy needs. This may change in the future as improved technologies are developed.

Commercial nuclear energy is relatively new, having been available for only about 50 years. The principles of nuclear energy and nuclear reactors are explained in Chapter 10. Advantages and disadvantages of nuclear energy will be described and the issue of available resources addressed. The questions of dealing with radioactive waste and reprocessing of spent fuel and the possibility of incidents and accidents as well as other safety issues are also considered in this chapter. We finish the chapter with a consideration of controlled thermonuclear fusion, which offers a truly exciting prospect as a future energy source.

Electricity is an energy vector more and more widely used. This is reflected by the fact that the demand for electricity increases at a larger rate than the demand for primary energy. Chapter 11 is devoted to this important energy vector and to the specific problems associated with producing and distributing it, the main one being that demand must be balanced by production in real time.

Storing energy is an important issue. This is the subject of Chapter 12. As far as electricity is concerned, it is important to be able to store electricity in very large quantities at off-peak hours to use it at peak hours. This allows smoothing the energy production and decreasing the installed power capacities which are usually dimensioned to meet peak demands. Intermittent renewable energies also demand methods of electricity storage. For heating or cooling purposes thermal energy storage is also an important issue. Being able to store heat in the summer to use it in the winter or cold in the winter for use in the summer would allow great progress in thermal energy management.

Transportation (Chapter 13) and housing (Chapter 14) consume a large part of the total energy used today. Transportation is necessary for trade as well as for many other activities. Presently it relies mostly on oil-derived products (gasoline, diesel oil, jet fuel) which are more and more expensive and will become scarcer in the future. We are not very far from having one billion road vehicles in the world. Transportation and housing are connected since most people must travel from home to the work place, shopping place, and so on.

Housing requires a lot of thermal energy. It is used to heat or cool buildings and produce hot water. It would be relatively easy to save quite a lot of energy in this domain. Methods by which this could be accomplished are presented.

Chapter 14 treats the production, transport, and use of hydrogen in various energy applications. Hydrogen is a very appealing energy vector for the future. Much consideration has been given to using it for road transportation in fuel cell vehicles. Unfortunately the physical properties of hydrogen make this difficult in the short term. Many problems remain to be solved and hydrogen vehicles will probably not be used at a large scale for several decades. However, there is a great interest in obtaining large supplies of hydrogen for use in petrochemistry and to exploit all of the carbon atoms contained in the ligno-cellulosic biomass in order to produce second-generation biofuels.

CHRISTIAN NGÔ
JOSEPH B. NATOWITZ

ACKNOWLEDGMENTS

Energy is a broad domain. C. N. would like to thank all of the people from different specializations, scientists, engineers, economists, industrialists, and others, from whom he has learned so much during the last decade. He is especially indebted to the CEA (Commissariat à l'Énergie Atomique) for all the fruitful years he spent working there and exchanging information with his colleagues. J. B. N. thanks C. N. for educating him on so many aspects of the energy supply problem.

C. N.
J. B. N.

CHAPTER 1

We Need Energy

Energy is a thermodynamic quantity equivalent to the capacity of a physical system to produce work or heat. It is essential to life. If we live better than our primitive ancestors, it is because we use more energy to do work, to produce heat, and to move people and goods. Energy can exist in various forms (chemical, mechanical, electrical, light, etc.). It is in the process of transforming energy from one form to another that we are able to harness part of it for our own use.

BASIC NATURE OF ENERGY

Energy is related to a fundamental symmetry of nature: the invariance of the physical laws under translation in time. In simple words this means that any experiment reproduced at a later time under the same conditions should give the same results. This symmetry law leads to the conservation of the physical quantity which is energy. There are also other symmetries which lead to important conservation laws. Space invariance with respect to translation or rotation leads respectively to conservation laws for momentum and angular momentum. This means that if we translate or rotate an experimental arrangement we will get the same experimental results. Conservation of energy, momentum, and angular momentum are of basic importance and govern the processes occurring in the universe.

1.1. GENERALITIES

1.1.1. Primary and Secondary Energy

All of the energy sources that we use, except geothermal and nuclear energies, are derived initially from solar energy (Figure 1.1). The fossil fuels that we use

Our Energy Future: Resources, Alternatives, and the Environment
By Christian Ngô and Joseph B. Natowitz
Copyright © 2009 John Wiley & Sons, Inc.

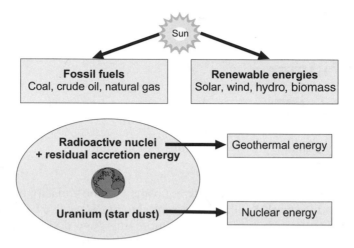

Figure 1.1. Origin of different sources of energy used by humans.

today—coal, oil, and natural gas—are derived from organisms (primarily ocean plankton) that grew over several hundreds of millions of years, storing the solar energy which reached the earth's surface. Renewable energies—hydro, biomass, and wind—are also directly or indirectly derived from the energy of our sun. Solar energy, though technically not renewable, is normally classified as such because it is effectively inexhaustible on any practical timescale.

Nuclear energy is derived from uranium nuclei contained in the earth. This element was formed in heavy stars and was scattered in space when those stars died. Uranium nuclei were present in the dust from which the solar system was formed about 4.5 billion years ago. The earth formed by accretion of such dust and some thermal energy due to this process still remains. However, most of the thermal energy contained in the earth comes from the decay of radio-active nuclei present in the earth and initially produced in stars.

It is useful to distinguish between primary and secondary energy sources. Primary energy sources correspond to those that exist prior to any human-induced modification. This includes fuels extracted from the ground (coal, crude oil, or natural gas) or energy captured from or stored in natural sources (solar radiation, wind, biomass, etc.). Secondary energy sources are obtained from the transformation of primary sources. Gasoline or diesel fuel from crude oil and charcoal from wood are examples of secondary sources.

We can also distinguish between nonrenewable and renewable energies. Nonrenewable energies are in finite quantities on the earth. Like uranium, which comes from the dust of stars, they could have been present at the earth's formation (about 4.5 billion years ago) or, like fossil fuels (coals, natural gas, crude oil, oil shale, etc.), they could have been synthesized several hundred million years ago. In contrast to the nonrenewable energies, renewable energies will be available as long as the earth and the sun exist, which is estimated to be about 5 billion years.

1.1.2. Energy Units

The *joule* is the standard energy unit in the international system. Defined as $1\,kg\cdot m^2/s^2$, it is a very small quantity of energy compared the amounts we use in daily life. For that reason we will frequently use another unit widely used in the energy domain: the *kilowatt-hour* and its multiples:

$$1\,kWh = 3.6 \times 10^6\,J = 3.6\,MJ$$

Prefixes defining multiples of any physical quantity are shown in Table 1.1.

For measurements of heat energy, the *calorie* (cal) or its multiple, the kilocalorie, is an older unit which is sometimes still used. One calorie is the quantity of heat necessary to increase the temperature of $1\,g$ of water by $1\,^\circ C$:

$$1\,cal = 4.18\,J \qquad 1\,kcal = 1000\,cal = 1.16\,Wh$$

The *British thermal unit* (Btu), also still used on occasion, is defined as the amount of heat necessary to raise 1 pound (lb) of water through $1\,^\circ F$ ($1\,Btu = 1055.06\,J$).

Use of another unit, derived from the international system, the *gigajoule* ($1\,GJ = 10^9\,J$), is increasing and is supported by the International Organization for Standardization (ISO). From time to time we will also use this unit.

Two units sometimes used in the United States are the *quad* ($1\,quad = 10^{15}\,Btu$) and the *therm* ($1\,therm = 10^5\,Btu$).

A much older unit, the *horsepower* (HP) is still sometimes employed also. It was introduced at a time when animals were the primary source of energy used to work in the fields. By definition $1\,HP = 746\,W$. In fact, this original evaluation of the power of a horse was quite optimistic and corresponds more closely to the power of 3 horses.

TABLE 1.1. Multiple Prefixes

Prefix	Multiplicative Factor	Symbol	Prefix	Multiplicative Factor	Symbol
Deca	10^1	da	Deci	10^{-1}	d
Hecto	10^2	h	Centi	10^{-2}	c
Kilo	10^3	k	Milli	10^{-3}	m
Mega	10^6	M	Micro	10^{-6}	μ
Giga	10^9	G	Nano	10^{-9}	n
Tera	10^{12}	T	Pico	10^{-12}	p
Pecta	10^{15}	P	Femto	10^{-15}	f
Exa	10^{18}	E	Atto	10^{-18}	a

TABLE 1.2. Conversion between Selected Units

	MJ	kcal	toe	Btu	kWh
MJ	1	238.8	2.388×10^{-5}	947.8	0.2778
kcal	4.1868×10^{-3}	1	10^{7}	3.968	1.163×10^{-3}
toe	4.1868×10^{4}	10^{7}	1	3.968×10^{7}	11630
Btu	1.0551×10^{-3}	0.252	2.52×10^{-8}	1	2.931×10^{-4}
kWh	3.6	0.86	8.6×10^{-5}	3412	1

Another unit sometimes employed for very large amounts of energy is the *ton of oil equivalent* (toe). It corresponds to 10 Gcal or 4.1868×10^{10} joules (http://www.economicexpert.com/a/Joule.htm). This is the (accepted) amount of energy that would be produced by burning 1 ton of crude oil. This unit is often used in energy statistics.

The unit toe was defined to answer the following question: Given an energy source, how much oil would be required to produce the same amount of energy? Thus it provides a means for making rough comparisons of the amounts of energy available from different energy sources. This in fact depends upon the nature of the energy produced. It will not be the same for electricity as for heat. It will also depend on the system used to produce the energy as some systems are more efficient than others. Furthermore, the energy content of a ton of oil can vary slightly depending on where the oil comes from. The value quoted above has been adopted by convention. Nevertheless, this unit is useful to compare different energy sources.

Conversion equivalents between some common units are shown in Table 1.2.

Some equivalence values of energy sources are given in Table 1.3. The energy content depends very much of the nature of the source. In the case of fossil fuel it may vary depending upon origin. For example, the gross calorific value of natural gas is equal to about 52.6 MJ/kg if it comes from Norway but only 45.2 MJ/kg if it comes from the Netherlands.

1.1.3. Power

Power is defined as an amount of energy delivered per unit of time. The standard unit is the watt, which corresponds to 1 J/s. In practice the kilowatt and the megawatt ($1 kW = 10^{3} W$ and $1 MW = 10^{6} W$) are often used. Power and energy should not be confused. In particular one should not confuse 1 kW (of power) with 1 kwh (of energy). One kilowatt-hour corresponds to the energy of a device which has a power of 1 kW (e.g., an electric iron) working for a period of 1 h. A 1-kW device which is not functioning does not consume energy.

TABLE 1.3. Net Calorific Value[a] in Toe of Some Energy Sources

Energy Source	Amount	GJ	toe
Hard coal	1 t	26	0.62
Coal coke	1 t	28	0.67
Lignite briquettes	1 t	32	0.76
Lignite and recovery products	1 t	17	0.4
Crude oil	1 t	42	1
Liquefied petroleum gas (LPG)	1 t	46	1.1
Automotive gasoline and jet fuel	1 t	44	1.05
Petroleum coke	1 t	32	0.76
Ethanol	1 t	26.8	0.64
Biodiesel	1 t	36.8	0.876
Natural gas	1 Nm^3	≈34.9	0.077
	1 MWh (GCV)	3.24	
Wood	Stere (1 m^3)	6.17	0.15
Electricity (nuclear)	1 MWh	3.6	0.26
Electricity (geothermal)	1 MWh	3.6	0.86
Electricity (other)	1 MWh	3.6	0.086
Hydrogen (1 kg H$_2$ ≈ 11.13 Nm^3 H$_2$)	1 t	120.1	2.86

Note: For natural gas the energy is indicated in gross calorific value (GCV). 1 Nm^3, which means one normal cubic meter, is measured at 0 °C and 760 mm Hg. One has 1 Nm^3 = 0.946 Sm^3, where the standard cubic meter is defined at 25 °C and 760 mm Hg.
[a]Also called low calorific value.

POWER AND ENERGY

One liter of gasoline contains about 10 kWh of energy. Assume we have 40 liters of gasoline at our disposal. The total amount of energy contained is $E = 400$ kWh. Using this gasoline in a car would allow us to drive about 250 miles. At a constant speed of 62.5 mi/h, it would take $t = 4$ h to do that. The power developed would be $E/t = 100$ kW. If on the other hand we burn this gasoline in 30 s, the power of the process 48 MW.

1.1.4. Energy and First Law of Thermodynamics

It is a basic law of nature that energy is conserved. In other words energy can neither be created nor destroyed: It can only change form. The first law of thermodynamics applied to the internal energy of a system (which applies to equilibrium states) is just a statement of energy conservation in heat and work conversion processes. The internal energy (U) is a state function, which means that in any thermodynamical transformation the change of internal energy depends only upon the initial and final states of the system under consideration and not on the way in which the transformation is carried out, that is, the "path."

The first law of thermodynamics relates the change of internal energy ΔU to the work W done on the system and the heat Q transferred into the system:

$$\Delta U = W + Q$$

We should note that the convention used in this evaluation is to treat work done on the system and heat put into the system as positive. Work done by the system and heat removed from the system are designated as negative. In other words, the equation above means that the change in internal energy between two equilibrium states is equal to the difference of heat transfer (Q) *into* the system and work (W) done *by* the system. Work corresponds to an organized energy while heat is completely disorganized energy since this energy is shared among all the microscopic degrees of freedom of the system. Transforming disorganized energy into organized energy is not an easy task. The reverse operation is much easier. This explains why we never get a 100% yield when extracting work from a heat source.

1.1.5. Entropy and Second Law of Thermodynamics

The first law of thermodynamics tells us whether a process (A ↔ B) is energetically possible but it does not tell us the direction (A → B or B → A) in which the process can occur spontaneously. In order to answer this question, we have to consider the second law of thermodynamics, which addresses the concept of entropy, a second state function. At the microscopic level, entropy is a quantity related to disorder. The higher is the disorder of a system, the larger its entropy. The unit of entropy is joules per kelvin.

The second law of thermodynamics tells us that the entropy of an isolated system can either spontaneously increase or can remain the same: $\Delta S \geq 0$ for an isolated system. This means, at the microscopic level, that disorder either increases or remains the same.

Thus, the first law of thermodynamics tells us that the total energy of the universe remains constant while the second law tells us that the quality of the energy constantly decreases. The second law tells us about the direction of irreversible processes. For example, we know from experience that, for an isolated system made of two bodies at different temperature, heat goes spontaneously from the high-temperature body to the low-temperature one and not in the reverse direction.

> At the microscopic level, entropy can be expressed as follows:
>
> $$S = k \ln \Omega$$
>
> where k is the Boltzmann constant ($k = 1.38 \times 10^{-23}$ J/K) and $\ln \Omega$ is the natural logarithm of the number of microscopic states, Ω, available to the system.

In thermodynamics there are several ways of expressing the second law of thermodynamics. One, due to Clausius, is the following: *There is no process in which the only result is to transfer heat from a cold source to a hot one.* It is possible to transfer heat from a cold sink to a hot source, but one needs to provide external work to make this occur. This is the operating principle of refrigerators or heat pumps. A second formulation goes a little further. It is due to Kelvin and Planck: *There is no process in which it is possible to produce work using a constant-temperature heat source.*

During the nineteenth century, steam engines were used to produce work. It was observed that a large part of the energy needed for this purpose was lost in the form of heat. Sadi Carnot, a French physicist, formulated a principle which allowed calculation of the maximum yield for the heat engines which were used at that time. This principle applies generally to any closed system producing work by using two heat sources at different temperatures.

Designate the temperature of the hot source as T_H and the temperature of the cold one as T_C. According to the Carnot principle, the maximum theoretical yield η for producing work in a reversible cycle operating between two heat sources at different temperature is given by

$$\eta = \frac{T_H - T_C}{T_H} = 1 - \frac{T_C}{T_H}$$

1.1.6. Exergy

Contrary to what the name suggests, thermodynamics deals with equilibrium phenomena. However, real processes are often nonequilibrium ones. Here, by equilibrium we refer to the equilibrium of a system with its environment. To better characterize real processes a new quantity, *exergy*, has been introduced. The exergy content of a system indicates its distance from thermodynamic equilibrium. The higher the exergy content, the farther from thermodynamic equilibrium is the system and the greater is the possibility to do work. Quantitatively the exergy is the maximum amount of work that can be done during the process of bringing the system into equilibrium with a heat bath (a reservoir at constant temperature). With the same original energy content, it is possible to produce more work if we use a high-temperature source than a low-temperature one.

Assume the following notation: U, V, S, and n are the internal energy, the volume, the entropy, and the molecular or atomic concentration of the system. The values of these quantities when thermodynamic equilibrium with the environment exists are U_{eq}, V_{eq}, S_{eq}, and n_{eq}, P_0, T_0, and μ_0 are the pressure, the temperature, and the chemical potential of the environment. Using these quantities the exergy E_x can be defined as

$$E_x = U - U_{eq} + P_0(V - V_{eq}) - T_0(S - S_{eq}) - \mu_0(n - n_{eq})$$

For the system there is a driving force toward equilibrium. At constant pressure and chemical potential, the exergy is just the classical free energy of equilibrium thermodynamics. In an irreversible process moving toward equilibrium, the total energy is conserved but the exergy is not conserved. It decreases as the entropy increases.

1.1.7. Going Back to the Past

Since early times when our ancestors used their own muscles or those of slaves and animals to perform work and improve their living conditions, the quest for new energy sources has been one of the main driving forces. In practice, most of mankind's energy history has been dominated by renewable energies. This started with the mastery of fire about 500,000 years ago. Making a fire allowed our ancestors to produce heat and light and to cook their meals. Wood was the energy source that was mainly used. It is still widely used today, especially in underdeveloped countries where it is sometimes the only readily available energy source.

Around 3500 B.C., Egyptians used the power of the wind to move boats. The harnessing of this new energy source allowed them to travel greater distances and promote trade with other lands. About 640 B.C., the power of wind was probably also used to grind grain by the Persians, who built windmills in the area which is now Iran. Solar energy was harnessed about 500 B.C. by the Greeks, who developed homes to better use the incoming heat from the sun. Around 85 B.C. geothermal energy was harnessed by the Romans, who used hot springs to heat baths. About the same time running water was also exploited by the Greeks, who used waterwheels to grind grain.

In antiquity, because of an abundant supply of labor, there was not great pressure for development of new energy sources. During the Middle Ages this was less true and large-scale use of renewable resources such as water or wind developed rapidly. By the eleventh century water mills became very common in countries such as England and France, which had good water resources. The extracted energy was used to grind grain, press olives, operate hammers or the bellows of forges, and so on. Windmills were developed mostly in dry countries like Spain and the Netherlands, where they were used to pump water. In the Netherlands this allowed the retrieval of land from the sea.

The industrial revolution, based on mechanization, started around 1750. In this era the steam engine played a key role because it provided power that did not depend upon the flow of rivers or movement of the wind. The first steam engines were used to pump water from coal mines, allowing miners to dig deeper and get more coal. Since the industrial revolution energy development has moved swiftly. Fossil fuels (coal, oil, gas) have been increasingly exploited and have become essential to modern society. More recently, nuclear energy has been mastered and used to produce large amounts of electricity.

1.1.8. Humans and Energy

A human needs energy to live. This energy is derived from food. Our basal metabolism requires about 2.7 kWh of energy per day. This corresponds to a power of 110 W. This is quite a small power considering the work done. With this small amount of energy all the organs are able to function, and it is also possible to carry out some limited activities. Interestingly, humans are actually more energy efficient than man-made devices.

> Humans and living species in general are very efficient energetic systems. As illustration of that we can compare the amount of energy emitted from the sun divided by its mass (emitted energy per unit of mass) to the energy of the basal metabolism of a human divided by his mass. We find that the latter is more than 7000 times larger than the sun's energy density.

Pregnancy is a particularly energy consuming human activity. It lasts nine months and it requires about 90 kWh of extra energy on the average. This corresponds to a daily extra energy of about 330 Wh or a little bit more than 10–15% of the average total energy needed. This explains why pregnant women need more food.

> A 1.5-ton car driving at 100 km/h ($\approx$28 m/s) has a kinetic energy of about 580 kJ. Estimations show that a person hitting a nondeformable obstacle has a high probability of being killed if the car's kinetic energy is larger than about 700 J. This is a small amount and shows that if only a small part of the initial kinetic energy is transferred to the body of a passenger in a car accident the passenger can die. Today's cars are designed in such a way that the materials they are made from deform and absorb a large part of the kinetic energy in a collision.

1.2. ALWAYS MORE!

The use of energy allows humans to be more efficient and to improve their way of life. Throughout history, humans have searched for better energy sources and better ways to harvest energy. A rough estimation of mankind's energy consumption through different periods of history is displayed in Figure 1.2.

The first energy source used by our remote ancestors (before humans mastered fire) was food. It was hard to find food, and it is estimated that the average food consumption provided an energy of 2 kcal per day. After fire was discovered and wood could be used for cooking and heating, a larger amount of energy (about 2.5 times more) was used. Agricultural activities again increased the energy needs and the average total energy consumption nearly doubled. About 5000 years ago, primitive agricultural humans used animals to assist them in this work. By the end of the Middle Ages in Western Europe, advanced agricultural humans added the power of wind, water, and small amounts of coal. Transportation of goods was also developing and required more energy. Between 1400 and 1820, a French citizen's average wealth doubled primarily through the use of renewable resources. For comparison, a doubling of wealth in the second part of the twentieth century took only 25 years due to the more concentrated forms of energies. During the industrial revolution the energy consumption of industrial man rose by a factor of 3. The steam engine consumed large amounts of energy but also produced a lot of

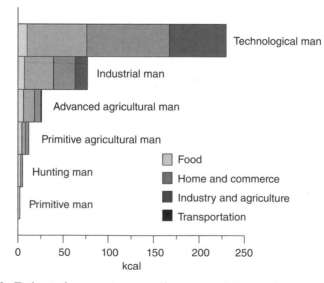

Figure 1.2. Estimated energy consumption per person per day over the ages. Data from E. Cook, *Scientific American*, 1971, and http://www.wou.edu/las/physci/GS361/ electricity%20generation/HistoricalPerspectives.htm.

work. The advent of the use of the fossil fuels stored in the earth allowed a quick development of mankind's wealth.

Since the 1970s technological man might be defined as an average U.S. citizen: This person consumes more than 100 times as much energy as the primitive human. Electricity accounts for almost a quarter of this energy consumption and large quantities are used for transportation means, for industrial purposes, and for housing.

1.2.1. Why Do We Need More Energy?

It is estimated that the cumulative global population since the appearance of *Homo sapiens sapiens* has been about 80 billion people. Starting about a century ago the rate of increase of the population became very steep. Each day there is a net increase of about 200,000 people more on the planet (difference between babies who are born and people who die). These new inhabitants need energy to live and this leads naturally to a continuous increase of primary energy consumption. The 1 billion wealthiest people in the world consume 66% of the food and 12 times more oil per capita than people of underdeveloped countries.

Energy consumption increases over the ages for two main reasons. The first is that the population increases. Figure 1.3 shows the evolution of the global

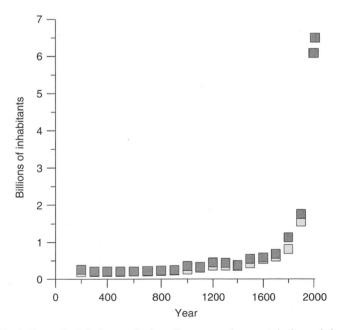

Figure 1.3. Evolution of global population. Because of uncertainties minimum and maximum estimates are indicated by open and closed symbols, respectively. Data from www.wikipedia.com.

population over the last 2000 years. Figure 1.4 shows the dates for which successive population increases of one billion inhabitants have been reached.

> Before the French Revolution, more than 200 years ago, the energy consumption per capita in France was about 14 times less than today. Since the French population was about half of today's population, the total energy used in the country was about 28 times less at that time. The increase over more than two centuries is large but actually corresponds to an increase of only 1.3% per capita per year and 1.75% per year for the whole country. At the same time the life expectancy has increased from about 28 years at that time to 80 years today.

The second reason why energy consumption increases is that the majority of people living on the earth live in countries that are still developing. There are currently 2.8 billion people living on less than $2 per day and about 1 billion who live on less than $1 per day. For a basis of comparison to energy costs, 1 kWh generated by an off-grid photovoltaic system (cells plus battery) costs around $1.5. The only effective way for people in developing countries

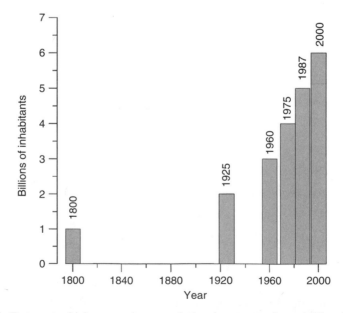

Figure 1.4. Dates at which successive population increases of one billion inhabitants have been reached. Data from http://villemin.gerard.free.fr/Economie/Populati.htm.

to increase their standard of living is to use more energy to develop their agricultural, industrial, and trading activities.

Life expectancy is strongly correlated with the amount of energy used. In Figure 1.5 the average lifetime expectancy is shown as a function of the energy consumption per capita. This is a mean curve which incorporates data from a number of different countries. The main message to be taken from the average trend shown in Figure 1.5 is that a minimal energy consumption is needed to reach a good life expectancy. People with little access to energy have short life expectancies. People having insufficient access to energy generally also have insufficient access to food, medicine, potable water, and so on. Data for most of the countries fall close to this curve, but there are a few exceptions. Some are shown in Figure 1.5. South Africa and Zambia have low life expectancies relative to the mean. This is primarily due to the AIDS epidemic. For Russia life expectancy is lowered by widespread alcoholism.

Figure 1.5 shows also that, above a certain threshold in energy consumption, about 3 toe per capita per year, the life expectancy levels off, indicating that in terms of lifetime expectancy there is no extra advantage of consuming more energy.

The increase in the world population and the increase of the standard of living in developing countries lead to an increase of global energy consumption which averages about 2% per year. Sustained at this level this would lead to a multiplication of our energy consumption needs by 7 times between 2000 and 2100. This is clearly unsustainable as far as fossil resources are concerned. To keep improving our standard of living and allow developing countries to

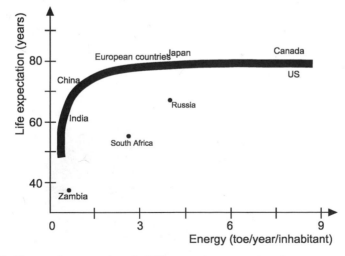

Figure 1.5. Curve of average trend of life expectancy versus primary energy consumption. Data from United Nations Development Program (2003) and B. Barré, *Atlas des energies*, Autrement, 2007.

reach a similar standard of living, we have to develop alternative sources and learn how to use energy differently.

1.2.2. Energy Sources We Use

The different energy sources described at the beginning of this chapter are not equally used. Figure 1.6 shows the contributions of different energy sources to the world's total energy consumption in 2005. Only recently in the history of humankind have fossil fuels and nuclear energy, both concentrated sources of energy, been used extensively. Fossil fuels have been used for about two centuries and nuclear energy has been used (to produce electricity) for only half a century. Their use allowed a rapid development of industrial and technological civilizations.

In 2005 energy derived from fossil fuels (crude oil, natural, and coal) provided about 80% of the total energy used. Our modern world is extremely dependent upon fossil fuels. In the near future it will probably become increasingly more difficult to meet our needs in this way.

In Figure 1.7, a sketch is shown depicting the contribution of various energy sources to primary energy consumption between the years 1800 and 2000. The "others" category includes renewable energies. These were heavily used in 1800. After the advent of the industrial revolution the share attributable to renewable energies progressively decreased. Today it accounts for just a little more than 12% of the total energy consumption. During the nineteenth century, coal progressively increased in importance, particularly at the beginning of the industrial revolution. The peak of coal's share was reached in the first quarter of the twentieth century. The relative contribution of coal decreased and then plateaued after midcentury. Today there is a greatly renewed interest in coal because proven coal reserves are much larger than those of oil or gas. Around the middle of the twentieth century oil began to

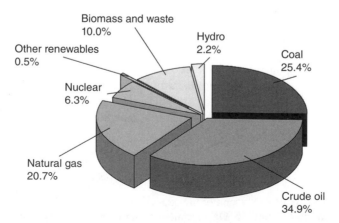

Figure 1.6. Sources of primary energy consumed, 2005. Data from www.iea.org.

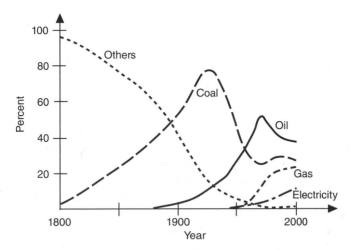

Figure 1.7. Percentage of different energy sources of total primary energy consumption, 1800–2000. Data from www.cea.fr.

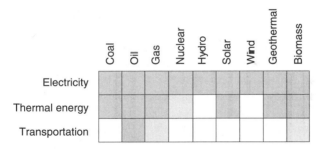

Figure 1.8. Possible uses of energy sources. Dark squares indicate that the application is possible and white ones that it is not possible. Light grey squares indicate limitations of use.

be a very important energy source because of its convenience. It is a liquid with a high energy density particularly suitable for transportation applications. For a long time natural gas was not used but flared. Fortunately it was realized that it is a very good energy source and it is now widely used due to the development of combined-cycle gas turbines that provide high efficiency for production of electricity.

The main uses of energy are to produce electricity, to produce heat (or cold), and for transportation. Not all sources of energy are well suited to meeting the needs for these applications, as is summarized in Figure 1.8.

Electricity is not an energy source but an energy vector. It is employed to transport energy from one point to another for use in many electric appliances. Electricity is being increasingly used in modern societies. The rate of growth of consumption for electricity is currently greater than the rate of growth of total energy consumption. Electricity can be produced by any source of energy. This possibility makes electricity a very convenient energy vector and more and more systems are now powered by electricity.

All sources of energy cannot easily be employed to produce thermal energy directly. For example, it is not possible to produce heat efficiently with falling water. However, first producing electricity with a turbine and using this electricity in an electrical heater can generate heat. This is an indirect heat production method. Wind also cannot be used to produce heat directly. A nuclear reactor produces heat, a part of which is used to produce electricity. Each time 1 kWh of electricity is produced 2 kWh of heat is released into the environment. Heat produced by nuclear power can be used directly, but currently this is not done except in very specific cases.

Modern transportation is mostly based on oil as the energy source. While electric-powered vehicles exist, trains for example, they account for only a small amount of the energy used in transportation. In France, trains use about 7 TWh of electricity per year. This is a little less than the quantity of electricity produced by a single nuclear plant. For comparison, the power lost in the grid by the Joule effect is about 12 TWh per year in France.

Natural gas and biomass (through the use of biofuels) can be used for transportation but so far are used in limited amounts. Biofuels can provide a limited part of the demand but cannot completely replace oil. It is not possible, with the existing cultivatable lands, to produce enough biofuels to meet the demand. Natural gas-powered vehicles can also be used. They have the advantage of being less polluting than gasoline or diesel fuel. However, even compressed, a gas is less convenient than liquid fuels. Natural gas can meet a part of the energy demand for transportation but not all. Furthermore, in the longer term, natural gas suffers from the same problem of limited reserves as does crude oil.

In Figure 1.9 a rough estimate of the usage distribution of the world's primary energy is displayed. Thermal energy (heating or cooling buildings, domestic hot-water production, and so on) accounts for the largest share. This means that the domain of thermal energy is the priority domain to be investigated in order to reduce our oil or gas energy consumption. It is also the domain where large effects in terms of reduction of pollution can be made. It is interesting to note that electricity accounts for only a small part of the initial primary energy consumption. The initial energy content of fuels used for transportation is about twice this amount. Just a small part of this primary energy is actually transferred to the wheels in the case of road transportation.

A large amount (about 40%) of primary energy is lost before reaching the end user. Much of this loss is due to physical laws which dictate maximum values for the efficiencies of thermodynamic processes. For example, the efficiency of a heat engine functioning with two heat sources at different temperatures is limited by the Carnot principle. On top of that there are also losses

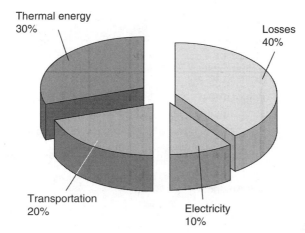

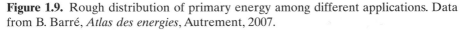

Figure 1.9. Rough distribution of primary energy among different applications. Data from B. Barré, *Atlas des energies*, Autrement, 2007.

due to the fact that there is usually a difference between an ideal system and a real operating device.

1.2.3. Security of Supply

Security of the energy supply is an essential concern for any country. In Figure 1.10 energy production and energy consumption (all energy sources are considered) are shown for different regions of the world. If the production is equal to the consumption, the region is energy sufficient. We see that there are regions which do not produce enough energy and others that produce too much energy for their own needs. North America, Asia, and Europe do not produce enough energy and they must import energy. The figure for Europe includes the former USSR and in particular Russia, which has large energy resources, including a lot of natural gas. This brings Europe's production and consumption close to each other but does not reflect the real political situation. Indeed, the remaining part of Europe imports a lot of energy from Russia and elsewhere.

Table 1.4 shows the level of the European Union's dependence on external sources for its oil and gas. It is large and expected to increase in the next decade.

Figure 1.11 shows the distribution of the different energy sources used to meet primary energy consumption in the United States. Oil supplied 40.5% of the primary energy. The U.S. oil production was not capable of meeting the whole demand. Therefore it was necessary to import oil. The origin of the oil imports is shown in Figure 1.12. Dependence on Middle East oil is not as large as most might think. However, in the future the share derived from that source

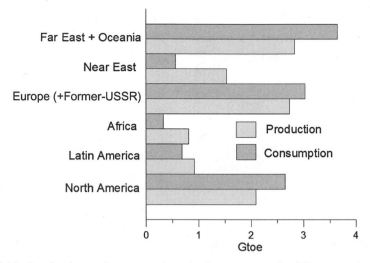

Figure 1.10. Production and consumption of primary energy in different regions of the world, 2006. Data from Pétrole, *Elements de statistiques*, Comité professionnel du pétrole, 2006.

TABLE 1.4. Level of Dependence of European Union on External Energy Sources of Oil and Gas

	Level of Dependence (%)	
	2004	2020 (Forecast)
Oil	80	90
Gas	54	70

Source: Data from J. P Favennec, *Géopolitique de l'énergie*, Technip, 2007.

is expected to increase because most of the petroleum reserves are in this area of the world.

In Table 1.5 the daily production of oil in barrels (bbl) is shown for different countries of the Middle East in 2007. Saudi Arabia dominated oil production in this area, producing more than 10 Mbbl/day.

The regions to which crude oil produced in the Middle East is exported are shown in Figure 1.13. A large part is exported to Asia.

Because oil and gas are such important energy sources, a lot of effort is devoted to finding new sources. The number of oil and gas wells drilled between 1998 and 2006 is shown in Figure 1.14. The number almost doubled in that period, reflecting some urgency in the effort to discover new resources.

In Figure 1.15 are shown percentages of the different energy sources employed in Russia in 2005. Natural gas was the dominant source of energy because of the large resources of natural gas in that country. The world's largest known reserves of natural gas are found in the former USSR.

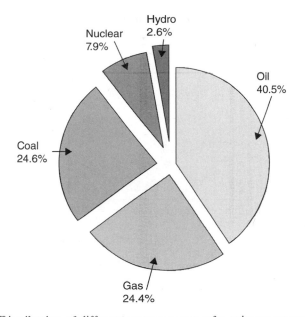

Figure 1.11. Distribution of different energy sources for primary energy consumption in United States, 2005. Data from J. P. Favennec, *Géopolitique de l'énergie*, Technip, 2007.

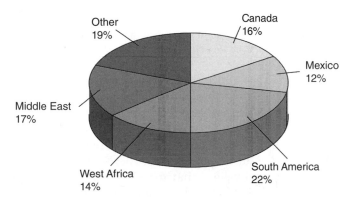

Figure 1.12. Origin of different oil imports done by United States, 2005. Data from J. P. Favennec, *Géopolitique de l'énergie*, Technip, 2007.

TABLE 1.5. Daily Production in Selected Middle East Countries, 2007

Country	Production (Mbbl/day)
Saudi Arabia	10.9
Iran	4.4
Iraq	2.1
Kuwait	2.6
United Arab Emirates	2.9
Qatar	1.2
Oman	0.7
Yemen	0.3

Source: Data from BP, *Statistical review of world energy*, www. bp.com, 2008.

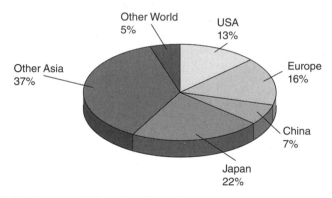

Figure 1.13. Regions to which crude oil produced in the Middle East is exported. Data from J. P. Favennec, *Géopolitique de l'énergie*, Technip, 2007.

Natural gas is also exported from Russia in large quantities. The regions to which it is exported are shown in Figure 1.16 as percentages of total exports. We see, as noted above, that the European Union is very strongly dependent upon Russian gas.

In Figure 1.17 the evolution of the different sources used to produce electricity in France is shown starting from 1950. There has been a strong increase in the demand for electricity since that time, and new means for production of electricity were needed to meet that increasing demand. Until the 1970s, electricity was mostly produced by hydro and fossil fuel plants. In order to ensure safety of its energy supply, France strongly developed nuclear power after the first oil shock. After the decision was taken to build the nuclear plants, it took a long time to construct them. This explains why a noticeable amount of electricity produced by nuclear power appeared only in the 1980s. The production of electricity by nuclear power has increased the level of energy

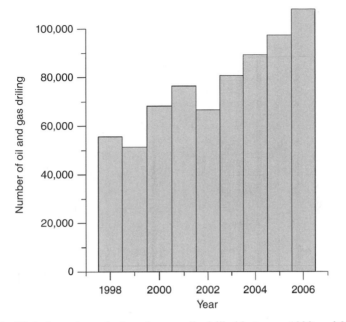

Figure 1.14. Global number of oil and gas wells drilled between 1998 and 2006. Data from Pétrole, *Elements de statistiques*, Comité professionnel du pétrole, 2006.

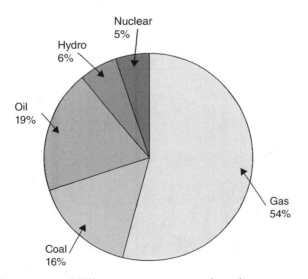

Figure 1.15. Percentages of different energy sources for primary energy consumption in Russia, 2005. Data from J. P. Favennec, *Géopolitique de l'énergie*, Technip, 2007.

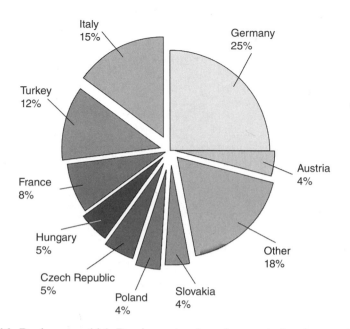

Figure 1.16. Regions to which Russian natural gas is exported and percentages with respect to total exported, 2005. Data from J. P. Favennec, *Géopolitique de l'énergie*, Technip, 2007.

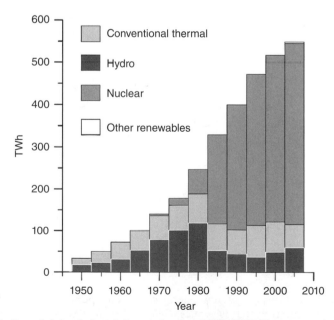

Figure 1.17. French inland electricity production, 1950–2005. The conventional thermal, hydro, nuclear, and other renewables contributions are shown on a grey scale. Data from *Energy Handbook*, CEA, 2006.

independence of the country ($\approx 50\%$). At present 90% of France's electricity is produced without CO_2 emissions (nuclear plus hydro).

1.2.4. Environmental Concerns

Progress in the energy domain has always been necessary to meet expanding demand. The main problems associated with keeping pace with the demand are that global population increases and energy consumption per capita also increases.

Our use of energy has an impact on our environment. Since a huge amount of energy is consumed in the world, the impact can be large. This impact can be on a local or a global scale. Nitrogen oxides emitted by the exhaust pipe of a car pollute at the local scale while the CO_2 emitted has an effect at the global scale. Pollution is not a new phenomenon. Chronicled already around the fifth century by Lao Tsu in China who described the impact of human activities on the environment, pollution already existed much earlier. Towns and cities were especially polluted. The energy supply in the Middle Ages was dominated by wood. This resource began to become increasingly scarce and the replacement of wood by coal allowed the industrial revolution. There were less people during the nineteenth century compared to today, but pollution in towns was greater because coal was widely used and no special care was taken against pollution. For a long time coal was not a welcome energy source. For example, in England, Edward I published strong regulations restricting the use of coal in London close to the royal palace: "whosoever shall be found guilty of burning coal shall suffer the loss of his head." Later the use of coal was regulated on a larger scale by Richard II. The famous London smog was already described in a publication around 1650. As late as 1952, between the December 5 and 9, about 4000 people died in London during the "great smog."

The age of the automobile ushered in an era of significant pollution in many other urban areas. Due to regulation and technological advances, emissions from cars have been significantly reduced. Except for CO_2 a reduction factor of 100 is often achieved for pollutants coming out of the exhaust pipes of vehicles. Environmental concerns are now being taken more seriously by society, and practices which were common a few decades ago are no longer as accepted today. The environmental impact of different energy sources will be discussed in the individual chapters dealing with those sources.

CHAPTER 2

Oil and Natural Gas

Petroleum, or crude oil, and natural gas are organic hydrocarbons formed from atoms of carbon and hydrogen. Produced during a process that lasted millions of years, they are derived primarily from ocean biomass (plankton) that grew due to an abundance of solar energy reaching the earth's surface. As a result, these hydrocarbons can be viewed as storage vehicles for solar energy and their combustion as a means of liberating that stored energy. Together with coal, these energy sources are critical to the functioning of modern civilization.

Oil and natural gas have been known and used since ancient times. For example, 5000 years ago bitumen[1] which seeped from the ground was of major importance in the Mesopotamian civilization. It was used to join together bricks made of clay or baked clay, to fire potteries or bricks, and in waterproofing of buildings, ships, canals, and so on. Burning oil was employed as a weapon by the Persians against the Greeks in the fifth century B.C. Before it became extensively used as an energy source, the main use of petroleum in ancient times was as a medicine or a liniment.

In some regions natural gas also leaked from the ground. Spontaneous fires, probably ignited by lightning, were interpreted as being of holy origin. In the region of Iraq, "eternal fires" were reported as early as 100–125 B.C. Leakage of natural gas from the ground was also observed by Indians in North America, mostly along the western side of the Appalachian Highlands.

In China natural gas has been used for cooking since the tenth century when it was transported inside bamboo pipes. In Europe natural gas was originally considered to be a scientific curiosity. Some scientists produced it by distillation of coal and used it for lighting and for inflating hot-air balloons. In 1885, Robert Bunsen developed a burner that easily mixed air and natural gas to produce heat. This provided an efficient and controlled means for burning gas to produce heat for warming buildings and cooking food. The natural gas

---

[1]Bitumen is a solid, semisolid, or viscous colloidal hydrocarbon also called asphalt. It is used mainly for the construction of roads and for roofing material.

Our Energy Future: Resources, Alternatives, and the Environment
By Christian Ngô and Joseph B. Natowitz
Copyright © 2009 John Wiley & Sons, Inc.

industry was really born in the United States during the nineteenth century when the gas was used mostly as a fuel for lamps. The first gas pipeline was built in 1870. It was 40 km long and the pipes were made from hollowed trunks of pine trees. Two years later metal pipelines were used to transport natural gas. Today, the length of the natural gas pipeline network in the United States is more than four times the distance from the earth to the moon.

The heating value of a fuel is the amount of energy released during the combustion of a given amount of it. The "higher heating value" (called also gross calorific value) is measured by bringing all the combustion products to the initial precombustion conditions (in particular to the initial temperature). If water vapor is formed during the combustion, the heat of vaporization of water is then recovered. The "lower heating value" (net calorific value) is the higher heating value minus this heat of vaporization. The higher the hydrogen content of a fuel, the larger the difference between the higher and lower heating values. In most practical combustions, the heat of vaporization is lost and the lower heating values apply. For coal and oil the difference between the gross calorific value and the net calorific value is approximately 5%. It reaches 9–10% in the case of natural gas. For natural gas, harnessing the heat of vaporization provides a large increase in the yield of useful energy.

2.1. GENESIS OF OIL AND NATURAL GAS

Oil and natural gas are produced from deceased sea organisms (zooplankton, phytoplankton, shellfish, algae, animals, and so on) buried in sediments of sand and mud. Only a small fraction (less than 1%) of the original dead living organisms is found in the sediment. The remaining part has been recycled within the biosphere. As time passed, progressive layers of organic material and sediment were deposited. Because of the pressure exerted by the upper layers, the mixture of organic matter and sediments in the lower layers were progressively converted into sedimentary rock. These sediments remained on the ocean floor for millions of years. In the absence of oxygen the organic materials trapped inside the sediments were transformed into oil and gas.

Two major phenomena govern oil and gas formation: *subsidence* and *compaction*. Subsidence corresponds to a downward motion of the earth's surface. The progressive accumulation of sediments in estuaries and deltas forms sedimentary basins. Under the increasing weight of the sediments the bottom of the sedimentary basin subsides allowing new sediments to accumulate above. At the same time compaction of the sediment under the pressure of the upper layers takes place. As both the pressure and temperature increase, water is expelled from the sediments. Physical, chemical, and biological processes collectively known as diagenesis take place and lead to the formation of rocks. These rocks, containing transformed organic, matter are usually called *source rocks*. At higher temperatures and pressures a metamorphic recrystallization of the rocks may occur.

In order that the organic matter contained in sediments can be transformed into carbon-rich compounds and not recycled into the biosphere, the medium must be oxygen free. In this case the organic matter is transformed into an organic material called *kerogen*. Kerogen contains heavy molecules having molecular masses more than 1000 times that of hydrogen. Except for the lighter fraction, known as bitumen, it is mostly insoluble in common organic solvents. There are several types of kerogen and, under specific conditions of pressure and temperature, fine-grain sedimentary rock containing kerogen can produce oil and gas. Under increasing pressure and temperature the kerogen is cooked and transformed progressively into crude oil and gas. A temperature of at least 60 °C is necessary to initiate the transformation into oil and a temperature of 120 °C to start the transformation into gas. On average, a temperature above 100 °C is sufficient to transform kerogen into hydrocarbons on a geologic time scale. The higher the temperature, the more likely that the transformation leads to the formation of natural gas. This is schematically illustrated in Figure 2.1 showing the amount of crude oil and natural gas typically produced as a function of depth. The "oil window" corresponds to the conditions under which the probability to form oil is large. This probability peaks around 2500 m depth. The gas formation probability corresponds to higher temperatures and therefore to deeper locations, ≈3500–4000 m deep.

The earth's surface consists of two layers. The outer one, the *lithosphere*, includes the crust and the upper part of the mantle of our planet. The *asthenosphere* is located beneath the lithosphere. It consists of a low-viscosity and low-shear-strength solid which flows like a liquid but on geologic timescales. The lithosphere is broken up into pieces called tectonic plates. There are seven major *tectonic plates* and several minor ones on the earth. Plate tectonic theory explains the large-scale motion of the earth's surface and in particular continental drift. Tectonic plates drift slowly on the asthenosphere at a velocity of a few centimeters per year. Although small, such drift velocities correspond to large distances at the geologic timescale. For example, a drift velocity of 2 cm/year over a billion years corresponds to a distance of 20,000 km.

Sedimentary basins look like a layered millefeuille pastry with layers of thickness ranging from a few hundred meters to about 20 km. These layers rest on platforms of crystalline rocks (metamorphic rocks, granite rocks, and so on). The motion of tectonic plates causes the earth's crust and thus the

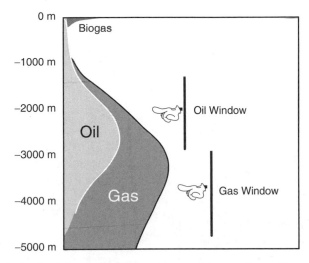

Figure 2.1. Schematic of depth at which source rock is transformed into hydrocarbons. Near the surface, biogas can be formed in the surface sediments. In this region oil and gas have little chance to be formed. The oil window, which is the range where oil has a high probability of formation, is located at a greater depth. Natural gas formation becomes more probable at even lower depths where the temperatures are higher. In the gas window oil is transformed into natural gas. When an oil or gas reservoir is formed, it may then shift to different depths and locations. From P. R. Bauquis and E. Bauquis, *Pétrole et gaz naturel*, Hirle, 2004.

sedimentary basins to move and deform. Because of tectonic plate drift and the fact that the level of the sea has changed over the ages, some of these sedimentary basins can now be found inland. Under the weight and pressure of upper sediment layers a migration process starts and the initial hydrocarbons move over several kilometers vertically and many tens, sometimes hundreds, of kilometers laterally. As the pressure rises, hydrocarbons are slowly expelled from the source rock: This process is called expulsion. Oil and gas are then mixed with salt water present in the porous surrounding medium. Since gas is lighter than oil and oil lighter than salt water, the gas is on top and water on the bottom in this mixture (Figure 2.2). Sometimes impermeable traps are created. These are geologic structures that are able to contain reservoirs of oil and gas. They are basically of two types: structural traps formed by deformation and stratigraphic traps resulting, for example, from the sealing off of the top of the reservoir by a nonporous formation. Fractures in the crust may occur, because of earthquakes, for example, allowing oil and gas to move to the surface. In this case gas may be lost into the atmosphere while the oil can evaporate, leaving just the heavier components such as bitumen.

An oil or gas field is a large accumulation of oil or gas. The nature of the crude oil or natural gas varies from field to field. Some fields contain a heavy black viscous crude oil while others provide a pale oil which flows like water.

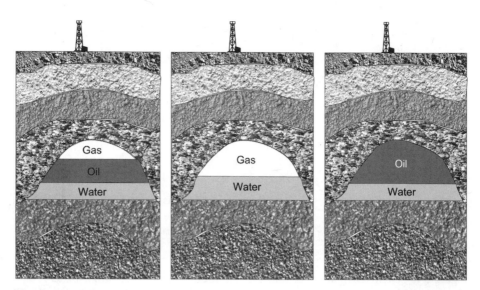

Figure 2.2. Schematic showing that it is possible to find deposits containing gas, oil, and water; gas and water; or oil and water. Water is denser than oil and gas is lighter than oil. Usually, these reservoirs are located at different depths (gas is formed at greater depths than oil).

Different qualities of natural gas also exist. These different qualities of crude oil and natural gas are valued differently. The more expensive are those which are more easily transformed into useful fuels.

A reservoir looks more like a sponge than a pocket of free hydrocarbons. The hydrocarbons are confined in very small cavities inside the source rock. Natural gas is relatively easily removed from the reservoir. As soon as the pressure in the reservoir is higher than the atmospheric pressure, gas flows out. The recovery factor is therefore very high. Recovering all of the crude oil contained in this spongelike rock would be very difficult. On average only 30–40% of the deposit is recovered. Improving this recovery factor would increase the exploitable crude oil reserves.

2.2. RECOVERING OIL AND GAS

The best reservoirs in terms of oil or gas exploitation are those which have a good permeability and porosity. Permeability is a measure of how well a liquid or a gas diffuses through a rock. Permeability is measured on a scale ranging from 1 (lowest) to more than 13,000 (highest). Porosity is a measure of the volume of the spaces within the rock compared to the total volume of the rock. A sponge has a large porosity while a solid rock has a negligible porosity. The average porosity of reservoirs ranges between 7 and 40%.

About 140 million years ago, a rift occurred separating Africa from South America. Water, first lake water and then seawater, flowed into these rifts, leading to favorable conditions for oil and gas formation. Many offshore oil and gas fields now exist in the western part of Africa (e.g., Nigeria, Angola, Congo) and in the eastern part of South America (e.g., Brazil).

In earlier days oil was much more valued than natural gas. The latter was often dispelled by being burned in a flare stack. This was wasteful but better than releasing natural gas into the atmosphere since the greenhouse effect of methane is greater than that of carbon dioxide. Currently, natural gas is a valued resource and the world demand for natural gas is increasing faster than the demand for oil. Energy companies are therefore eager to find new gas reservoirs.

Finding hydrocarbons is a hard job. Before a well which may cost millions of dollars is drilled, it is important to be convinced that the probability of finding oil or gas is large. The initial explorations start with extensive geophysical surveys. These rely on conventional geology, aerial photography, satellite imagery, magnetic field measurements, and an ever-improving understanding of the physics of rock formation. Unfortunately many of these techniques are less useful for exploring deep underwater oil deposits. Fortunately, sophisticated seismic techniques have been developed. Seismic measurements employ the propagation and reflection of sound waves. Sound waves are created just beneath the land surface or on the ocean's surface. They propagate down through the layers of rocks and are scattered. Scattered waves are detected with small receivers placed on the ground or at the sea surface and give information on the interfaces which are encountered by the incident waves. The data from the receivers are analyzed using powerful computers, and it is possible to get precise information on the reservoir and to build a three-dimensional picture of it. With seismic measurement it is even possible to distinguish between oil, natural gas, and salt water. The large progress recently made in seismic techniques has greatly increased the probability of finding hydrocarbons in drilling operations. Drilling techniques for oil or gas have also improved. Tungsten or diamond drills are employed. In soft rocks the drilling speed can reach 100 m/h but is much slower in hard rocks. Powerful improvements in rotary drilling make it possible to drill in three dimensions and not only vertically.

Once an oil field has been discovered, recovering the maximum possible amount of oil or gas contained in the reservoir becomes paramount. The natural pressure of the reservoir usually allows recovery of only a very small part of the deposit, typically about 20–40%. The recovery factor can be raised by injecting water, gas that increases the pressure in the reservoir, or chemicals that reduce the oil's resistance to flow. Enhanced recovery techniques can increase the range of recovery factors to 30–60% and perhaps even more in the future.

There are three major enhanced oil recovery techniques that allow an increase of the recovery factors to 30–60%: The more frequently used technique is thermal recovery, which uses hot steam to lower the oil's viscosity and improve its ability to flow. This technique represents over 50% of the U.S. applications in this domain. Gas injection uses nitrogen or carbon dioxide to increase the pressure in the reservoir and push the oil out. In the United States, this technique is employed nearly 50% of the time. Chemical injection represents less than 1% of the U.S. enhanced oil recovery methods. It consists of injection of detergent compounds that decrease the oil's surface tension of polymers that increase oil flow.

Producing oil or gas from offshore reservoirs is more difficult than on land and requires large investments. Drilling rigs can be supported by legs that stand on the ocean floor if the depth of water is less than about 120 m. For larger depths the rigs have to float and be anchored at a precise position whatever the weather conditions. Deep offshore technologies have improved greatly since the early 1970s when offshore production in the North Sea took place at depths near 200 m. Today, it is possible to exploit reservoirs located at depths reaching 2000 m (Gulf of Mexico, Angola, west coast of Africa).

The world's largest offshore drilling rig is Troll, located in the Norwegian part of the North Sea. In 1999, it provided 15 Gm^3 of natural gas to France (about 40% of its consumption). The gas is transported by an 840-km underwater pipeline connecting the Norwegian reservoirs of Sleipner and Troll to France. Natural gas in the pipeline moves at a speed near 30 km/h and must be recompressed every 80–120 km in order to maintain that rate of flow.

2.3. PEAK OIL

It took millions of years for nature to synthesize the existing reservoirs of hydrocarbons. At any given time, only part of this oil can be recovered at an acceptable cost using the techniques then available. This recoverable part constitutes the "known oil reserves." Oil reserves are sometimes classified in three main categories in the oil industry. The first one refers to the "proven reserves," which are the reserves of crude oil that are reasonably certain and exploitable using the current technology and at the current oil prices. Proven reserves include developed and known undeveloped ones. In terms of probability, proven reserves have a 90% chance to be exploited. The second category, referred as "probable reserves," includes those which are believed to have a 50% chance of being produced using the current technology (or a slightly improved one) and within the current economic paradigm. "Possible reserves" belong to the third category. They have a 10% chance of being exploited in the future and need very favorable circumstances for that to occur. Political or commercial reasons may lead to pressures to underestimate or overestimate the oil reserves in some countries. Consequently an accurate estimate of oil reserves is very difficult to produce.

The useful reserve depends on the price that the consumer is ready to pay. Some oil resources may be interesting to exploit only if the price of oil is above a certain value. Currently the energy resource often referred to as conventional oil is being used at a rate that can theoretically deplete this resource within a half a century. Knowing when conventional crude oil production will decline because exploitable reserves are running out is a key question for formulation of national energy policies.

The current yearly consumption of petroleum is larger than the amount that is newly discovered each year. This situation has existed since the 1980s. Presently, about 80% of oil being recovered comes from deposits discovered before 1973. Some analysts believe that we have discovered 90% of the conventional crude and that production will soon begin to decline. This is based on an extension of studies originally undertaken in the 1950s by King Hubbert, a geophysicist working at the Shell oil company. His estimation was based on the pattern of discovery and depletion of crude oil, two events similar in nature but shifted in time. In 1956 he predicted that the oil production in the United States would peak in 1969 and decline thereafter. This prediction proved to be very accurate since in reality production declined starting in 1970. The basic idea of King Hubbert's model is that the exploitation of any finite resource follows a logistic curve as shown in Figure 2.3. The maximum of the curve (the peak) is reached when the resource is half exploited.

Applying this model to the world's oil deposits is complicated because there are several unknowns, as discussed above. Because of these uncertainties, there is debate over oil reserves and there are basically two opposite schools of thought, the pessimists and the optimists. The pessimists think that the oil peak will soon be reached ($\approx$2010 or even before). The optimists say that the oil

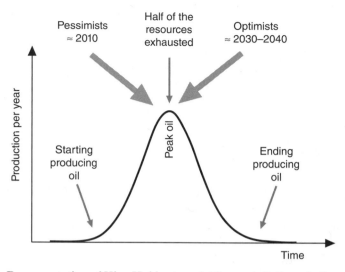

Figure 2.3. Representation of King Hubbert model for exploitation of oil reserves. The curve is a logistic distribution which looks like a normal distribution in shape but with more extended tailing.

peak will not occur before 2030–2040. They claim, in particular, that new oil discoveries can be made if people search in unexplored areas. Interestingly, many of these optimists are mostly associated with oil companies while many of those belonging to the pessimistic community are persons retired from the oil industry.

By 1997, about 110 Gt of oil had been extracted from the earth. The global reserves quoted in the BP statistical review at the end of 2006 amounted to 164.5 Gt (the annual world production of oil in 2006 reached 3.9 Gt). Other estimates exist but are of the same order. These estimates imply that mankind is close to having extracted half of the earth's conventional crude oil. At an annual consumption rate of 3.9 Gt, equal to that of 2006, current reserves would be expected to last 42 years. Actually, this inference is quite uncertain since new nonconventional sources of oil can be used and the recovery factor of oil in the reservoir can be improved. Nevertheless, some experts predict that the amount of crude oil extracted around 2050 will be about 3.5 Gt, very comparable to current production. At that time, the total primary energy demand will probably have been multiplied by a factor of 2 or so and we would need an additional 3.5 Gt of oil equivalent in other forms of energy to compensate for the lack of petroleum. This is a huge amount that cannot be completely compensated for by renewable energies, nuclear energy, or energy use efficiency alone. A combination of the three will be required to meet our needs.

In any case, whether the oil peak occurs in 2010 or 30 years later is not the critical point. The important point is that even the optimists expect the peak to occur within decades. Therefore we will soon be faced with a progressive shortage of crude oil and we must change the way we use oil because the price of this resource will progressively increase and its availability will diminish.

Similarly we can also expect a gas peak to occur. It will be reached after peak oil, probably around the middle of the century. It could come earlier if the consumption of natural gas increases strongly to partially compensate for diminishing supplies of reasonably priced oil.

2.4. RESERVES

By the end of the century it is virtually certain that oil and gas will be in short supply. The part of these resources that remains will probably be reserved for applications in production of industrial chemicals and pharmaceuticals and some liquid fuels. The questions of how much crude oil and natural gas reserves currently exist and where are important issues in addressing the problems of safety and security of supply.

2.4.1. Crude Oil Reserves

At the end of 2006, 61.5% of the known crude oil reserves were in the Middle East, as can be seen in Table 2.1. Europe and Eurasia have 12% of the reserves, mostly located in the Russian Federation (6.6%) and in Kazakhstan (3.3%). Africa has 9.7%, mostly in Libya (3.4%) and Nigeria (3%). In South America, oil is mostly found in Venezuela (6.6% of the global reserves). North America has only 5% (2.5% in the United States, 1.4% in Canada, and 1.1% in Mexico). Major oil companies like Exxon, Shell, BP, and Total control only 15% of the total crude oil production. The remaining part is the property of many smaller entities.

In Figure 2.4, the seven countries of the world having the largest crude oil reserves are displayed. Not surprisingly, five of them are in the Middle East. This means that the rest of the world will become increasingly dependent upon this region of the world for petroleum resources.

TABLE 2.1. Share of Proved World Reserves of Crude Oil among Different Regions, End of 2006

Regions	Proved Reserves (Gt)	Percent
North America	7.8	5
South and Central America	14.8	8.6
Europe and Eurasia	19.7	12
Middle East	101.2	61.5
Africa	15.5	9.7
Asia Pacific	5.4	3.4
Total world	164.5	100

Source: Data are from the BP *Statistical review of world energy*, 2007. A consequence of rounding errors is that the sum of the percentages is not exactly equal to 100%.

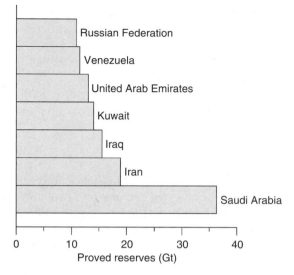

Figure 2.4. Countries with largest proved crude oil reserves. Data from BP *Statistical review of world energy*, 2007.

2.4.2. Natural Gas Reserves

In 2006 the world's natural gas consumption was equal to $2.87\,Tm^3$ (or $2.6\,Gtep$). Currently there is plenty of natural gas and the amount discovered each year is larger than the amount consumed. The proved reserves of natural gas by the end of 2006 amounted to $180.5\,Tm^3$, or $162.4\,Gtep$. The ultimate reserves estimated under today's economical and technological conditions are about $400\,Tm^3$.

Using the global proved reserve value and assuming the consumption rate of 2006, one can deduce that there remains ~63 years of natural gas. For the same reason as for crude oil, this figure has to be taken with great care. The annual consumption may increase and other reserves may be exploited in the future. It is possible to transform gas into synthetic oil, which might be interesting for transportation applications. Indeed, a liquid fuel has a greater volume energy density than natural gas and is more convenient for powering vehicles.

In Table 2.2 the shares of proved natural gas reserves for different regions of the world are shown. Most of them are concentrated in Eurasia and the Middle East, mostly in Iran (15.5%) and Qatar (14%). In Africa, most of the reserves are found in Nigeria (2.9%) and Algeria (2.5%). In South America, there are good reserves in Venezuela (2.4%). In North America, the United States has 3% of the global reserves of natural gas.

In Figure 2.5 we show the seven individual countries with the largest proved reserves of natural gas. Most of the reserves are concentrated in the Russian Federation, the top country in terms of natural gas reserves with 26.3%. Iran has 15.5% and Qatar 14%. In Africa, most of the reserves are found in Nigeria

TABLE 2.2. Shares of Proved World Reserves of Natural Gas for Different Regions, End of 2006

Regions	Proved Reserves (Tm^3, or $10^{12} m^3$)	Percent
North America	8	4.4
South and Central America	6.9	3.8
Europe and Eurasia	64.1	35.3
Middle East	73.5	40.5
Africa	14.2	7.8
Asia Pacific	14.8	8.2
Total world	181.5	100

Source: Data from BP *Statistical review of world energy*, 2007.

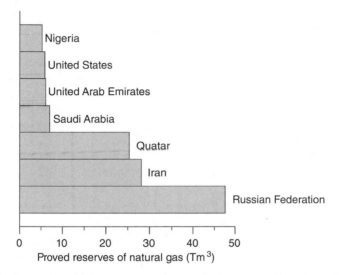

Figure 2.5. Countries with largest proved natural gas reserves. Data from BP *Statistical review of world energy*, 2007.

(2.9%) and Algeria (2.5%). In South America, there are good reserves in Venezuela (2.4%). In North America, the United States has 3% of the global reserves of natural gas. Currently, North America is self-sufficient to a large extent and Canada is a large exporter to the United States. Europe depends very much on the former Soviet Union countries and on Algeria for its natural gas.

Natural gas can be transported by pipelines or in the liquid form—liquid natural gas (LNG). Both methods are expensive and a long time is required to pay back the initial capital investment. Pipelines are less expensive for transportation over short distances (typically less than 3000 km). Natural gas is liquid below −161 °C and its volume is decreased by a factor of 614 compared to the gaseous state at normal temperature and pressure. Liquid natural gas is produced in a liquefaction train, which is a series of operations leading from the gaseous state to the liquid state. This requires an energy equivalent to between 13 and 16% of the energy contained in the LNG. The LNG is transported in specially designed ships or tanks. Large investments are required for the liquefaction plant and the ships. At the point of arrival the LNG is returned to the gaseous state, usually using heat from the sea (about 0.25 kWh/kg is needed). The advantage of LNG is that it is more flexible than pipeline transportation. It is easy to change the destination of an LNG ship. In 2004, 680 Gm^3 of natural gas (25% of the global production) was transported around the world, mostly by gas pipeline (73%), but more and more in the form of LNG. In 2004, the volume of natural gas transported by tankers in the form of LNG represented 27% of the global exchanges. It went from 5 Gm^3 in 1974 to about 180 Gm^3 in 2004. The LNG represents 6.5% of the world production. Transporting LNG is about four to five times more energy consuming than transporting the same energy in the form of oil.

In 2004, natural gas supplied 21% of the world's primary energy supply. In North America it provided 23.5% of the energy and in Europe 23%. Natural gas is mostly used in three sectors: residential and commercial applications, industry, and power generation.

Storage of natural gas is an important issue. The amount of gas stored in a country may represent at least 30 days of demand. Additional strategic storage is also a means to get a better security of supply.

2.5. PROPERTIES OF HYDROCARBONS

In Table 2.3 some properties of different oil products are indicated.

In Table 2.4 we give the main properties of natural gas. Its main drawback is that it has a small volume energy density and occupies a large volume compared to oil. This makes it less convenient for transportation applications.

There are different types of crude oils. Those containing light hydrocarbon molecules are more easily exploited than those containing very heavy ones. The average crude oil contains about 1.5 hydrogen atoms per carbon atom. This is because there are not only hydrocarbons but also other molecules containing oxygen, nitrogen, and sulfur. Heavy petroleum contains more of these molecules than light ones. Oil refining allows extraction of the right molecules for a given use.

Natural gas contains mainly methane (CH_4) but also ethane (C_2H_6), propane (C_3H_8), and butane (C_4H_{10}). The numbers of hydrogen atoms per carbon atom are 4, 3, 2.67, and 2.5, respectively. Natural gas contains also nitrogen, carbon dioxide, and hydrogen sulfide. Natural gas provided to the consumer is purified and contains mainly methane.

TABLE 2.3. Typical Properties and Calorific Values for Selected Petroleum Products

Product	Density		GCV			NCV		
	kg/L	L/kg	MJ/kg	KWh/kg	KWh/L	MJ/kg	kWh/kg	kWh/L
Aviation gasoline	0.71	1.41	47.4	13.2	9.3	45.0	12.5	8.8
Motor gasoline	0.74	1.35	47.1	13.1	9.7	44.8	12.4	9.2
Kerosene jet fuel	0.80	1.25	46.9	13.0	10.5	44.6	12.4	9.9
Gas/ diesel oil	0.84	1.18	45.7	12.7	10.7	43.4	12.1	10.2
Fuel oil	0.94	1.06	42.8	11.9	11.2	40.7	11.3	10.7

Note: GCV = gross calorific value or higher heating value; NCV = net calorific value or lower heating value.
Source: Data from the International Energy Agency, Paris.

TABLE 2.4. Gross Calorific Values and Net Calorific Values for Selected Natural Gas Products

Product	Density (kg/m^3)	GCV (MJ/kg)	NCV (kJ/kg)	GCV (kJ/Nm^3)	NCV (kJ/Nm^3)
Methane	0.72	55.6	50.0	39.9	35.9
Natural gas[a]	≈0.7–0.9	67.85	61.3	42.1	38.1
Ethane	1.26 (NTP)	51.9	47.5	69.9	64.1
Propane	1.88	50.4	46.4	101.3	93.1
n-Butane	2.5	49.5	45.8	133.6	123.1

Note: Data from www.imteag.com/2-2005-06.pdf. All densities except ethane are given at standard temperature and pressure (STP), 0 °C, and 1 atm. The density of ethane is given at normal temperature (20 °C) and pressure (1 atmosphere), NTP. Ethane data from www.engineeringtoolbox.com/gas-density-d_158.html.
[a]This may have different compositions.

Petroleum is refined in order to separate and/or transform the different types of molecules that it contains into useful petroleum products such as gasoline for cars, fuel for oil-fired heaters, jet fuel for planes, and so on. In terms of volume, 1 barrel of crude oil gives 1.15 barrels of petroleum products. This comes from the fact that additives are used during the refining process to produce the final products.

Figure 2.6 shows the percentage share of different products derived from one barrel of California crude oil. We see that, in this particular refining process, half of the barrel of crude oil is converted into gasoline. In the European Union, where diesel oil is extensively used, refining is tuned to lead to different proportions of final products and about the same amounts of

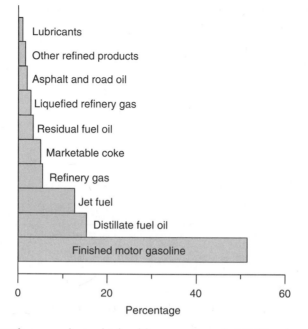

Figure 2.6. Petroleum products obtained from one barrel of California crude oil, based on 2004 data. From http://www.energy.ca.gov/gasoline/whats_in_barrel_oil.html, California Energy Commission, Fuels Office, PIIRA database.

gasoline and diesel oil are obtained. In France the refining process is adjusted to yield more diesel than oil because of the large number of diesel cars.

Consider a few-order-of-magnitude comparisons for the combustion of commercial fuel and that of natural gas. The energy contained in $1\,m^3$ of natural gas is roughly the same as that contained in 1 liter of fuel. Burning $1\,m^3$ of natural gas produces about $2\,kg$ of CO_2 whereas burning 1 liter of fuel produces $2.7\,kg$ of CO_2. This is because oil has a larger fractional content of carbon than does methane. On the other hand, more water vapor is produced when burning $1\,m^3$ of natural gas ($1.68\,kg$) than 1 liter of fuel ($0.9\,kg$). Finally, burning $1\,m^3$ of natural gas or 1 liter of fuel requires about $10\,m^3$ of air. Generally more air is needed than indicated by the reaction stoichiometry (about 20%). Using less air than the stoichiometry requires may produce carbon monoxide, CO, which is a poisonous gas. A CO concentration of 0.2% leads to death in less than half an hour.

2.6. OIL FIELDS

Regions with many petroleum wells are called oil fields. Because oil reservoirs can extend over large areas, many oil wells are drilled to exploit a given field.

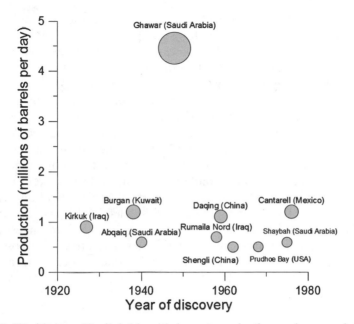

Figure 2.7. World's top 10 oil fields with largest production and years of discovery. Data from Simmons & Company International and J. L. Wingert, *La vie après le pétrole*, Autrement, 2005.

There are more than 40,000 oil fields around the world but most of them are small ones. Some are on land, others offshore. About 40% of the global petroleum reserves are located in more than 900 giant oil and gas fields clustered in 27 regions located mainly in the Persian Gulf and in the Western Siberian basin. These contain about 500 million oil barrel equivalents.

The oil fields with the largest daily production are shown in Figure 2.7. The Ghawar field in Saudi Arabia is the largest one. The bad news is that, despite the fact that we now have more sensitive and sophisticated techniques to find petroleum on land and offshore, the 10 largest oil fields were all discovered before 1980. Each year since 1980, oil demand has surpassed the amounts discovered, indicating that we shall probably reach peak oil soon.

2.7. PRICES

Finding oil is a risky task. Oil or gas is found in about one in five new drilling attempts. If one includes extensions of already exploited oil fields, this average figure is closer to one in three. Exploration costs amount on the average to $1–$2 per barrel. The production cost is about $2–$3 per barrel in good oil fields found in the Middle East and can reach $10–$15 in less accessible areas (e.g., the North Sea) or older oil fields (offshore fields in the Far North).

Transportation costs vary according to the distance the oil must be transported. It costs typically $1–$1.5 to transport petroleum from the Middle East to Europe. At the end of 2007, the CIF (initial cost, insurance cost, and freight cost) was about $5 a barrel for oil from a good Middle Eastern oil field. In April 2008, the market price for a barrel of oil surpassed $115 per barrel and reached $147.5 on July 11, 2008. The price of oil varies, of course, with the quality of the crude oil. The price of a barrel of oil typically reported in the media refers to certain standard quality oils: Brent in Europe, West Texas Intermediate in the United States, and Dubai in the Middle East.

The production cost of oil is quite different from the market price. Consequently, there is a considerable profit margin. The market price suffers many fluctuations depending on several factors, but political and economic considerations are those having the largest impact. In the past, the low initial cost of Middle Eastern oil has deterred oil companies from investing in oil fields with significantly larger production costs. Therefore, it is possible that before being faced with a crude oil shortage due to the finite amount of reserves, the world might be faced with a shortage due to a lack of sufficient investment in higher cost fields.

The "barrel," usually noted bbl, is the common unit of volume used for oil. This measure was first adopted in the United States during the exploitation of the first oil wells in Pennsylvania when petroleum contained in wooden barrels was transported by horse-drawn carts. Between 1859 and 1861 two types of barrels were used: whisky or beer barrels (36 gallons) or herring barrels (42 gallons = 159 liters). In 1861, the herring barrel became the standard for the oil industry. The first ship transporting oil carried 4000 barrels of oil. This is negligible compared to today's big tankers, which transport volumes of more than 2 million barrels.

Useful conversions: 1 barrel = 159 liters; 1 ton of oil = 7.3 barrel and 1 bbl/day = 50 t/year.

In Figure 2.8 the evolution of the price of oil between 1861 and 1999 is shown in both the currency of the day and 1999 dollars, that is, adjusted for inflation. We see that, adjusted for inflation, the price of a barrel of oil was essentially constant over decades between 1880 and 1970. Although the price of oil is determined by a number of factors, the increase and strong fluctuations observed after 1970 correspond to various world events: the oil crisis of 1973 (OPEC), the Iranian revolution in 1979, and the Gulf War in 1990. The effect of the last was milder and shorter than that of the two preceding events. As this is written the war in Iraq, which began in 2003, continues. Oil prices

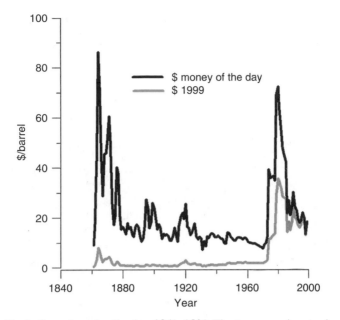

Figure 2.8. Evolution of crude oil price, 1861–1999. The top curve is actual price in U.S. dollars of the day. The bottom curve corresponds to 1999 dollars. Data between 1861 and 1994 correspond to the U.S. average. Between 1945 and 1985 it is for Arabian light oil posted at Ras Tanura and between 1986 and 1999 it represents the Brent spot price. Data from www.eia.doe.gov/pub/international/iealf/BPCrudeOilPrices.xls.

reached about $150 per barrel in 2008 but has since declined. (Note that there is sparkling mineral water in France sold at ≈$150 per barrel in supermarkets.)

In Figure 2.9 we show the evolution of the annual averaged crude oil price between 1983 and 2007. The steep rise observed from 2000 on indicates that we are now entering a period where petroleum is likely to remain expensive.

The price history of natural gas shows an evolution similar to that of oil, as can be seen in Figure 2.10. However, prices of natural gas are much more dependent on the place to which it is delivered since the cost of transport is a more significant part of the total cost.

2.8. CONSUMPTION

The world's total consumption of primary energy in 2004 was 11.2 Gtoe [International Energy Agency (IEA)]. The consumption of crude oil was 3.9 Gtoe and of natural gas was 2.3 Gtoe. That of coal was in between: 2.8 Gtoe. Consequently, fossil fuels represented around 80% of the global primary energy consumption. In 1990, the world's primary energy consumption was

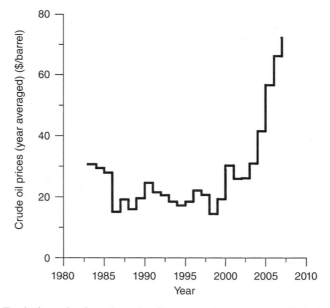

Figure 2.9. Evolution of price of crude oil averaged over a year, 1983–2007. The data are relative to light sweet crude oil at Cushing, Oklahoma, and prices from the New York Mercantile exchange. Data from www.eia.doe.gov/emeu/international/oilprice. html.

8.7 Gtoe with the following shares: oil (3.2 Gtoe), natural gas (1.7 Gtoe), and coal (2.2 Gtoe). Again fossil fuels provided 80%. Since the total energy consumption increased by 28% over a decade and a half and the share of fossil fuel did not change, the total increase in fossil fuel consumption has also been about 28%. This is an average value and the increase of fossil fuel consumption is not the same in all countries. In Table 2.5 we show the percentage increase of oil consumption over the past 20 years for some selected countries. In that period developed countries saw very little increase in their oil consumption (United States, Japan) or even decreased it (France). However, developing countries greatly increased their oil consumption. One of the leading reasons is that a large share of the manufactured objects that people of developed countries are now buying are made in the developing countries. To meet that demand the developing countries need energy and are relying on fossil fuel to meet their need. Other notable consequences of this shift in the manufacturing base are the increases of jobs and of pollution in the developing countries.

The per capita consumption of oil in 1971 and 2005 is shown in Figure 2.11 for selected countries. We note again the same tendency as in Table 2.5. The large increase in the world's oil consumption is occurring in developing countries.

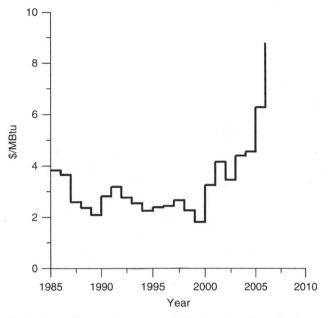

Figure 2.10. Evolution of annual average CIF of natural gas in European Union, 1985–2006: 1 MBtu = 1055 MJ = 0.293 MWh. Data from Heren Energy Ltd. and *Natural Gas Week*.

TABLE 2.5. Increase in Oil Consumption in 20 Years for Selected Countries

Country	Increase in 20 Years (%)
South Korea	306
India	240
China	192
Brazil	88
United States	16
Japan	12
France	−12

Figures 2.12 and 2.13 show the breakdown of primary energy sources in 1973 and 2005 for the United States, Germany, and France. For these three countries we see that between 1973 and 2005 the share of oil in total energy consumption decreased. Meanwhile there was an increase of natural gas consumption, except in France, where nuclear energy (classified in other) produced an increasing share of France's electricity.

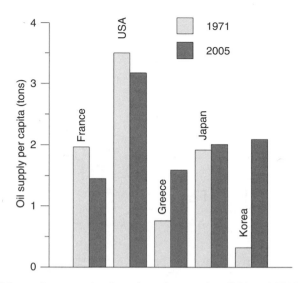

Figure 2.11. Oil supply per capita for selected countries, 1971 and 2005. Data from *Oil Information*, International Energy Agency/Organization for Economic Co-operation and Development (IEA/OECD), Paris, 2006.

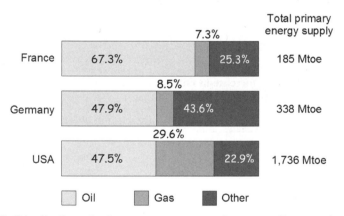

Figure 2.12. Distribution of primary energy sources between oil, gas, and other, 1973. Data from *Oil Information*, IEA/OECD, Paris, 2006.

2.9. ELECTRICITY GENERATION

Oil is mainly used in transportation and in the petrochemical industry and less frequently to produce electricity. In contrast, natural gas is extensively used to produce electricity. Its main advantage is that it has a lower environmental impact than coal. The second advantage is that new technologies (combined-cycle gas turbines) have been developed in gas-fired power plants allowing these plants to reach operating efficiencies between 55 and 60%.

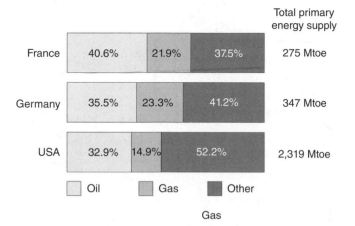

Figure 2.13. Distribution of primary energy sources between oil, gas, and other, 2005. Data from *Oil Information*, IEA/OECD, Paris, 2006.

A decade ago electricity produced in gas-fired power plants was less expensive than that from coal-powered plants. This is not currently the case. Today, natural gas is typically three times more expensive than coal. This is because the price of gas is correlated, with some delay, with the price of oil and has increased considerably during the last few years. However, a gas-fired power plant emits only about half as much CO_2 as a coal-fired plant, and other emissions, dust particles, for example, are much lower. Since less CO_2 is emitted into the atmosphere for the same amount of power produced, the comparison of the cost between coal-fired power plants and gas-fired ones may change if high carbon taxes are introduced as a means of pollution control.

SINGLE-CYCLE GAS TURBINE

The simplest way to use natural gas is with a single-cycle gas turbine. The principle is shown in Figure 2.14. Compressed air and natural gas are burned in a combustion chamber at almost constant pressure. The exhaust gases of the combustion chamber, having an initial temperature around 1500 °C, enter the turbine at temperatures between 1200 and 1450 °C and expand before they are emitted into the atmosphere at a temperature near 600 °C. The heat content is usually lost. However, it is possible to recover part of the heat using a heat exchanger. The efficiency of a 40-MW$_e$ single gas turbine is about 40%. It is a little bit less for 200–300-MW$_e$ turbines (38%).

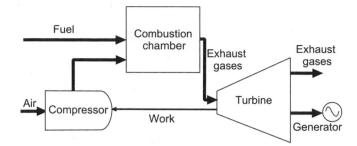

Figure 2.14. Principle of single-cycle gas turbine.

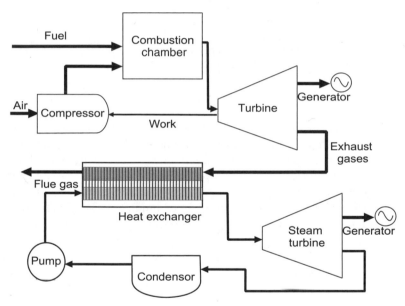

Figure 2.15. Principle of combined-cycle gas turbine.

COMBINED-CYCLE GAS TURBINE

More power can be obtained with the combined-cycle gas turbine tech-nology, which consists of two thermodynamic cycles. The first is the single gas cycle discussed above for the single-cycle gas turbine. The second one is a steam cycle. The principle of the simplest combined-cycle gas turbine is displayed in Figure 2.15. The exhaust gases of the first turbine, which have temperatures of 560–640 °C, go through a heat exchanger. Steam is produced and used to drive a steam turbine. Overall yields between 54 and 60% can be obtained.

Single-cycle gas turbines have a very short startup time (between 15 and 30 mn) and quick load change capabilities. They can be used to meet base-load, cycling, and peak-load demands. Combined-cycle gas turbines are less flexible because the steam cycle operates under higher pressure and heavier duty materials are required. Care must be taken to avoid thermal stresses. As a result, an interruption in power generation can introduce delays in restarting. If operation is interrupted for less than 8 h, full power can be regained in 40–50 min. Reaching full power from a cold state takes 10–16 h. Faster starts are possible if the steam turbine is bypassed, but including this capability increases the investment and maintenance costs. During operation, it is possible to alter the power output 5% per minute. There are several variants in design which can increase the flexibility, combining two single turbines with a steam turbine, for example.

Because air is ~80% nitrogen and 20% oxygen, the high operational temperatures of gas turbines lead to production of nitrogen oxides. Nitrogen oxides can lead to nitric acid, which contributes to acid rain. The amount of nitrogen oxides emitted depends on many parameters, among them flame temperature, fuel-to-air ratio, combustion pressure, and duration of combustion. Combustors usually have fuel-to-air ratios of the order of 1 or even a bit smaller. The greatest emissions occur in the case of stoichiometric combustion. Several methods have been developed to minimize nitrogen oxide emissions.

Other technologies employing natural gas may also be employed. Aeroderivative gas turbines used in aviation can be used for small power outputs (below 50 MW). Cogeneration, producing electricity and heat at the same time, or cocombustion, with biofuels, for example, may also be employed.

Gas-fired power plants may be constructed relatively quickly (two or three years) and require smaller investments than other means of power production such as coal, nuclear, or renewable energies. The investment cost is typically of the order of 400–800 €/kWh$_e$. That for coal-fired power plants is 1000–1500 €/kWh$_e$. Sizes of individual units can also be small (a few hundred megawatts or smaller). The investment, operation, and maintenance costs of gas-fired power plants account for only about 20% of the cost of the electricity produced. A much larger part of the cost, $\approx$70–80%, is the fuel itself. Consequently, the economic viability of gas-fired power plants is very sensitive to the price of natural gas. While gas-fired power plants were very competitive economically at the end of the twentieth century, they are currently less so when compared to coal-fired plants.

2.10. IMPACT ON ENVIRONMENT

The use of oil and natural gas has several impacts on our environment. Since both contain carbon atoms in their molecules, CO_2 is emitted during their combustion. Generating 1 kWh of electricity with oil produces about 700–800 g of CO_2 whereas using natural gas produces about 480-780 g of CO_2, depending on the specific technology used. If the appropriate technologies are employed, there is a clear advantage for natural gas compared to oil as far as CO_2 emissions are concerned. In addition, natural gas produces practically no SO_2 because it contains almost no sulfur. The emissions of nitrogen oxides are also lower than those for coal or oil.

A 1000-MW_e oil-fired plant generating electricity 6600 h/year produces 6.6 TWh of electricity. For that it needs 1.5 Mt of fuel and 4.8 Mt of oxygen. It emits on the average 4.7 Mt of CO_2, 90,000 t of SO_2, 6000 t of NO_2, and 1500 t of ashes.

However, although burning natural gas can emit between 40 and 50% less CO_2 than coal and 25–30% less than oil, it is estimated that leakage during the transport and utilization of natural gas amounts to about 2–4% of the total natural gas consumed in the world. This corresponds to a release of about 25–50 Mt of CH_4 per year into the atmosphere. Because CH_4 is about 23 times more efficient than CO_2 as a greenhouse gas, the effect is equivalent to an emission of a volume of CO_2 comparable to half or more of the total volume of natural gas used in the world. The environmental advantage of natural gas usage is clearly diminished when this is taken into account.

Methane is also present in coal mines (firedamp or coal bed methane) and there are leaks of methane from them. In the 1990s the global level of emissions from open-cast coal mines was estimated to be about 4 Mt/year and that from underground mines to be 44 Mt/year. There are great uncertainties regarding the level of these emissions.

It is interesting to compare the emissions associated with fossil fuel exploitation to those from naturally occurring sources or due to other human activities. Methane is emitted from natural wet lands (e.g., swamps, peat bogs, shallow lakes, temporarily flooded areas) as well as from cultivated areas such as rice paddies. Emissions range from 1 mg to 1 g of CH_4 per square meter per day depending on the nature of the region and the climate conditions. Globally, estimates are that between 40 and 160 Mt of CH_4 per year are released from natural wetlands and between 60 and 140 Mt per year from paddy fields. The amount of methane emitted from the sea (there are large concentrations in the Gulf of Mexico and on the Black Sea) are much smaller (of the order of 3.5 Mt/year). Animals, especially ruminants, also emit CH_4 (about 74 Mt/year for domestic animals, 70% of this by cattle, and between 2 and 6 Mt/year from wild animals). Termites also make a contribution (around 20 Mt/year). Open garbage deposits emit between 30 and 70 Mt of methane per year. Each year fires affect areas of about $20 \times 10^6 \, km^2$ around the world and between 1 and 2 Gt of dry biomass is burned. Between 20 and 110 Mt/year of methane is emitted. Globally, the quantity of methane emitted by human activities is 1.7 times larger than that emitted by natural processes.

Leakage of petroleum or petroleum products during transport is also an environmental concern. About half of the world's shipping capacity is used to transport fossil fuels and a huge amount of oil is transported by sea (1.6 Gt in 2000). Releases of petroleum into the ocean can have a severe impact on the environment. The events that have the largest impact are large oil spills from tankers, combined carriers, and barges. Some of these oil spills come from accidents, collisions, or grounding, while others occur during operations such as loading or discharging in ports or at oil terminals. Over the last 30 years, the number of large spills (>700 tons) has significantly decreased, as can be seen in Figure 2.16. Since 1967 there have been 10 oil spills with a size greater than 100 Mt and 10 between 50 and 100 Mt. The five largest oil spills since 1967 are listed in Table 2.6.

TABLE 2.6. Five Largest Oil Spills Since 1967

Year	Name of Ship	Location	Spill Size (Mt)
1979	*Atlantic Empress*	Off Tobago, West Indies	287
1991	*ABT Summer*	700 nautical miles off Angola	260
1983	*Castillo de Bellver*	Saldanha Bay, South Africa	252
1978	*Amoco Cadiz*	Brittany, France	223
1991	*Haven*	Genoa, Italy	144

Source: Data from The International Tanker Owners Pollution Federation Limited (ITOPF), www.itopf.com.

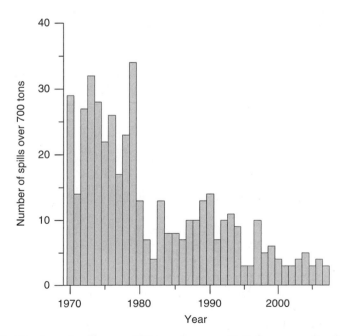

Figure 2.16. Number of spills over 700 tons by year. Data from www.itopf.com.

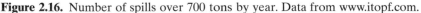

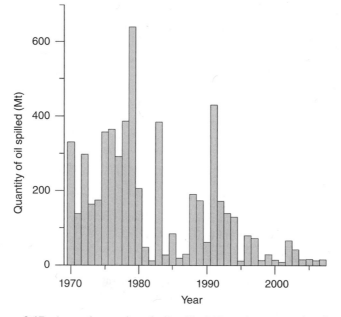

Figure 2.17. Annual quantity of oil spilled. Data from www.itopf.com.

Although serious accidents do occur, 91% of operational oil spills involve a release of less than 7 tons of oil and the quantity of oil accidentally released into the ocean has strongly decreased in the last few years (see Figure 2.17). Pollution due to spills leads to only about 5% of oil pollution in the oceans. Most of the pollution comes from used engine oil released into the waterways and from routine ship maintenance. Illegal releases by ships seeking to avoid costs associated with disposal operations in ports are common.

The use of natural gas may also lead to serious accidents due to explosions. For example, there were two such accidents in the winter of 2003–2004. The first occurred in China in a drilling well. Methane and H_2S were released into the atmosphere: Dozens of people died and hundreds were seriously injured. The second occurred in Algeria (in Skida). An explosion occurred in an LNG terminal killing 27 people and injuring many more.

2.11. UNCONVENTIONAL OIL AND GAS RESOURCES

The world is used to cheap "conventional" oil, but this resource is progressively declining. By the end of this century conventional oil will not be available in the same quantities as today. Nevertheless there are other nonconventional sources of oil and oil may be synthesized from other materials containing carbon and hydrogen. So far these alternatives are not economically competitive, but the rise of the price of conventional oil will make some of them much more interesting in the future.

2.11.1. Oil Shale

Oil shale is a fine-grained sedimentary rock containing a large amount of kerogen. (Actually the name is misleading since oil shale does not contain oil and it is usually not shale.) Oil shale can be considered as an immature oil and gas deposit because it did not go through the "oil window" of heat to produce crude oil naturally. If oil shale is heated to sufficiently high temperatures, nonconventional oil and shale gas can be produced.

So far relatively little exploration has been undertaken to identify new oil shale locations and determine the amount of the reserves. In 2005 global reserves of oil shale were estimated to be about 400 billion tons, which corresponds to around 3 billion barrels of recoverable oil. Deposits exist in, for example, the United States, Russia, Brazil, and China. The United States has a little more than 60% of these reserves and three countries—the United States, Russia, and Brazil—have 86% of the recoverable reserves. The largest known deposit of oil shale is in the western United States in the Green River Formation in Wyoming, Utah, and Colorado. It is estimated that there is a deposit of about 1.8 trillion barrels. Although there are large uncertainties, the amount of oil shale reserves in the world is potentially much larger than that of conventional oil. More than 30 countries have oil shale deposits which may be economically recoverable in the near future. In most cases it is currently not economically competitive to exploit this resource. It is usually assumed that interesting oil shale deposits should provide more than 40 liters of oil shale per ton of rock. Lowest production costs correspond to about $60 per barrel of oil and are starting to be economically interesting.

By the seventeenth century oil shale had already been exploited in several European countries. In Sweden, alum shale was used to extract potassium aluminum sulfate for tanning leather and fixing color. Later, in the 1800s, alum shale was used to produce hydrocarbons. The production was halted only in 1966 because it was no longer competitive with crude oil. Also in Europe, a French oil shale deposit in Autun was exploited as early as 1839. Between 1980 and 1991, an oil shale deposit in Colorado (United States) produced 4.5 million barrels of oil. The yield was on the average 0.8 barrel of oil per ton of rock.

Oil shale contains other interesting commodities such as uranium, vanadium, and zinc. Processing also leads to other useful by-products, specialty carbon fibers and carbon black, for example. Recovering these decreases the net cost of exploiting an oil shale deposit.

There are two main technologies for extraction of oil shale deposits. In the first the rocks are fractured in place and heated to recover gases and liquids. In the second the oil shale is mined and transported to another location where it is heated to 450 °C and hydrogen is added. Both processes need large quantities of water and also large quantities of energy, which often makes the exploitation of oil shale uncompetitive at present. A sustained high price of oil is needed in order to make this resource competitive on a large scale.

In 1999, about 350,000 tons of oil were produced from oil shale in Brazil, Estonia, and Australia. At this date the proved amount of oil shale in place was equal to about 3.5 trillion tons, from which it is expected that between 75 and 95 billion tons of synthetic oil can be derived. The average efficiency of oil recovery from shale varies within broad limits: from around 10 kg of oil per ton in South Africa to 126 kg of oil per ton in the Ukraine.

Today, there are still a few thermal power plants using oil shale in Estonia (3 MW), Israel (12.5 MW), China (12 MW), and Germany (10 MW). Other countries have phased out their oil shale power plants because they have become uneconomical.

2.11.2. Tar Sands

Tar sands, also called bituminous sands, oil sands, or extra heavy oil, are a mixture of sand or clay, water, and extremely heavy crude oil. Over time, the lighter fraction of the oil escaped and the remaining fraction has been partly degraded by bacteria, leaving a very heavy hydrocarbon product. Viewed in terms of convenience of exploitation tar sands may be considered to be overly mature in the same sense that oil shale deposits may be viewed as immature.

Tar sands are found in many locations, notably in Canada, Venezuela, the United States, Russia, and some Middle Eastern countries. The largest deposits are found in Canada, which has 3.5 times more tar sand resources than Russia and 62 times more than the United States, and in Venezuela. In Canada, the deposits are in Athabasca. In Venezuela, they are in the Orinoco, but they are buried more deeply, are less viscous than those of Athabasca, and are more difficult to exploit. Assuming a 10% recovery factor and not including those of Venezuela, the accessible resources are estimated to be of the order of 350 billion barrels of oil. Canada exploits tar sands at a significant level in the region of Alberta. Tar sands account for about 40% of Canada's oil production. Some part of this production is exported to the United States. With the increasing price of crude oil, tar sands become more and more economically competitive. The extraction techniques for the tar sands are different depending upon their nature. For example, Canadian tar sands are wetted with water in the extraction process while tar sands in the United States are hydrocarbon wetted.

If a tar sand deposit is close to the surface, it can be exploited by open-pit mining techniques. After transportation to the extraction plant, bitumen is separated from the sand using hot water. Roughly three-quarters of the bitumen contained in the tar sand is recovered. About 1 barrel of oil may be obtained from about 2 tons of tar sand.

If the tar sands are not near the surface, in situ techniques can be used to extract bitumen from tar sands. Both steam injection and solvent injection are employed, although steam injection is favored. Oxygen is also injected to burn part of the resource and provides heat. As with oil shale, these techniques require large amounts of water and energy. Consequently they are less economically interesting than conventional crude oil. However, as this latter resource becomes scarcer, tar sands will become fully competitive.

2.11.3. Coal Bed Methane

Coal bed methane is methane found in coal seams. It is formed by either biological processes (microbial action) or thermal processes occurring at large depths where the temperature is high. Not all coal seams contain coal bed methane. Coal bed methane is held in the coal by the surrounding water pressure. Because of that, there are coal seam aquifers. Coal bed methane travels with groundwater in coal seams and is extracted when water is pumped from the seam in order to decrease the pressure and liberate the methane. The pumping of large amounts of water may have an environmental impact on streams, springs, and wells. However, pumping and using coal bed methane generate fewer pollutants than similar operations with oil or coal. Coal bed methane is used in the same way as traditional natural gas. It meets 7% of the total natural gas demand in the United States. There are resources in other regions: Wyoming, Montana, Colorado, and Utah. It is estimated that the Rocky Mountain region has a deposit of about 0.9–1.6 trillion cubic meters of coal bed methane.

In principle it is practical to exploit coal bed methane if one can recover at least 1.5–2 m^3 of gas per ton of coal. To be economically competitive, one should have a coal seam thicker than about 6 m. The largest reserves of coal bed methane are found in Canada, Russia, and China. Estimates are that the global coal bed methane reserves are between 110 and 200 Tm^3. Estimates made on the same basis indicate reserves in Canada of 17–92 Tm^3, Russia 17–80 Tm^3, China 30–35 Tm^3, Australia 8–14 Tm^3, and the United States 4–11 Tm^3 (data from the University of Montana).

2.11.4. Methane Hydrates

Methane hydrate is a white crystalline solid. Also called methane clathrate or methane ice it is a material in which methane molecules are trapped inside a crystalline cage structure (hence the name "clathrate") similar to ice. On the average methane hydrates contain about 13% methane, corresponding to one molecule of methane for a bit less than six molecules of water. The average density is $0.9\,g/cm^3$. Methane hydrate is stable at low temperatures and high pressures. It can be found in permafrost soils, regions where the ground is permanently frozen, and oceanic sediments at depths greater than 300 m where the pressure is large. Methane hydrates give stability to sediments that might otherwise collapse. Most of the carbon trapped in biomass is contained in methane hydrates and fossil fuels. Under normal conditions of temperature and pressure methane hydrates melt and the methane that is released can easily be ignited with a lighted match.

There are many estimates of the quantities of methane hydrate trapped inside the permafrost or in ocean sediments, but these estimates have very great uncertainties. Globally, it is estimated that there is between 0.5×10^{12} and 2.5×10^{12} tons (between 10^{15} and $5 \times 10^{15}\,m^3$) of methane hydrates. These resources are larger than those of natural gas but smaller than all known other fossil fuel reserves.

Harnessing methane from methane hydrates is not a simple task. Only a small part of the deposit is expected to be recoverable and currently no accepted industrial method exists for that recovery. So far methane hydrates have proved to be sources of difficulties during extraction of offshore crude oil, sometimes blocking drilling tubes. Several techniques are under development to monitor the release of methane during drilling operations: increasing the temperature, decreasing the pressure, injecting methanol to shift the dissociation boundary in the methane hydrate phase diagram, and injecting CO_2, which releases CH_4 while stabilizing the remaining hydrate.

Scientists have found evidence that in the past natural releases of methane due to the decomposition of methane hydrates have occurred with dramatic consequences on the earth's climate. Such an event occurred offshore of Norway 8100 years ago. An important current environmental question is whether human activities could trigger a major decomposition of methane hydrates and a release of large quantities of methane into the atmosphere, amplifying the greenhouse effect in an irreversible manner.

Methane has a long dwell time in the atmosphere (about a decade) and is normally oxidized into water and carbon dioxide. However, there have been reducing periods where methane was not destroyed as rapidly. This was the case during the anoxic ocean events that occurred in the Jurassic and Cretaceous periods, for example. The climate was very warm and high concentrations of carbon dioxide in the atmosphere produced a super–greenhouse effect. These periods were probably characterized by an up-welling of ocean water rich in hydrogen sulfide, a toxic compound that may have caused the extinctions of many plants and animals. On the positive side, these conditions were however especially good for oil and gas formation, and most of the deposits of these fuels were probably produced during these anoxic periods.

> The dissociation of methane hydrate into methane and water is an endothermic process requiring between 300 and 500 kJ of heat per kilogram of methane hydrate. About 100 g of methane can be extracted from 1 kg of methane hydrate. Burning 1 kg of methane releases 46,000 kJ of energy. Consequently to recover 1 kWh it costs 0.1 kWh of heat.

2.12. CONCLUSION

Oil and natural gas provide more than half of the global primary energy consumed. Oil is the most convenient form of energy storage because it has a large volume and mass energy density. Furthermore, it is a liquid that can be transported easily. Oil is essential for modern transportation where it provides about 97% of the fuel used to propel road vehicles, ships, and planes. Due to high-yield technologies, natural gas is more convenient for electricity production and provision of heat and hot water for buildings.

The burning of both these energy sources produces CO_2 and contributes to the greenhouse effect with a corresponding impact on global warming. In this respect natural gas is superior to oil because of a larger hydrogen content, but this is mitigated somewhat by releases of methane into the atmosphere associated with leakage during natural gas transport.

Oil and gas exist in finite quantities. They were formed over hundreds of millions of years but are likely to be consumed in only a few centuries. The availability of cheap oil will progressively decline. This will be followed by a similar situation for natural gas. Oil can be produced from nonconventional sources such as oil shale and tar sands as well as from other carbon-containing energy sources, natural gas, coal, and biomass. Nevertheless a significantly higher price must be expected.

The wealth of our civilization owes much to fossil fuels. Oil has played the major role and we have become acutely dependent upon conventional oil, particularly as far as transportation is concerned. That dependence will have to change in the near future. The world will have to adopt different ways to produce and use energy.

CHAPTER 3

Coal: Fossil Fuel of the Future

The use of coal as an energy source has a long history. The Chinese have mined and exploited coal as a fuel since 10,000 B.C. They also used it for smelting of ores. This resource was also used in several parts of the world during the Bronze Age, between 2000 and 1000 B.C. In more recent times, in the eleventh century, coal became heavily used in China because significant deforestation caused by the extensive use of charcoal which is obtained from wood had resulted in fuel shortages. From then until the eighteenth century, China remained the world's largest producer and consumer of coal.

Although archaeologists recently discovered that coal was already being used as a fuel by German hunters 120,000 years ago, the extensive use of coal came much later in Europe than in China. It was probably around the twelfth century that poor people started to use coal, which could be easily found near the surface in several countries in Europe. At that time wood was becoming more and more expensive because it was being extensively used as a fuel and for building houses and ships. At the beginning, coal was used with some reluctance because people found this energy source to be dirty and to smell bad. Because the coals often contained high sulfur content, their burning produced the odors of sulfur compounds. Such odors were associated in the popular mind with witchcraft and heresy. The French Academy of Science recommended that coal not be used. In London burning coal was even forbidden. However, by the eighteenth century, because of increasing shortages of wood, Europeans made the choice to use coal rather than to let the economy decline. Thus coal played a major role in the steel industry during the first part of the eighteenth century and became the fossil fuel which allowed the industrial revolution at the end of the eighteenth century. It became the leading primary energy source in the late nineteenth century, progressively replacing renewable energies, and remained in the lead until the 1950s, when oil took first place. In 2005, coal still provided 25% of the world's primary energy supply, ranking it behind oil (35%) but ahead of natural gas (21%). Because

Our Energy Future: Resources, Alternatives, and the Environment
By Christian Ngô and Joseph B. Natowitz
Copyright © 2009 John Wiley & Sons, Inc.

reserves of coal are large compared to those of oil and natural gas, coal is an important energy source for the future.

3.1. GENESIS OF COAL

Coal is a stratified sedimentary rock composed of more than 50% carbon. Because of the means of formation coal has both organic and inorganic content. The organic content is mostly carbon but also includes smaller amounts of hydrogen, nitrogen, and oxygen. The inorganic mineral content can be 9–30% by weight. Coal may also contain sulfur and other potentially dangerous elements: for example, arsenic, beryllium, cadmium, mercury, uranium, and thorium. It can also contain a small amount of asbestos.

For coal to be formed, peat has to be buried and preserved. Peat is formed from the deposition of dead organic material. Most coal ($\approx$90% of the world reserves) has been formed from peat which decomposed in an oxygen-poor environment after being buried in swamps or peat bogs. It forms in a mire that may initially be located close to the sea (estuaries, deltas, and lagoons) or isolated from the sea. Coal formed in these conditions is called humic coal. Other kinds of coal, called sapropelic coal, formed from spore or aquatic plants in an aquatic environment. Occurring only in deposits of relatively small size, sapropelic coals constitute about 10% of the world reserves.

As in oil and gas formation, the organic matter buried in the sediments undergoes a series of complex transformations which occurs in two steps. The first step is a biochemical degradation assisted by organisms (bacteria and fungi). Tropical environments, in which high rates of precipitation compensate for high rates of evaporation, are more efficient for this process. The second step is a physicochemical decomposition, a "coalification" process occurring because of the burial environment. The combination of pressure and temperature in this environment produces a thermal cracking. Water is squeezed out and carbon dioxide is released. Later hydrogen-rich volatile compounds escape. Above a temperature of the order of 110 °C, this thermal cracking produces coal.

Factors such as climate and pH of the water available have an impact on the characteristics of coal. The greater the degree of coalification, the better the quality of the coal. The degree of coalification is specified by the "rank" of the coal, an indicator of its maturity. This is an important determiner of its quality. Peat has the lowest rank and anthracite the highest. The carbon content of coal varies between about 55% for lignite and 90% for anthracite. The typical hydrogen content in coal is about 4%, about half that of crude oil. Coal also contains volatile matter which is liberated by heating at high temperature in the absence of air. This volatile matter includes hydrocarbons, aromatic hydrocarbons, and some sulfur.

The oldest humic coals were formed during the Devonian Era, about 400 million years ago. Large deposits also formed during the Carboniferous Era, about 350 million years ago, and the Cretaceous period, about 140 million years ago. Sapropelic coals are even older. Some between 1.8 and 3 billion years old have been found in Siberia. Others, found in Australia, were formed during the Cambrian period (from 540 to 490 million years ago).

3.2. RANK OF COALS

The value of coal and its use as an energy source depend very much on its degree of maturity, that is, its degree of coalification. The "rank" of a coal was introduced as a means of specifying that. Rank is obtained by measuring the moisture content, the gross calorific value, and optical properties.

Coal is a complex heterogeneous material composed of many distinct organic subunits called macerals. Macerals are the smallest entities which can be seen using a microscope. They are a mixture of compounds with specific chemical and physical properties. The rank of a coal is determined by the average characteristics of the macerals which constitute that coal.

Macerals can be classified into three groups:

- The vitrinite group originates from coalification of woody plant material such as trunks, roots, branches, stems, and so on. This is the most abundant group, constituting 50–90% of North American coals. The macerals are derived from the cell wall material of plants containing the polymers cellulose and lignin. For bituminous and black coals, the vitritine reflectance is used as a parameter to determine the rank.
- The liptinite group is derived from more decay resistant parts of plants: spores, resinous and waxy parts, cuticles, and so on. These coals comprise 5–15% of North American coals. These macerals are abundant in the Appalachian region.
- The inertinite group contains material which has been altered and degraded prior to coalification. Inertinite macerals are found in fossil charcoal and fusinite, for example. Between 5 and 40% of most North American coal is inertinite material. This group is found in Appalachian coals and is also common (50–70%) in some western Canadian coals.

Several different systems have been introduced around the world to classify the different types of coals. Those of the International Organization for Standardization (ISO), the American Society for Testing and Materials (ASTM), the British Standards Institution (BSI), and the Australian Standard (AS) are the most frequently used classifications. Here we shall also use that of the International Energy Agency (IEA).

Table 3.1 shows some properties of families of coal arranged by increasing rank.

TABLE 3.1. Percentage of Carbon, Gross Calorific Value, Percentage of Inherent Moisture, and Percentage of Vitrinite Reflectance for Different Types of Coal

Rank	Percent Carbon (Dry and Ash Free)	Gross Calorific Energy (MJ/kg)	Inherent Moisture	Percent Vitrinite Reflectance (Random Measurement)
Peat	≈60	≈15	≈75	0.20
Brown coal	≈71	≈23	≈30	0.40
Subbituminous coal	≈80	≈33.5	≈5	0.60
High-volatile bituminous coal	≈86	≈35.6	≈3	0.97
Medium-volatile bituminous coal	≈90	≈36	<1	1.47
Low-volatile bituminous coal	≈91	≈36.4	<1	1.85
Semianthracite	≈92	≈36	≈1	2.65
Anthracite	≈95	≈35.2	≈2	6.55

Source: Data from C. F. K. Diessel, *Coal-bearing Depositional Systems*, Springer-Verlag, 1992; L. Gammidge, available: www.newcastle.edu.au. For comparison, the gross calorific value of dry wood varies between 14.4 and 17.4 MJ/kg, that of charcoal is about 30 MJ/kg, and that of lignite is about 16 MJ/kg.

The best coals, like anthracite, are usually called "hard" coal while those like lignite are called "soft" coal. There can be a factor of 2 in the calorific value between the best and the worst coals and a corresponding difference of cost on the market.

3.3. CLASSIFICATION OF COALS

For statistical purposes, coals are classified in different categories. The classification may vary with country.

The technical classification used by the IEA depends on the quality of the coal:

- Lignite or brown coal is a nonagglomerating coal with a gross calorific value smaller than 17.4 MJ/kg and volatile matter larger than 31% by weight. Typically lignite has a gross calorific value between about 5.5 and 14.3 MJ/kg. Oil shale that is directly burned is also included in this category.

- Subbituminous coal is a nonagglomerating coal with a gross calorific value larger than 14.3 MJ/kg but smaller than 17.4 MJ/kg and volatile matter larger than 31% by weight.

- Bituminous coal and anthracite are nonagglomerating coals having a gross calorific value larger than 17.4 MJ/kg. Bituminous coals have a gross calorific value between about 18.8 and 29.3 MJ/kg. Anthracites have high gross calorific values, of the order of 30 MJ/kg.

Subbituminous and bituminous coals account for 82% of the world reserves. Anthracites constitute only 1% of the world reserves. The apportionment among different ranks is summarized in Figure 3.1.

To simplify statistical reporting, two main categories of coal are often employed: brown coal and hard coal. Two qualities of coal are also defined: coke coal, which corresponds to a quality of coal allowing the production of coke which can be used in a blast furnace, and steam coal, suitable for use in steam boilers. Steam coal used in power plants is often subbituminous coal with a calorific value in the range 5.2–9.1 kWh/kg.

Lignite belongs to the brown coal category. Subbituminous coal either is reported in the brown coal category or is included in steam coal depending on the country. In the United States, France, or Japan, for example, subbituminous coal is classified as steam coal whereas in Germany it is put in the brown coal category. Bituminous coal and anthracite are considered to be steam coal.

With these definitions, hard coal includes coking coal and steam coal. Brown coal contains mostly lignite and, in some countries subbituminous coal. In Figure 3.2 we summarize the properties of different types of coal.

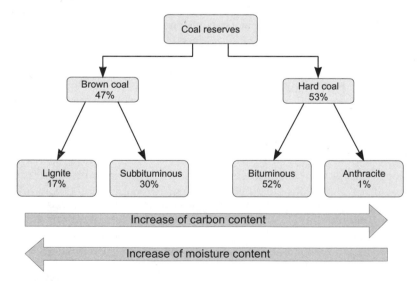

Figure 3.1. World share of coal reserves in different ranks (US classification). Data from *Panorama*, Institut Français du Pétrole, 2008, www.ifp.fr.

The estimation of the total coal resources in production or in consumption is obtained after conversion to a common energy unit, the ton of coal equivalent (tce):

$$1\,tce = 29.3\,GJ = 8141\,kWh = 27.78\,MBtu = 7,000,000\,kcal = 0.7\,toe$$

$$1\,ton\ of\ hard\ coal \approx 0.8\,tce \approx 0.56\,toe$$

$$1\,ton\ of\ lignite \approx 0.3\,tce \approx 0.21\,toe$$

Total coal is the sum of hard coal and brown coal.

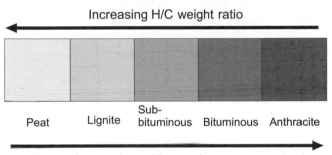

Figure 3.2. Different types of coal (peat has been included). From peat to anthracite, the energy content increases and the hydrogen-to-carbon (H/C) ratio decreases. This ratio, which is typically around 1 for peat falls below 0.5 for anthracite.

3.4. PEAT

Peat is a young organic material compared to coal. Made of plant matter which has not been completely transformed into coal, it is mixed with deposited minerals. The dry and ash-free part of peat has a calorific value between 20 and 22 MJ/kg.

The estimation of peat resources is imprecise. By convention land is called peatland if the depth of the peat is larger than 20 cm on drained land or 30 cm on undrained land. In Figure 3.3, the estimated distribution of peatland area (megahectares) in the world is indicated. One-quarter of Europe's peatland area has already been depleted.

Peat is used:

• In agriculture, for example, as a growing medium, to improve the quality of soil, as a compost ingredient

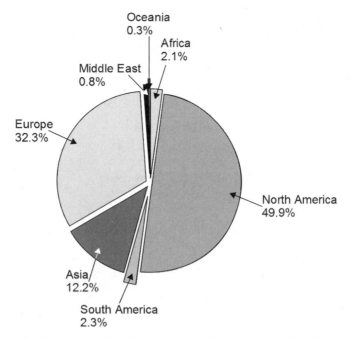

Figure 3.3. Distribution peatland. Data from World Energy Council (WEC), www. worldenergy.org.

• As an energy source for electricity and heat generation
• In organic chemistry (e.g., as a source of resins, waxes)

The main producers and consumers of peat are Ireland, Finland, Belarus, Russia, Sweden, and the Ukraine. In Figure 3.4 numbers for the production and consumption of peat are shown for different regions of the world in 2004. In that year, 13.4 Mt of fuel peat was produced and 17.3 Mt of fuel peat was used in the world. The difference came from already extracted stocks. The annual production varies because of weather conditions.

Peat may be viewed as a (slowly) renewable fuel. In that respect it is different from other fossil fuels. As far as its impact on the greenhouse effect is concerned, peat falls between fossil fuels and renewable energies.

Peat is used in three main forms:

• Sod peat—slabs of peat used for household fuel
• Milled peat—granulated peat produced on a large scale and used in power stations or to prepare briquettes
• Briquettes—small blocks of dried, highly compressed peat used for household fuel

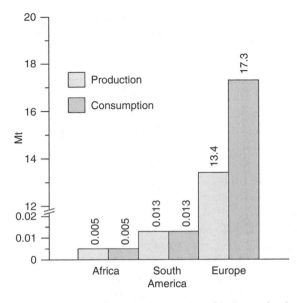

Figure 3.4. Production (light grey) and consumption (dark grey) of peat (in Mt) in different areas of the world, 2004. Note the break in the ordinate scale. Data from WEC, www.worldenergy.org.

3.5. USE OF COAL

Coal is primarily used as a fuel. More than three-quarters of the demand is for production of electrical power. It is also used as a household fuel for cooking and heating and as a feedstock in carbochemistry to produce dyes, textile fibers, fertilizers, and so on. In addition, coal is essential to the steel industry. Producing 1000 kg of steel requires 600 kg of coal. Coke coal is produced in a carbonization process in which blended coal is heated in a coke oven.

The price of coal and its use depend on its properties. Price varies according to the coal's origin. High-grade coal is more expensive than low-grade coal. In the average spot market of March 7, 2008, the price of the Central Appalachian coal (12,500 Btu/t) was $84.3/t, the Illinois Basin coal (11,800 Btu/t) cost $46.8/t, and the Powder River Basin coal (8800 Btu/t) cost $14.35/t.

In the energy sector, coal is mostly used for electricity production (7351 TWh in 2005) but also for producing heat (1196 TWh in 2005). Assuming an 80% efficiency for an oil-fired furnace and a 70% efficiency for a coal-fired furnace, heat production using coal is cheaper than that using oil by a factor of 2.5–3.

3.6. COAL RESERVES

Estimated coal reserves correspond to the amounts of coal considered to be economically extractable using current technologies. This means, in particular, that at present only coal seams located at depths up to a maximum of 1800 m and having thicknesses larger than about 35–80 cm are considered.

As with petroleum and natural gas, these resources can be classified as proven, indicated, or inferred. Proven resources include only coal deposits that are known from direct measurement and analysis. Indicated and inferred reserves are those which are extensions of known deposits, undiscovered ones in known mining areas, or estimated ones with a given probability based on geologic studies. This leads essentially to two main categories of reserves:

- Proven reserves that can easily be recovered in the future
- Estimated "additionally recoverable" reserves that are reasonably believed to be recoverable in the future

Figure 3.5 gives the proved recoverable reserves of three grades of coal—bituminous plus anthracite, subbituminous, and lignite—at the end of 2005. About half of the proved resources are good grade coal.

Some countries have larger coal reserves than others. In Figure 3.6 the 10 countries with the world's largest proved coal reserves are shown. Since the importance of coal as an energy source will increase, countries with large coal reserves can be expected to play a very important role in meeting the world's future energy needs.

The geographic distribution of the world's total recoverable reserves of coal is shown in Figure 3.7. These reserves are more widely distributed than those of oil and gas. Most of the reserves (83%) are in North America, Europe, and Asia. In contrast to petroleum and natural gas, little coal exists in the Middle East.

Figure 3.8 shows the distribution of the proved recoverable bituminous and anthracite resources among regions of the world. These comprise the top grade of coal. Asia, North America, and Europe (the Russian Federation) have 78% of these reserves. Significant resources are found in Africa (South Africa) and Oceania (Australia).

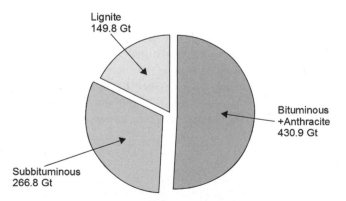

Figure 3.5. Proved recoverable reserves at end 2005. Data from WEC, www. worldenergy.org.

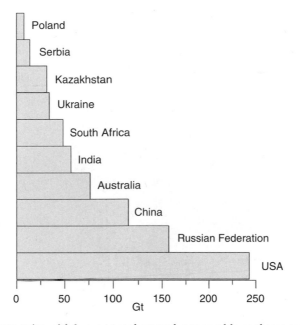

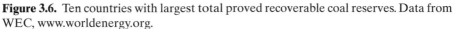

Figure 3.6. Ten countries with largest total proved recoverable coal reserves. Data from WEC, www.worldenergy.org.

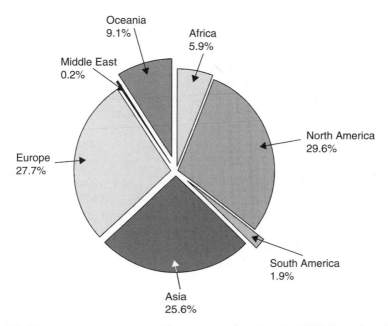

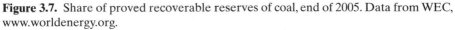

Figure 3.7. Share of proved recoverable reserves of coal, end of 2005. Data from WEC, www.worldenergy.org.

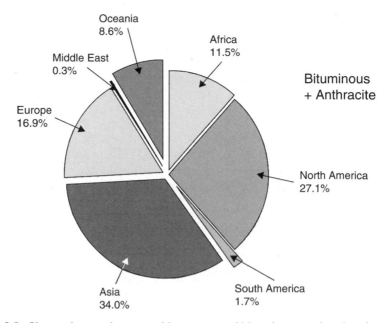

Figure 3.8. Share of proved recoverable reserves of bituminous and anthracite, end of 2005. Data from WEC, www.worldenergy.org.

Subbituminous coal is coal of a lower grade. The distribution of its proved recoverable resources is shown in Figure 3.9. Europe and North America are the major areas where subbituminous coal can be found (82%).

Lignite is a poor-grade coal that is also one of the most extensively polluting fuels. Practically all the proved reserves are found in Europe, Oceania, Asia, and North America (Figure 3.10). There are only very small amounts in Africa (3 Mt in Central Africa Republic) and South America (24 Mt in Ecuador).

A conventional 1000-MW coal-fired power plant without any depollution system and operating 6000 h a year produces 1500 tons of particulates, 5 million tons of CO_2, 40,000 tons of SO_2, and 20,000 tons of NO_x. Radioactive nuclei are also emitted as well as some toxic mineral elements. It has been estimated that, in 1982, coal power plants in the United States liberated around 800 tons of uranium and 1970 tons of thorium. At the world level, about 3640 tons of uranium and 8960 tons of thorium were released. This has however no significant impact on human health and is small compared to natural radioactivity. The average radioactivity of coal is about 156,000 Bq/t.

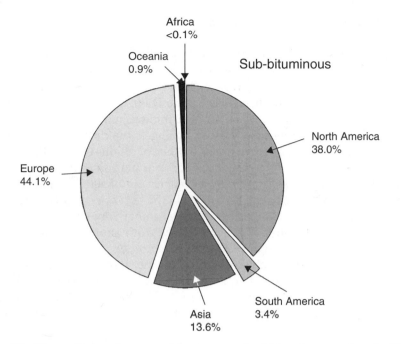

Figure 3.9. Share of proved recoverable reserves of subbituminous coal, end of 2005. Data from WEC, www.worldenergy.org.

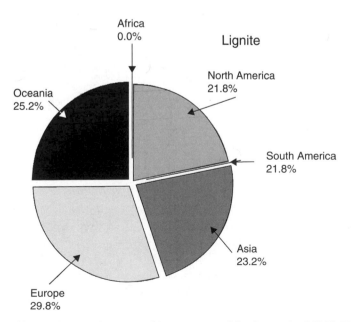

Figure 3.10. Share of proved recoverable reserves of lignite, end of 2005. Data from WEC, www.worldenergy.org.

3.7. PRODUCTION AND CONSUMPTION

Coal mining depends on the nature of the coal deposit (depth and quality of the seams) and the technology used to harvest the coal. The depth of coal seams depends very much on region (less than 100 m in the Appalachians, 400 m in France, and more than 700 m in the Ruhr in Germany). Coal-mining processes are usually different in surface mining and underground mining. Investment and operating costs vary according to the difficulty of extraction.

The productivity of a miner depends on the conditions under which the coal is extracted. It is about 300 tons of coal per year per miner in China (about 1 ton of coal per day), 4000 tons in South Africa, and 8500 tons in the United States. It is possible to reach higher values in the best open pits. In some places in the United States 15,000 tons per year per miner is possible.

The cost of extraction varies significantly; from about $10/t for the most competitive coal mines to about $200/t for the most expensive ones. More than half of the proved recoverable resources can be harnessed at a price lower than $100/t. Since 1 t of oil is equivalent to 1.83 t of raw coal, a $100/t cost of extraction is the energy equivalent of a $25 barrel of oil. Coal is therefore very competitive economically when compared to oil, but its price is also increasing.

Figure 3.11 shows the coal production and consumption figures for different regions of the world. Production and consumption are well balanced, reflecting

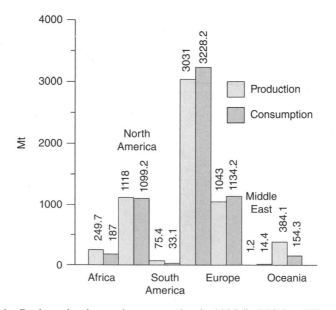

Figure 3.11. Coal production and consumption in 2005 (in Mt) for different regions of the world. Data from WEC, www.worldenergy.org.

the fact that trade in coal is mostly on a regional scale. While the international trade in coal is expanding, it still involves only a small part of the world's total coal production (of the order of 15%). Coal is an energy source which is mainly used for domestic markets. The top-10 coal-exporting nations are shown in Figure 3.12 and the top-10 importers are presented in Figure 3.13. China, which was self-sufficient until the recent past, is now importing a small part of the coal that it uses.

The average increase of energy consumption in China has been 12.5% per year since 2002. This is much larger than in the preceding decade (3.5% per year between 1992 and 2002). China is now the world's second largest consumer of primary energy, 1.7 Gtoe in 2006. However, the energy use per capita is still low, 1.1 toe in 2003.

Coal is the major energy source used in China and the large coal deposits existing in China are critical for its energy independence. Between 2000 and 2006 coal consumption in China doubled. China uses more coal than the United States, Europe, and Japan combined (almost 1.2 Gtoe in 2006). In 2006, 2.38 Gt of coal (almost 40% of the world's production) was mined in China. China is no longer self-sufficient in coal and has started to import slight quantities (2%). As a consequence the price of coal on the international market rose to $130/ton CIF at the beginning of 2007. However it has dropped to ~$80/ton in 2009.

Transportation of coal is expensive and the price of this resource depends very much on the distance between the place where it is produced and the place where it is used. Historically, industries using coal as an energy source or for other purposes were located close to the production mines. Presently this is not necessarily the case. Sometimes coal has to be transported over large distances within a country. This may co-opt a large part of a country's transportation capabilities. This is true in China where the transport capacity is presently saturated. Investment and time will be required to build adequate rail infrastructure from the northwest part of the country, where coal is produced, to the eastern and southern parts, where it is primarily consumed. The same is true in India where coal is mostly mined in the eastern part of the country and consumed in the northern or south, western part. Because of frequent rail transport problems, countries are often obliged to import coal to satisfy demands in particular regions.

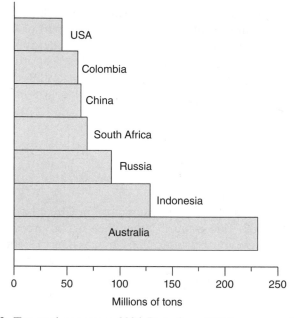

Figure 3.12. Top coal exporters, 2006. Data from WEC, www.worldenergy.org.

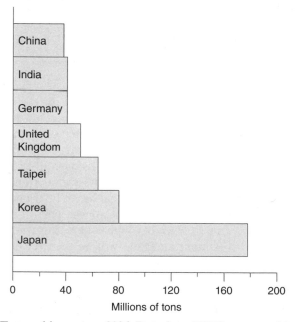

Figure 3.13. Top coal importers, 2006. Data from WEC, www.worldenergy.org.

India ranks fourth in the world for coal resources: 252.3 Gt at the beginning of 2007. Of these 95.9 Gt are proved resources, 119.8 Gt are indicated resources, and 37.7 Gt are inferred resources. The proved resources (essentially bituminous coal) represent more than 10% of the world reserves. In India as in China, coal is the dominant source used to meet primary energy demands. Power plants use 75% of the domestic production of coal and 69% of Indian electricity comes from this resource. India, number 3 in global coal consumption, used 238 Mtoe in 2006. The primary energy consumption per capita was only 0.5 toe in 2003 but increased 6.7% per year between 2003 and 2006. India is no longer self-sufficient in coal and needs to import (more than 60 Mt/year). Coking coal, in particular, is needed and imported from Australia.

The higher the calorific value of a coal, the higher is the price. The price at final destination is different from the price at the extraction site. Two types of prices are usually used: price FOB and price CIF. The price FOB (freight on board) is the price at the boarding port. The seller pays for the transportation to the port of shipment and the loading cost. The buyer pays freight, insurance, unloading, and transportation costs to the final destination. The price CIF is the price at the arrival port. The selling price includes the original price of the coal, the freight or transport cost, and the insurance. For steam coal mined in the United States, for example, the CIF cost is about 20% larger than the FOB cost.

Even though the price of coal has doubled between 2003 and 2007, coal remains the cheapest fossil fuel. As this is written, coal is almost five times cheaper than oil and three times cheaper than natural gas for the same energy content.

3.8. ELECTRICITY PRODUCTION

Worldwide coal makes a major contribution to electricity generation. This is illustrated for different regions and countries in Table 3.2 and Figure 3.14 using data from 2005 and 2006. The overall global contribution of coal to power generation was 40.2%, up from 38% in 2003. This supported a global power production of 16,600 TWh. China produces more than three-quarters of its electricity from coal. In contrast, in the Middle East electricity is mainly produced using oil and gas (91%) and coal accounts for only 5.5% of electricity

TABLE 3.2. Contribution of Coal to Production of Electricity in Selected Regions and Countries, 2005

Region or Country	Total Electricity Production (TWh)	Electricity from Coal (TWh)	Contribution of Coal (%)
World	18,307	7,351	40.2
Total OECD	10,459	3,947	37.7
Total non-OECD	7,848	3,404	43.4
OCDE North America	5,149	2,293	44.5
European Union	3,311	1,001	30.2
China	2,536	1,996	78.7
Africa	566	251	44.4

Note: OECD, Organisation for Economic Co-operation and Development.

Source: Data from IEA, www.iea.org.

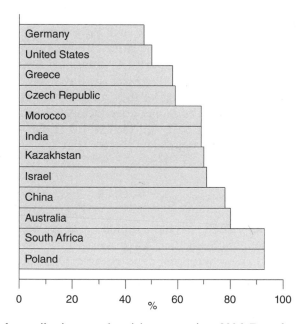

Figure 3.14. Coal contributions to electricity generation, 2006. Data from IEA, www.iea.org.

generation. In the European Union, the contribution of coal is a bit smaller than in other regions because around 30% of electricity is provided by nuclear power.

Unfortunately, coal is the worst fossil fuel as far as pollution is concerned. Replacing coal-fired by gas-fired technology would decrease the amount of

CO_2 emitted into the atmosphere, but in the long run natural gas supplies will diminish. Coal will become even more important in the future because of the large available coal reserves which exist. Learning to use it in the cleanest way possible should be given high priority.

3.9. COAL COMBUSTION FOR POWER GENERATION

Pulverized coal combustion is the major technology used to produce electricity. The advantage of this technology is the ready availability of a wide range of coals that can be used as fuel. Coal powder is burned at temperatures of 1500–1700 °C if bituminous coal is used and 1300–1600 °C with low-rank coals. Several configurations of burners in the combustion chamber can be used. The heat generated in the burner is transferred through a heat exchanger to produce superheated steam. This high-temperature and high-pressure steam is expanded in a steam turbine coupled to a power generator. The solid residues of the combustion consist of 80–90% fine fly ashes. Most of them flow with the flue gases as particulate matter which can be captured.

More than 90% of coal-fired power plants use pulverized coal combustion technology. It has been used for more than 60 years. The average investment cost is about $1500/$kW_e$ but this varies by country. In China, for example, the investment cost is $800–$1000/kW_e. The most sophisticated coal-fired power plants, including sophisticated depolluting systems, can reach costs of $1800/$kW_e$.

Pulverized coal combustion plants can have powers between 50 and 1300 MW_e, but most new ones have a power around 300 MW_e. Plants have been continuously improved to increase their efficiencies and decrease the cost of electricity. The thermal yield of older plants was of the order of 30% or less. Now, with large combustion units, a thermal yield of the order of 35–36% can be reached routinely and pulverized combustion power plants can have a yield up to 41% when they are operated in optimal conditions with high-quality coal.

3.9.1. Advanced Pulverized Coal Combustion

Advanced pulverized coal combustion employs so-called supercritical or ultra-supercritical conditions to run the power plants. This development has been possible due to the availability of new materials able to perform under very demanding conditions. A pulverized coal combustion power plant is said to be supercritical if the pressure is above 250 bars and the temperature is larger than 565 °C. Efficiencies of 43–47% can be obtained. It is called ultra-supercritical if the pressure is larger than 300 bars and the temperature above 585 °C. Between 1985 and 1999, 19.4-GW_e-capacity supercritical coal-fired power plant was installed in the OECD countries. The capital required to build an advanced pulverized coal combustion power plant is higher than that of a conventional one. However, since the efficiency is increased, the operating cost is decreased by more than 15%.

3.9.2. Fluidized-Bed Combustion at Atmospheric Pressure

Fluidized-bed combustion at atmospheric pressure is useful to burn low-grade coal. Particles around 3 mm in size are fed into the combustion chamber. Two types of beds are used: circulating beds and bubbling beds. They operate at temperatures ranging from 850 to 950 °C, which leads to less NO_X emissions than pulverized coal combustion.

Atmospheric pressure circulating fluidized-bed combustion units typically are 250–300 MW$_e$ in size, but higher powers, up to 600 MW$_e$, are also possible. The thermal efficiency is about 3–4% less than that realized in pulverized coal combustion power plants. Atmospheric pressure bubbling fluidized combustion units have a smaller size, up to 25 MWe, but larger powers are also possible. Bubbling beds use a low fluidizing velocity compared to circulating beds. The overall thermal efficiency is around 30%.

3.9.3. Pressurized Fluidized-Bed Combustion

The efficiency of fluidized-bed combustion units can be improved by pressurized fluidized-bed combustion, which is a particularly useful technology to burn high-ash coal or low-grade coal. In pressurized fluidized-bed combustion, air is injected at a pressure between 12 and 25 bars and combustion takes place between 800 and 900 °C. The thermal efficiency of pressurized fluidized-bed combustion power plants is over 40%, similar to that of a supercritical pulverized coal combustion power plant. It will probably be possible to achieve 45% in the future. Pressurized fluidized-bed combustion power plants built as demonstration units are small (about 80 MW$_e$), but two larger units (250 and 360 MW$_e$) have started operation in Japan. Their larger size was dictated by the power needs of the gas turbines.

3.10. COMBINED HEAT AND POWER GENERATION

In combined heat and power generation, or "cogeneration," electricity and heat are produced at the same time. As a consequence the total efficiency of the unit is increased since the total delivered energy (power plus heat) for a given amount of fuel is larger. In order to carry out cogeneration efficiently, the turbine must be specially designed and flexible enough to produce electricity alone or both electricity and heat. Combined heat and power generation is interesting to provide heat for industrial processes or in cold regions such as Scandinavia or Eastern Europe where winters are long.

Despite its potential, cogeneration is used in less than 10% of the power-generating facilities in many countries and is sometimes much rarer. Currently cogeneration is used more frequently with natural gas than with coal. An interesting example of a large cogeneration unit is the one in Avedore, Denmark. It was conceived as a multifuel plant capable of burning coal, natural gas, or biomass as a fuel. In 1996, the government of Denmark banned the use of coal. Avedore switched entirely to gas and biomass and Now uses renewable biomass as its main fuel. Avedore reaches a global efficiency (heat and power) of 55% at 200 MW$_e$ in combined heat and power mode or 42% at 300 MW$_e$ if only electricity is produced.

3.11. INTEGRATED GASIFICATION COMBINED-CYCLE POWER PLANTS

Integrated gasification combined cycle (IGCC) is an emerging technology able to produce electricity with a high efficiency and lower emission of pollutants than conventional coal-fired power plants. The IGCC plants have high capital costs, about 30% higher than conventional coal-fired plants, and the operating costs are about 15% higher. They are presently at the demonstration stage. Several IGCC power plants, typically around 250 MW$_e$ in size, are operating in Europe and the United States. New projects are also under construction.

An IGCC plant consists of a gasification unit and a gas-fired combined cycle unit. Syngas (synthetic gas) is produced in the gasification unit. It is mainly a mixture of hydrogen and carbon monoxide. The syngas is used to power a combined cycle similar to the one described in the preceding chapter for natural gas. The exhaust gases of the turbine driven by the syngas are heat exchanged with water or steam to produce superheated steam to drive a steam turbine. The principle of an IGCC power plant is shown schematically in Figure 3.15.

Gasifiers are large pressure vessels operating at between 20 and 80 bars and at temperatures between 1300 and 1600 °C. For example, the gasifier of the Tampa, Florida, IGCC power plant is 40 m high and weighs 680 tons. Several technologies can be used for gasifiers: fixed beds (with lumps of coal), fluidized beds (with particles between 3 and 6 mm in size), and entrained flow (with pulverized coal). Between 60 and 70% of the power comes from the syngas gas turbine. Overall efficiencies over 40% can be obtained for electricity generation and the goal is to reach about 50% in the future. The 250-MW$_e$ Tampa plant and the 262-MW$_e$ Wabash River plant in Indiana have efficiencies of 41%. The 305-MW$_e$ IGCC power plant of Puertollano (Spain) reaches an efficiency of 45%.

The great advantages of an IGCC plant are that it can burn many fuels, including biomass, and the level of pollution is far below other coal-fired plants. They are also ready for CO_2 capture and sequestration when this process reaches the industrial stage. Even with these advantages, IGCC is not the preferred technology for new coal-fired power plants. If taxes for

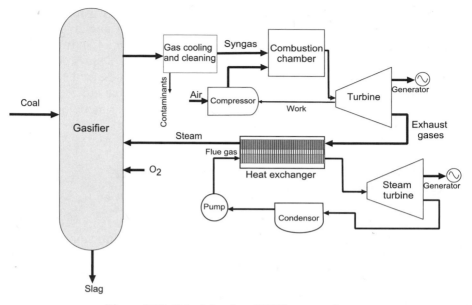

Figure 3.15. Principle of an IGCC power plant.

air pollution and CO_2 emission are imposed, this technology will become competitive.

3.12. COAL-TO-LIQUID TECHNOLOGIES

While coal reserves are greater than those of oil, liquid fuels are much more convenient to use than coal. Therefore, developing coal-to-liquid conversion techniques to produce synthetic fuels from coal is a very important technological goal.

There are two common approaches to the liquefaction of coal:

- A direct process using techniques originally developed in Germany between 1910 and 1927.
- The Fischer–Tropsch synthesis, an indirect process developed by two German chemists, Hans Fischer and Franz Tropsch, who started from a patent taken in 1913 by the BASF Company. Fischer and Tropsch produced the first motor fuel from coal in 1922, and this process was used in Germany during World War II to produce motor fuel from coal.

3.13. DIRECT COAL LIQUEFACTION

Direct coal liquefaction (CTL) is a process used with bituminous and subbituminous coal. The simplified principle is the following: Coal is reduced to a powder which is mixed with vacuum gasoil (a product obtained from vacuum distillation of oil) before undergoing hydrocracking in a reactor. Final products are separated and undergo a partial hydrotreatment. Vacuum diesel is sent back to the reactor while the synthesized diesel, mixed with naphtha, undergoes a hydrotreatment or hydrocracking to meet the specifications for motor fuel. The hydrogen necessary to the process is provided by coal gasification or by steam reforming of natural gas depending on the available resources. The hydrogen-producing units are usually located next to the coal liquefaction unit. With direct CTL technology, the typical products are 10% LPG, 20–30% naphtha, and 60–70% diesel oil.

The story of coal-to-liquid technology is an old one. It was first used on an industrial scale by Germany during World War II. Outputs of such processes are typically specified in volumes per stream day, a measurement used to denote the rate of oil or oil product flow while a fluid-processing unit is in continuous operation. Carrying out the direct liquefaction of Ruhr Valley coal the Germans produced about 120,000 barrels per stream day. This was almost sufficient to provide fuel to the entire Luftwaffe.

In 1955 the Sasol Company installed a 190,000-barrel-per-stream-day-capacity plant in South Africa in response to the international embargo imposed on the apartheid regime. The indirect coal liquefaction method was used with domestic coal to produce synthetic oil.

The CTL (coal-to-liquid), GTL (gas-to-liquid), and BTL (biomass-to-liquid) technologies, all based on the Fischer–Tropsch technology, are becoming ever more attractive economically due to large increases in the price of a barrel of oil.

3.14. INDIRECT COAL LIQUEFACTION

The principle of the Fischer–Tropsch coal-to-liquid method is shown in Figure 3.16.

The first step is to convert coal to syngas using pure oxygen. The syngas, which has a H_2/CO in the range 0.5–0.8 is transformed in a shift reactor to reach an H_2/CO ratio of 2, which is more suitable for the Fischer–Tropsch reaction. Once the impurities from the syngas are removed, the Fischer–Tropsch reaction is performed followed by an isomerization stage to get properties suitable for the specifications of the fuels.

Diesel fuel obtained this way is very pure. It contains only paraffinic molecules and has a high *cetane number*. The cetane number, a measure of the

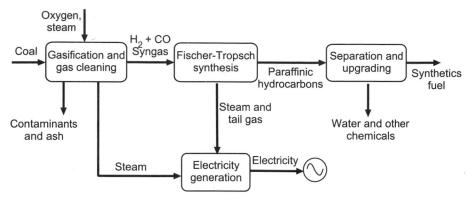

Figure 3.16. Principle of Fischer–Tropsch synthesis.

time delay between the start of injection and the start of combustion of diesel fuel, quantifies the combustion quality of the fuel. A diesel engine runs well when the cetane number of the fuel is in the range 40–55. In North America the cetane number is between 40 and 45 for regular diesel fuel and between 45 and 50 for premium fuel. In Europe the cetane number has been required to be larger than 51 since the year 2000.

The great advantage of indirect CTL is that it can be used for coals such as lignite which are not as good as bituminous or subbituminous coals. It can also be used with biomass. With the indirect CTL technology, the typical products are 20–30% naphtha, 25–35% kerosene, 35–45% diesel, and 0–5% fuel oil.

3.15. DIRECT OR INDIRECT CTL TECHNOLOGY?

If hydrogen is produced locally from coal, the amount of naphtha and diesel produced per ton of coal (moisture and ash free) is about: 3.5 bbl/ton for the direct CTL technology and 2.5 bbl/ton for the indirect CTL technology. Today there is one industrial coal liquefaction facility based on the indirect CTL technology. It is operated by Sasol in South Africa. In contrast, the Shenhua project in China, started in 2007, is based on direct CTL technology. The main difference between the two technologies is the quality of the diesel produced. Direct CTL technology produces a diesel fuel with a cetane number below the European specifications. An additional step is required to increase that number, but this increases investment and operating costs and decreases the yield. In contrast, indirect CTL technology gives a quality of diesel fuel which is higher than the specifications require. Blending outputs from the two processes would appear to be a practical move and projects of this type are planned.

The cost of the liquid fuel is correlated to the cost of coal. At a price of $20/t for coal, the CTL technology is economically competitive if the price of oil is larger than $70 a barrel. One can expect that more CTL facilities will be built in countries where coal is abundant and there are large demands for liquid motor fuel. The technology can also be used to obtain liquid fuel from "stranded gas." A *stranded gas* reserve is a natural gas field that has been verified but remains unused for economic reasons. About 40% of the world's gas reserves are classified as stranded.

In the 1920s to the 1950s the driving forces for developing CTL technology were military needs and economic security. In the 1970s to the 1980s, fears of an oil shortage and specific desires for energy independence, as in South Africa, spurred the development. In the present period, CTL technology is viewed as an important resource to meet the world's increasing demands for motor fuels. In the United States the military is currently taking a leadership role in promoting the development of liquefaction technologies for this purpose.

3.16. CARBON CAPTURE AND SEQUESTRATION

Fossil fuels are essential to our economic development and it will take a long time before we can significantly decrease our dependence on these fuels and meet a large part of our energy demands with other sources. Burning fossil fuels leads to carbon dioxide emissions, which increase the greenhouse effect. This may have serious consequences on the global climate of the earth. There are several ways to address this issue. An extreme one would be to stop all use of fossil fuels, but this is impractical because our civilization is so dependent on their use. A second extreme response would be to avoid all release of carbon dioxide while using fossil fuels. This means capturing and sequestering all of the carbon dioxide. This can only be possible in centralized plants because it is a complex process, requiring energy and large facilities. Sequestering CO_2 emitted by the exhaust pipe of a car would be very difficult to do in a practical way.

Practical control of CO_2 emissions requires parallel approaches in which:

- Fossil fuel consumption is reduced. This can be done by using carbon-free primary energy sources, renewable energies, or nuclear energy to replace fossil fuels where possible and by developing sensible technologies for fossil fuel consumption.
- Carbon dioxide emissions in large facilities that burn fossil fuels are captured and sequestered. This is discussed further below. The discussion is applicable to coal-fired power plants, which play a major role in electricity production, but also to natural gas- or oil-powered plants. It is also applicable to other industrial operations, cement production, for example.

Since 1750, about 1150 billion tons of CO_2 has been released into the atmosphere as a result of the use of fossil fuels and of cement production. Half of this amount has been emitted since the 1970s. In 2004, emissions of CO_2 were about 29 Gt (5.4% more than in 2003).

• CO_2 capture and sequestration is an important challenge. In order to be efficient against global warming, the storage of CO_2 must last for centuries (one could imagine releasing this stored CO_2 when the next ice age sets in). Figure 3.17 shows the relative importance of different sources of CO_2.

There are three steps in CO_2 capture and sequestration. The first is the capture, that is, separation of CO_2 from the other gases resulting from combustion. The second is transportation of the CO_2 to the place where it can be stored. The third is sequestration of the CO_2 so that it is stored safely.

3.16.1. Capture

The large quantity of CO_2 emitted by energy production facilities restricts the number of possible technologies to capture CO_2. The right technology depends

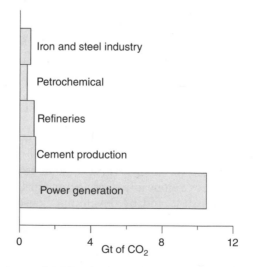

Figure 3.17. Emissions of CO_2 from stationary sources emitting more than 100,000 t/year. Data from greenhouse gas R&D program, IEA, Paris, and special report on CO_2 capture and storage, Intergovernmental Panel on Climate Change (IPCC), www.ipcc.ch.

on the nature of the power plant technology, which determines the characteristics of the gas stream from which CO_2 has to be separated. Currently three main technologies are employed:

- The first method is the postcombustion capture of CO_2 from the flue gases.
- The second is precombustion capture where the fuel is first converted to H_2 and CO_2 before combustion.
- The third is oxy-fuel combustion where fuel is burned in oxygen instead of air. In this case the flue gases contain a high concentration of CO_2.

The main steps in these three technologies are summarized in Figure 3.18.

It was in the 1970s that serious consideration was first given to separation of CO_2 from flue gas streams. The main reason was not concern over greenhouse gas emissions but rather a desire to obtain CO_2 for use in enhanced oil recovery and other applications. Apart from its use for oil recovery, CO_2 has many other applications such as the production of dry ice, urea, and beverages and the carbonation of brine. Several CO_2 capture plants were constructed in the United States in the late 1970s and early 1980s. Some of them are still in operation today. However, compared to the needs envisaged today, the scale of these plants is far too small.

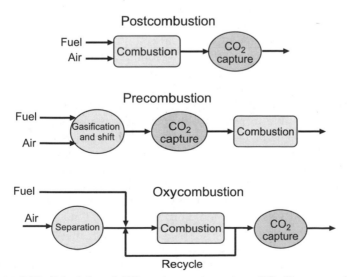

Figure 3.18. Principles of different ways to capture CO_2. From www.ifp.fr.

Postcombustion is capture technology, which has been thoroughly investigated. Derived from a process developed more than 60 years ago to remove acidic gas impurities (hydrogen sulfide and CO_2) during refining processes or gas field exploitation, it is based on chemical absorption. The flue gas stream is scrubbed by passing it through a basic solution, which can be monoethanolamine or a solution of ammonia. Carbon dioxide is dissolved in the solution and released when this solution is heated. The other gases, N_2, O_2, and steam, are released into the atmosphere. At the same time the solvent is regenerated.

The chemical process is

$$2R-NH_2 + CO_2 \rightarrow R-NH-CO_2^- + R-NH_3^+$$

Where $R = -CH_2CH_2OH$. This chemical reaction corresponds to the absorption process. The carbamate ion $R-NH-CO_2^-$ is stable but can be decomposed back into CO_2 by the following regeneration reaction:

$$R-NH-CO_2^- + R-NH_3^+ + \text{heat} \rightarrow 2R-NH_2 + CO_2$$

With this method, about 75–95% of the CO_2 is captured and it is possible to produce a fairly pure CO_2 stream. The capture operates at ordinary temperature and there is already some operating experience in industrial installations designed for other purposes. However, among the drawbacks is the fact that a large amount of heat is required in the regeneration process because the carbamate ion is fairly stable. This imposes a high-energy requirement for the whole process. There are also losses of solvent and corrosion problems to be dealt with. For an 80–85% capture of CO_2 the energy penalty leads to an 8–10% efficiency loss in the best cases and the loss can be over 20%. Figure 3.19 illustrates, as a theoretical example, the effect of an energy penalty of 20% on a 48% thermal efficiency supercritical coal-fired power plant. The energy penalty can be reduced with further development. Environmental issues due to solvent degradation present another concern.

The main advantage of the postcombustion process is that it is an end-of-pipe treatment which can be implemented in any thermal power plant. Therefore it is possible to retrofit old coal and gas plants provided there is enough space to install the facility necessary for the process. However, retrofitting to old and inefficient power plants is not very interesting.

Precombustion can only be implemented in new power plants. It is based on coal gasification followed by a water gas shift. In the gasifier, coal is fed along with a controlled quantity of oxygen or air and steam. This produces syngas. The hot syngas is cooled and purified and its heat is extracted to produce a high-pressure steam. The clean syngas undergoes a water gas shift at high temperature over a catalyst. The shifted syngas is cooled, humidified, and expanded to produce electricity. Hydrogen sulfide and CO_2 are then removed from the shifted syngas in a special selexol unit and the final

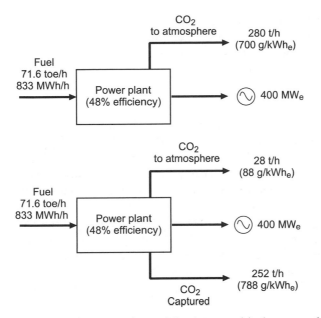

Figure 3.19. Effect of CO_2 capture in coal-fired supercritical power plant of 48% thermal efficiency and penalty of 20% for CO_2 capture.

gas is sent to a gas turbine or a combined-cycle turbine. Compared to methyl ethanol amine, which is a chemical solvent, selexol is a physical solvent (no chemical reaction), which means that recovering captured CO_2 needs less energy.

The precombustion technology is at an early stage of development but seems to be an interesting route to CO_2 capture. Compared to postcombustion technology, the loss of efficiency is lower. It is especially suited to IGCC power plants with the advantages and disadvantages discussed previously. There are however many developments and improvements still required. Gas turbines running with pure hydrogen have not reached an industrial stage yet. New IGCC power plants are usually qualified as "capture ready," but they have not generally been optimized for the precombustion process. Furthermore, application of IGCC technology in coal-fired power plants has some difficulties.

The third capture technology is oxy-fuel combustion. The idea is to carry out the combustion process under pure oxygen (energy is needed to separate oxygen from air in a previous stage). The advantage is that concentrations of CO_2 larger than 50% are obtained in the flue gases and there is no nitrogen in these. As can be seen in Figure 3.18, the flue gases are recirculated in the combustion chamber in order to moderate the temperatures. Carbon dioxide is captured by either the amine process or liquefaction based on chilling.

Oxy-fuel combustion technology is in the early stages of development and has only been tested on a small scale. It has the advantage of being usable in almost all conventional combustors. The required boilers are more compact and the volumes of flue gases which must be treated are smaller. If turbines are to be used, new designs are needed.

3.16.2. Transport

Transportation of the CO_2 from the place where it is captured to the place it can be stored will generally be required. As in the case of natural gas, CO_2 can be transported by pipelines (onshore or offshore) or by ship. The transportation cost depends on the distance traveled. For distances smaller than 500 km, a pipeline is the cheapest way to transport CO_2. Onshore pipelines are less expensive than offshore pipelines.

Because CO_2 is used for enhanced oil recovery, pipelines for CO_2 transport are a well-mastered technology. There already exist about 4000 km of pipelines for CO_2 transport, most of them in the United States. A single pipeline has a flow reaching 20 Mt/year. An example is the 325-km pipeline from the Great Plains Synfuels Plant at Beulah, North Dakota, to Weyburn, Canada, where the CO_2 is used for enhanced oil recovery.

Transportation by shipping becomes more competitive for distances larger than 1500–2000 km. The CO_2 tankers are similar to tankers for liquefied natural gas. The CO_2 is transported as a liquid at a pressure larger than 6 bars and a temperature smaller than −55 °C.

3.16.3. Sequestration

There are three main possibilities for CO_2 storage:

- In deep geologic formations.
- In the deep ocean, but this is not an acceptable solution because of possible impacts on seawater (e.g., ocean acidification).
- In the form of mineral carbonates, but this is too energy consuming. Nature does this over geologic timescales, but this method is not economically competitive on a human timescale.

At the storage site, CO_2 is injected into a geologic formation to a depth where it is in supercritical form. Carbon dioxide is a supercritical fluid at temperatures greater than 31.1 °C and pressures greater than 7.38 Mpa (73.8 bars).

For geologic formations there are three main possibilities:

- Deep saline aquifers which contain brines so that they cannot be used as drinking water.
- Depleted oil and gas fields.

- Unminable coal seams. In this case CO_2 injection can release some coal bed methane.

Deep aquifers present the best possibilities in terms of volume. They often have the advantage of existing near CO_2 production sites. Potentially between 400 and 10,000 Gt of CO_2 can be stored in these formations. Assuming that half of the emitted CO_2 is stored ($\approx$15 Gt), this corresponds to between about 30 and 700 years of present emissions. Currently these geologic formations remain incompletely characterized geologically.

Depleted oil and gas fields are usually well characterized geologically. Carbon dioxide injection has been used in declining oil fields to enhance oil recovery. There are however possible problems of corrosion. Carbon dioxide with water produces an acid which attacks cement. The capacities of storage are smaller than for deep aquifers, typically around 930 Gt of CO_2. This represents little more than about half a century of present emissions.

The capacity of unmined coal seams is smaller, around 30 Gt of CO_2 or two years of storage. There is enough room to store great amounts of CO_2 and deep aquifers are probably the best solution in the long term. The integrity of the storage has to be ensured. Stored CO_2 should not be allowed to leak. Surveillance and monitoring are therefore important issues. Surface leakage would create risks to human health. Carbon dioxide is not extremely toxic but a concentration above 3% may significantly affect people. Indeed, CO_2 is a local and cerebral vasodilator, may increase the blood pressure, may lead to mental confusion, and so on. At 4–5% concentration breathing rates increase to four times normal and blood pressure is significantly increased. At a concentration larger than 30% the death of people or animals occurs within minutes. Natural accidents involving the release of CO_2 have already occurred, for example, at Lake Nyos, Africa. The risk assessment and the social acceptance of CO_2 storage are important issues deserving great attention.

Lake Nyos, in Cameroon, is located in the throat of an old volcano. Dissolved CO_2 seeps from springs located beneath the lake. When the CO_2 saturation level is reached, bubbling occurs, triggering a release of large quantities of CO_2. In August 1986, this lake released a large quantity of CO_2 (about 100,000 tons). Since CO_2 is a dense gas relative to air, it flowed down the hill. People living above the lake were not affected, but among those living below the lake more than 1700 died. Thousands of cattle and many other birds and animals also died, even at distances over 25 km from the lake. This lake is now monitored and equipped to prevent such accidents. Other catastrophic natural releases have also occurred.

TABLE 3.3. IEA Demonstration Projects of Geologic Sequestration of CO_2

Project	Start Date	Location	Tons of CO_2 Injected per Year	Total Tons of CO_2 to be Injected
Sleipner	1996	North Sea	1 Mt	20 Mt
Weyburn	2000	Canada	1 Mt	20 Mt
Salah	2004	Algeria	1 Mt	18 Mt
Snohvit	2006	Barents Sea	0.7 Mt	
Gorgon	2008–2010	Western Australia	1 Mt (phase I)	125 Mt

Several geologic sequestration projects have already been started or are in preparation. Table 3.3 lists demonstration projects from the IEA's greenhouse gas program on CO_2 storage and capture.

Each of these projects allows a sequestration of about 1 Mt/year. Since several gigatons of CO_2 would need to be captured and sequestered each year, thousands of such sequestration facilities would be required. Little chance exists that such a development will occur before many decades have passed.

As has been pointed out by several scientists, CO_2 can also be directly fixed or transformed into a sequestered form by biologic means, for example, in photosynthesis, reforestation, and microbial fixation. The problem with these methods is the practical difficulty associated with generating a capability sufficient to absorb the quantities of CO_2 emitted by human activities.

3.16.4. Cost

Capture and compression is the most expensive step. Estimates are that costs would be between 30 and 60 €/t of CO_2. Transportation of the CO_2 from the place where it is captured to the place of sequestration would cost about 3.5 €/t per 100 km of distance. The costs of injection and storage would depend very much on the quantities handled. It is estimated to be about 20 € for 1 Mt/year but is reduced to 7 €/t for 10 Mt/year. All these costs are of course only indicative.

The energy penalty for CO_2 capture may be large depending on the technology. This directly affects the cost of electricity production. For example, in the case of a pulverized coal plant, energy penalties are between 22 and 29%. For an IGCC power plant it is lower ($\approx$12–20%). For an oxy-fuel power plant it may range between 15 and 43%. All these values for the energy penalty exclude transport and storage. Capture and sequestration will be a costly technology.

3.17. COAL PIT ACCIDENTS

Underground coal mining is the most dangerous activity in the energy domain. Accidents including explosions and roof cave-ins are common. Over the time that coal has been exploited as a fuel tens of thousands of miners have died in coal mine accidents. The difficulties associated with rescuing miners trapped in underground mines after an accident significantly increase the chances of fatalities. It is estimated that, worldwide, about 15,000 people die from coal mine accidents each year. The toll varies significantly with country, but every coal-mining country has seen deadly accidents.

China derives 74% of its energy from coal and official records indicate that, currently, more than 6000 miners die each year in Chinese mines, about 16 per day. Accidents are regular occurrences. For example, in October 2004, 148 miners died in the Daping Mine in Henan Province. A month later, in November 2004, 166 miners were killed in a gas explosion in the Chenjiashan mine in Shaanxi Province. Two months later, in February 2005, more than 214 pit workers died in an accident in the Fuxin mine in Liaoning Province.

In 2006, there were 47 coal-mining fatalities in the United States. This figure is much smaller than the fatalities in construction activities (1226) but comparable to those in air transportation (51).

In France, the worst coal-mining accident occurred on March 10, 1906, when (officially) 1099 pit workers, including many children, died in a coal dust explosion in Courrières. In fact, this number is very likely underestimated because of nondeclared workers. The accident was badly managed by the coal company. Many of the pit workers died, not because of the initial fire ignited in the mine, but as a consequence of bad decisions taken by the company which wanted to preserve the coal deposit and continue to exploit it. A major strike followed. One consequence of that strike was the introduction of the weekly day off in France.

A number of serious coal-mining accidents have been attributed to the presence of firedamp, a mixture of methane and air, inside the galleries. Great care is required not to ignite this gas. The worst coal-mining accident in the United States took place on December 6, 1907, at the Monongah mine in West Virginia. An explosion was caused by the ignition of firedamp, which in turn ignited coal dust, causing the deaths of 362 miners.

On April 26, 1942, the worst coal-mining disaster that ever occurred took place in Benxi, China, where a coal dust explosion killed 1549 miners.

3.18. ENVIRONMENTAL IMPACTS

Power generation from coal can have a severe impact on the level of greenhouse gas emissions and, more generally, on the environment. Since the number of coal-fired plants is expected to increase significantly, developing clean technologies—or at least cleaner technologies—at a reasonable economic cost should be given very high priority by those responsible for energy policy. Attempts to mitigate pollution effects from coal-fired technologies which must be directly addressed include:

- Increasing the thermal efficiency of coal-fired power plants in order to produce more electricity with the same amount of coal
- Reducing or eliminating pollutants such as nitrogen oxides, sulfur oxides, and particulate matter
- Reducing or eliminating CO_2 emissions

Many of today's conventional technologies can be improved or retrofitted in order to decrease atmospheric pollution. However, the capture and sequestration of CO_2, the ultimate technology in striving for zero CO_2 emissions, is the most costly and far from being broadly utilized at the industrial stage.

Figure 3.20 indicates the large differences in emission in CO_2 emission rates and thermal efficiencies realized by new generation (supercritical cycle)

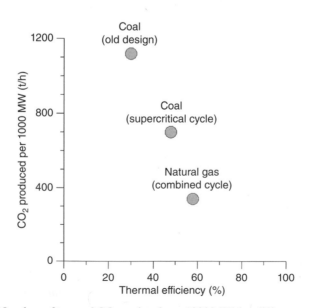

Figure 3.20. Number of tons of CO_2 emitted per 1000 MW for different types of power plants. Data from www.ifp.fr.

coal-fired power plants and those for plants of older design. It also shows that a combined-cycle technology using natural gas is much better in terms of both higher yields and lower emissions.

3.19. CONCLUSION

Fossil fuels will continue to play an important role in energy production for many years to come. The reserves of coal are far above those of oil and natural gas, which indicates that this fossil fuel will become increasingly important in the near future. Coal could be used to power electrical plants or to make synthetic fuels to power vehicles. Unfortunately coal is also the dirtiest fossil fuel and progress in development of clean coal technologies, increasing the thermal yields, and controlling emissions is essential. Carbon dioxide capture and sequestration is an appealing goal, but we are far from major application on an industrial scale. It will be many years before many capture and sequestration units are operational. In addition, the costs and safety issues surrounding CO_2 storage should not be underestimated. Finally, capture and sequestration of carbon dioxide may greatly increase energy costs. The economic viability of clean coal technologies may well depend on the level of any carbon taxes.

CHAPTER 4

Fossil Fuels and Greenhouse Effect

The average temperature of the universe is $-270\,°C$. Without any source of energy, Earth would also be at this temperature. However, the interior of Earth is hot, because of the initial heat accretion when our planet was formed and because natural radioactive isotopes in the earth continue to release energy. Through geothermic heat flow this internal energy diffuses slowly toward the surface. If this flow were the only energy source, the temperature of our planet would be $-243\,°C$ and the air would be a liquid. In fact we receive a flux of energy from the sun that is 4000 times larger than the geothermal flux. It is this solar radiation that makes our planet inhabitable.

The fraction of the sun's energy received by Earth depends on the effective surface area of Earth as seen from the sun. Since Earth will appear to be a disc of radius R_T, this effective area is $\pi(R_T)^2$. On average, the solar energy arriving at the top of Earth's atmosphere along a line perpendicular to the axis joining the centers of the sun and Earth, per unit of time and area, that is, the power per unit area as seen from the sun, is $1368\,W/m^2$. This quantity is called the *solar constant*. However, Earth is actually a sphere of total area $4\pi(R_T)^2$ and rotates with a regular period of 24 h. A given point on Earth passes successively through night and day, which means that it does not always receive sunlight. Thus the average power received at the top of Earth's atmosphere *per unit of total earth area*, obtained by multiplying the solar constant times the ratio $\pi(R_T)^2/4\pi(R_T)^2 = \frac{1}{4}$ is $342\,W/m^2$. The power received in a given location can, of course, differ notably from this value. For example, more energy per unit of area arrives at the equator than close to the poles. From this average power of $342\,W/m^2$ received on top of Earth, one-third ($107\,W/m^2$) is reflected toward space: by the clouds and the layers of the atmosphere on the one hand ($77\,W/m^2$) and by Earth's surface on the other hand ($30\,W/m^2$). The remaining part ($235\,W/m^2$) is transformed into heat. Of this, Earth's surface absorbs $168\,W/m^2$ and the atmosphere and clouds absorb $67\,W/m^2$ (Figure 4.1).

Our Energy Future: Resources, Alternatives, and the Environment
By Christian Ngô and Joseph B. Natowitz
Copyright © 2009 John Wiley & Sons, Inc.

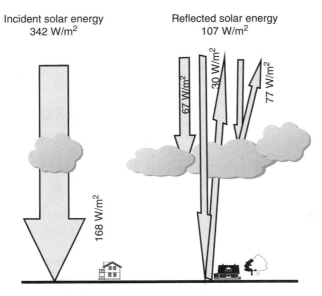

Incident solar energy
342 W/m²

Reflected solar energy
107 W/m²

67 W/m²

30 W/m²

77 W/m²

168 W/m²

Earth's absorption

Figure 4.1. Schematic indicating how incident solar energy arriving on top of the atmosphere is shared. Of the 342 W/m² incident power, 107 W/m² is reflected toward space, 67 W/m² is absorbed by the atmosphere, and 168 W/m² reaches the ground.

4.1. GREENHOUSE EFFECT

The energy absorbed by Earth's atmosphere is absolutely critical to life as we know it. Without this absorption the average surface temperature on our planet would be very cold—18 °C. Water would not normally exist in liquid form, and life could not have developed in the same way as it has. Fortunately, our atmosphere contains gases known as greenhouse gases, which, though present in small quantities, are largely transparent to the incident direct solar radiation but capable of absorbing the energy emitted from Earth's heated surface. This energy is emitted in all directions as longer wavelength infrared radiation. Absorption of Earth's infrared radiation increases the average temperature from −18° to +15 °C, making our planet a comfortable place to live. The result is very similar to that employed in greenhouses to grow plants. The greenhouse windows are transparent to visible solar radiation (so solar energy goes in) but absorbed energy reemitted from the soil or other objects located inside the greenhouse remains trapped and increases the internal temperature. Because the atmosphere serves both as a window to admit solar radiation and an absorber to trap part of the reemitted infrared radiation, the subsequent warming effect is known as the "greenhouse effect."

The first suggestion of the existence of the greenhouse effect was made in 1827 by Joseph Fourier, a French mathematician and physicist. He advanced the idea that Earth's temperature was much greater than that of free space because the atmosphere acted to confine additional heat, functioning much like a garden greenhouse. In 1895, the Swedish chemist Svante Arrhenius was the first to calculate quantitatively the effect that an increase of CO_2 concentration could have on atmospheric temperature. He predicted a global warming as a consequence of using fossil fuels and estimated that doubling the CO_2 concentration would increase Earth's average temperature by 5–6 °C.

Presently, there is growing concern that increasing greenhouse gas concentrations may adversely impact Earth's climate. A significant part of this increase in greenhouse gases results from human activities, notably the combustion of fossil fuels—coal, oil, and gas. Since these fossil fuels are essential to the present world economy and currently meet about 80% of the world's energy demand, a major reduction in their use is not feasible unless good energy source replacements are in place. In this chapter we quickly review the nature of the greenhouse gas emission problem and discuss current assessments of the impact that continued use of fossil fuels may have on the climate.

Both Earth's surface and the atmosphere radiate energy, but because their temperature is much lower than that of the sun, their emissions are in the far-infrared region. At equilibrium, there exists a stationary state in which the received power from the sun is equal to that emitted by Earth into space. Therefore, $342 W/m^2$ must be emitted. If this were not the case, the average temperature of our planet would increase or decrease until equilibrium is attained. As a consequence, $235 W/m^2$ of infrared radiation must be emitted toward space at the top of the atmosphere. This is added to the $107 W/m^2$ corresponding to the reflected incoming solar radiation. This $235 W/m^2$ of radiation has two origins. It consists of the fraction of infrared radiation coming from the ground which is not stopped by the atmosphere and the fraction of the infrared radiation emitted by the atmosphere in the direction of space (Figure 4.2). If the $235 W/m^2$ emitted from Earth's surface were lost, the mean temperature at the surface would be −18 °C.

Earth's surface, which is at an average temperature of 15 °C, emits $390 W/m^2$. The major part, 90% ($350 W/m^2$), is absorbed by the atmosphere and this is the greenhouse effect. The remainder ($40 W/m^2$) goes to space. The atmosphere, heated by the radiation of Earth's ground and the direct solar radiation ($67 W/m^2$), also emits in the infrared region depending upon its temperature. The atmosphere thus radiates $195 W/m^2$ toward space and $324 W/m^2$ toward the surface which absorbs this energy.

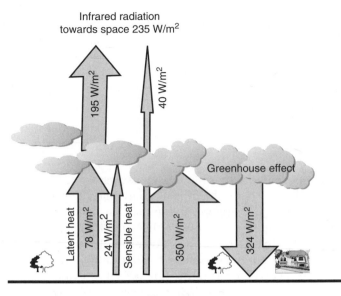

Earth's radiation

Figure 4.2. Schematic of energy emitted by Earth in equilibrium with the energy received from the sun: $235 W/m^2$ is radiated toward space to balance the incident radiation. Because of the greenhouse effect, Earth's surface receives $492 W/m^2$. It only reemits $390 W/m^2$. The difference is used to warm the air from the ground and in the water cycle. Actually, Earth's surface transfers its heat energy to the atmosphere partly by infrared radiation and partly as heat and as latent heat of evaporated water. The latter releases the energy as it condenses in the cold upper atmosphere.

With these data we see that the input energy flow balances the outgoing energy flow. However, there is no balance between Earth's ground and the atmosphere. Earth's surface receives $168 W/m^2$ from the sun plus $324 W/m^2$ from the atmosphere (in the form of infrared radiation), which equals $492 W/m^2$, whereas it emits only $390 W/m^2$ in the form of infrared radiation. The difference, $102 W/m^2$, is transferred by other mechanisms: conduction and evaporation. This corresponds to $24 W/m^2$. The evaporation process uses $78 W/m^2$.

On our neighboring planets in the solar system the greenhouse effect ranges from nonexistent to very important. The temperature on the planet Mercury, located closest to the sun at 58×10^6 km, ranges from $-183°$ to $427°C$. Mercury has an unstable, very thin atmosphere constantly renewed because it consists of atoms blasted off its surface by the solar wind. In contrast, Venus, located farther from the sun, 108×10^6 km, has an atmosphere that is more than 95% CO_2 and an atmospheric pressure almost one hundred times higher than that on

Earth. The greenhouse effect on Venus is so large that the temperature actually reaches 460 °C. It is higher by 60 °C than that existing on Mercury. Without this atmosphere the temperature on Venus would be a few degrees. Mars, located approximately 228×10^6 km from the sun (compared to 150×10^6 km for Earth) has a very thin atmosphere (0.6% that of Earth) and no greenhouse effect. There, the temperature at the equator is −73 °C.

Modeling the effect of increasing greenhouse gases on Earth's climate is a very complex problem. The problem is not only that the greenhouse effect is increased when some gases are emitted in the atmosphere but that there are many feedback mechanisms that may either amplify (positive feedback) or reduce (negative feedback) the effect. For example, approximately 71% of Earth's surface is covered by the oceans. Evaporation of liquid water requires a lot of energy (this corresponds to the latent heat of evaporation) and leads to a cooling of the water's surface. When the water vapor condenses, it transfers its latent heat of condensation to the atmosphere. The larger the concentration of greenhouse gases in the atmosphere, the more important is this mechanism of energy exchange between Earth's ground and the atmosphere. Another example of a positive feedback is the melting of snow or ice. An initially reflective area is replaced by an absorbing one. Consequently solar energy is absorbed more than before, leading to an increasing warming which induces more melting of ice and snow and so on. The problem is that there are more positive feedbacks than negative ones and the consequences of these phenomena are not well mastered in model calculations.

4.2. GREENHOUSE GASES

Interestingly, water vapor is the most important greenhouse gas, accounting for 55% of the natural greenhouse effect (Figure 4.3). Carbon dioxide (CO_2), the second most important greenhouse gas accounts for 39%. Other greenhouse gases present in small amounts in the atmosphere are methane (CH_4), nitrous oxide (N_2O), ozone (O_3), and halocarbons.

In our atmosphere water can exist in three phases: solid, liquid, and gas. Water vapor is produced naturally by evaporation (the water cycle), perspiration, and breathing. Its concentration in the atmosphere increases when the temperature increases, leading to a positive feedback. Water vapor represents, on average, approximately 0.4% in volume of Earth's atmosphere—about 0.1% in Siberia but 5% in equatorial maritime areas. It is responsible for about half of the greenhouse gas effect on Earth's average surface temperature. Condensation of all of the water vapor contained in the atmosphere would produce a 2.5-cm-thick layer of liquid over the entire surface of Earth. Since the average annual precipitation corresponds to about 1 m, this means that evaporated water molecules do not remain in the atmosphere very long before returning to Earth's surface.

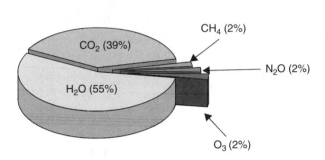

Figure 4.3. Share of contribution to natural greenhouse effect from gases present in atmosphere. Water vapor is the main contributor. Its global concentration in the atmosphere is of the order of 3–4%. Carbon dioxide has a much lower concentration (about 0.03–0.04%). The figures should be taken as order of magnitude because there is also a contribution from clouds which contain water vapor.

The burning of hydrocarbons liberates both CO_2 and H_2O. However, the quantity of water emitted by hydrocarbon combustion represents less than 0.003% of water evaporated naturally. Thus, although water vapor is an important greenhouse gas, the emission from human activities is negligible compared to the quantity of water evaporated from bodies of water, the ground, or vegetation. Some small amount of water vapor can be produced in the stratosphere due to the chemical destruction of CH_4 by ultraviolet (UV) rays. Airplanes emit both CO_2 and water vapor at high altitudes. Both can have a worse effect than if they are emitted at sea level.

As can be seen from Figure 4.4, a large part of *anthropogenic*[1] greenhouse gas emissions result from energy generation. Figure 4.5 shows that CO_2 emissions are strongly dominant. This CO_2 comes from fossil fuel use in transportation, electricity production, heating or cooling buildings, and so on. Atmospheric CO_2 holds a very small fraction of the carbon in the global carbon cycle but

---

[1] An effect is said to be anthropogenic if it results from human activities rather than naturally occurring environmental causes (i.e., without any human influence).

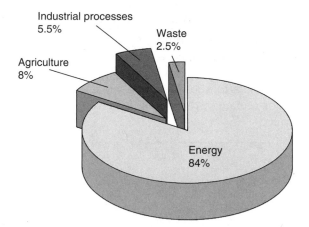

Figure 4.4. Relative contributions of various sources to world's anthropogenic greenhouse gas emissions, 2003. *Source*: *CO_2 Emissions from Fuel Combustion 2006*, IEA, Paris.

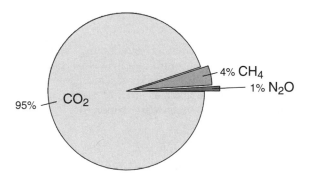

Figure 4.5. Relative proportions of world anthropogenic greenhouse gas emissions from energy generation, 2003. *Source*: *CO_2 Emissions from Fuel Combustion 2006*, IEA, Paris.

plays an important role in the greenhouse effect. The concentration of CO_2 in the atmosphere was equal to 280 ppmv (parts per million by volume; we use ppm in the following) in the preindustrial era.

Methane is produced mainly in anaerobic decomposition of organic matter issuing from biologic systems. Agriculture (e.g., paddy fields, ruminant mammals) but also wetlands or organic wastes are important sources of CH_4. Incomplete combustion of fossil fuels, leakage during the process of distribution of natural gas or oil, or gas releases during coal mining also contribute to increasing the concentration of CH_4 in the atmosphere. Methane molecules remain in the atmosphere for a shorter time than do CO_2 molecules. They react initially with hydroxyl radicals (OH) in the first step of a process which eventually results in the production of CO_2. The concentration of CH_4 in the atmosphere varied slowly between 580 and 730 ppb (parts per billion) over the 10,000 years prior to the preindustrial era.

Agricultural operations, fertilizers, and manure are the main source of N_2O. There are also other sources, such as wastewater treatment and waste or biomass combustion. Before the preindustrial era, the concentration of N_2O in the atmosphere was equal to 270 ppb by volume.

Ozone provides the third largest greenhouse effect contribution after CO_2 and CH_4. Ozone is present in both the troposphere and the upper stratosphere. The troposphere is the lower part of the atmosphere, in contact with the surface of the ground; its temperature decreases with altitude. The stratosphere is at an altitude ranging between 13 and 50 km from the ground. Its temperature increases with altitude. In the stratosphere, the "ozone layer" protects Earth against dangerous UV radiation. Because of an extensive emission of chlorofluorocarbons (CFCs) and other halocarbons into the atmosphere, the ozone layer has been depleted. With the Montreal protocol, settled in 1987, there has been a progressive prohibition of the use of CFCs, which were used primarily in refrigeration and air-conditioning circuits. Some of the products substituted for CFCs, though not damaging to the ozone layer, are greenhouse gases.

The ozone in the stratosphere is formed by the interaction of oxygen with UV rays. This ozone protects living beings from exposure to short-wavelength, high-energy UV radiation emitted from the sun. It absorbs practically all the UV rays with a wavelength shorter than 255 nm. Most of the ozone is produced above the equator. It then diffuses toward the poles where it accumulates. Its concentration reaches a marked maximum at altitudes between 20 and 25 km. The ozone concentration varies between day and night as well as according to season. In each hemisphere, the concentration is minimal at the end of winter and at the beginning of spring. The average concentration of stratospheric ozone reached a minimum in the early 2000s. It will be a few decades before earlier concentration levels can be restored.

The tropospheric ozone is synthesized in photochemical reactions involving precursors which are produced by combustion of fossil fuels or of the biomass: nitrogen oxides (NO_x), carbon monoxide (CO), and volatile organic compounds. It is a very reactive and toxic chemical product remaining only a few hours or a few days in the troposphere. It is a major contributor to air pollution.

The atmospheric dwell time of individual greenhouse gas molecules can vary greatly. On average, CO_2 molecules released into the atmosphere remain there for more than a century. In contrast, the dwell time of evaporated water molecules in the atmosphere is very short. Gases which remain for a long time are a problem because the effect of any reduction in their emission can only be realized at a later time.

A large amount of information and data on the greenhouse effect and its impact on climate can be found at the Intergovernmental Panel on Climate Change Internet site (www.ipcc.ch).

4.3. WEATHER AND CLIMATE

Although they are closely related to each other, the words *weather* and *climate* should not be confused. Weather is defined as the state of the atmosphere at a given time and place. This may be specified by variables such as barometric pressure, temperature, wind velocity, moisture, and so on. The word climate denotes the long-term trends of weather over long periods of time (typically 30 years or so).

Weather forecasting is the science of predicting the weather. In recent times, very sophisticated computerized theoretical models have been developed for weather forecasting on a short time scale (hours and days). It is now possible to make reasonably accurate weather forecasts for two- or three-day periods. Weather predictions for more extended periods are difficult because the atmosphere is a chaotic system. This means that a small perturbation can have an effect which is both significant and unpredictable (the so-called butterfly effect).

There is a pervasive impression that weather forecasts are usually wrong. In fact, in recent years, great progress has been made in this domain due to greatly improved observational capabilities (e.g., satellites), powerful computers, and a better understanding of the underlying physics. This can be illustrated in France where weather forecasting is rather difficult because of the geographical location and the widely varying topography over a small-distance scale. In 2000, 98% of the one-day forecasts, 93% of the two-day forecasts, and 70% of the three-day forecasts were correct. Over a one-year period, 76% of temperature forecasts and 89% of rainfall forecasts were correct.

Although it can often be extremely difficult to predict exactly the weather over a several-day period, projections of the longer term climate trends can be made with greater confidence. A metaphor can be used to help understand this difference between weather and climate predictions. For that let us consider the death of the inhabitants of a given country. It is not possible to predict exactly when a particular inhabitant will die. However, the average age of death can be specified with high accuracy (e.g., 80 years in France in 2007). A similar situation exists for weather and climate forecasts. Over long periods of time better predictions of some statistical parameters can be made for climate than for weather. It is possible to develop predictive models that can be tested on historical climate records and predict, using some hypotheses, the climate of the future. This naturally entails some uncertainty because a complete physical description of Earth's system (atmosphere, oceans, land, etc.) is very difficult to include.

4.4. NATURAL CHANGE OF CLIMATE

It takes one year for Earth to follow its entire orbit and one day to spin around its rotational axis. Variations in the orbital parameters of Earth have an influence on the climate because the incident radiation from the sun changes as Earth moves on its elliptic orbit. The best known variations are the seasonal changes occurring on an annual basis. The main parameter governing the seasons is the tilt of Earth's axis.

The Serbian Milutin Milankovitch related the longer term fluctuations in Earth's climate to orbital fluctuations of our planet. Earth's spin axis makes a precession motion with a period around 22,000 years and varies in degree of tilt from $21.5°$ to $24.5°$ with a period of about 41,000 years. The eccentricity of Earth's orbit cycle varies over a 100,000-year period. The so-called Milankovitch cycles correlate variations of the climate to the three major types of fluctuation in Earth's orbit. The onset and retreat of the great ice age of the past million of years can be explained within this framework.

Although the climate changes as a function of time, the problem with which we are faced today is that the change in the climate which may be induced by human activities will occur on a time scale so short that some living systems may have no time to adapt.

4.5. ANTHROPOGENIC EMISSIONS

Measurements have shown that the concentration of greenhouse gases has increased regularly since the start of the industrial era because of human activities. As we said above, greenhouse gas emissions due to human activities are known as anthropogenic emissions. About two-thirds of the anthropogenic emissions are coming from fossil fuel burning and one-third from land use changes. Carbon dioxide emissions due to fossil fuel burning are now on the average of 7 GtC/year (GtC means a billion of tons of carbon).[2] About half of the anthropogenic emissions are absorbed by natural processes (oceans and terrestrial biosphere). The remaining part remains in the atmosphere.

---

[2] Emission of CO_2 can be expressed in terms of either mass of CO_2 or mass of carbon. Since the molecular or atomic masses are in the ratio $44/12 = 3.67$, we have $1 \, tC = 3.67 \, tCO_2$.

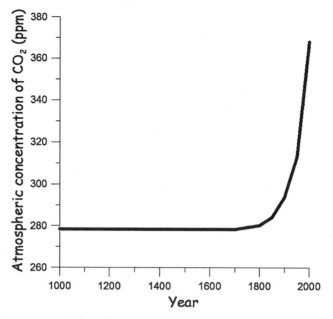

Figure 4.6. Average evolution of atmospheric concentration of CO_2 between 1800 and 2000. *Source*: Data from www.ipcc.ch.

The concentration of CO_2 in the atmosphere (Figure 4.6) was equal to 280 ppm in 1750 but has reached 379 ppm in 2005 (more than a 35% increase). This concentration was 367 ppm in 1999. In six years the increase has been equal to 12 ppm, which is a very large value compared to those determined from the historical preindustrial period record. In that period the CO_2 concentration has increased by only 20 ppm in 8000 years.

Temperatures at Earth's surface have been measured for a bit more than half a century. A global warming has been observed since the end of the last century. This occurred during two periods. From 1910–1940 there was a 0.35 °C increase in the average temperature. Later, during the time from the 1970s until the present, an increase of 0.55 °C has occurred. Between these two periods, from about 1940 to 1970, there was a small cooling of –0.1 °C. Globally, the average surface temperature has increased by 0.74 °C over the last 100 years. The last 25 years have been especially warm. Over the past 100 years, 11 of the 12 warmest years occurred in the period from 1995 to 2006.

Radiative forcing, a term often used by climatologists, is the change in power output per unit area of Earth which results from activities such as volcanic eruption or the anthropogenic emission of greenhouse gases. These changes are called "forcing" to distinguish them from the changes caused by internal parameters of the system. Measured at the top of the atmosphere, it is usually evaluated in watts per square meter. If it is positive Earth warms. If negative Earth cools. A radiative forcing of $1 W/m^2$ means that the greenhouse effect is increased by $1 W/m^2$. The climate may change under the influence of radiative forcing. The increase in the concentration of CO_2 since 1750 has induced a radiative forcing of $1.7 W/m^2$. In the decade between 1995 and 2005, the radiative forcing increased by 20%.

The contribution of methane to radiative forcing is smaller than that of CO_2. The atmospheric concentration of methane varied slowly between 580 and 730 ppb over the 10,000-year period prior to the preindustrial era. Since then it has more than doubled (Figure 4.7), increasing from approximately 750 ppb in 1750 to 1774 ppb in 2005. This increase in concentration leads to a radiative forcing of about $0.5 W/m^2$. The main part of CH_4 emissions is coming from sources other than fossil fuels (wetlands, ruminant animals, rice cultivation, biomass burning, etc.).

Atmospheric N_2O concentration has increased from approximately 270 ppb in 1750 to 319 ppb. This corresponds to an 18% increase (Figure 4.8). It had increased by only 10 ppb over the 11,500 years preceding the preindustrial era.

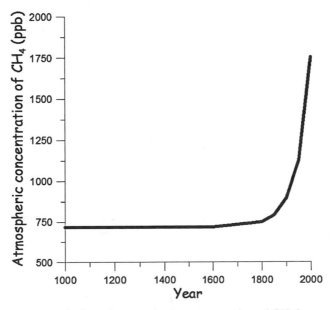

Figure 4.7. Average evolution of atmospheric concentration of CH_4 between 1800 and 2000. *Source*: Data from www.ipcc.ch.

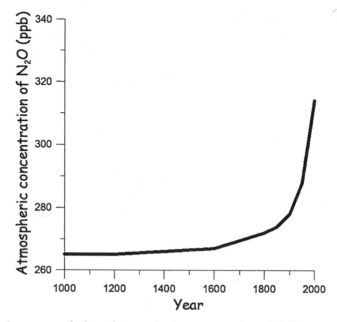

Figure 4.8. Average evolution of atmospheric concentration of N$_2$O between 1800 and 2000. *Source*: Data from www.ipcc.ch.

The associated radiative forcing is equal to 0.16 W/m^2. Anthropogenic N$_2$O comes mainly from agriculture and land use changes.

There are two significant contributions to radiative forcing from ozone. A positive one coming from tropospheric ozone produced partly and indirectly from fossil fuel uses (estimated to be +0.35 W/m^2) and a negative one coming from destruction of the stratospheric ozone (estimated to −0.05 W/m^2).

Artificial gases like CFCs or hydrochlorofluorocarbons (HCFCs) are entirely due to human activities (refrigerators, air-conditioning systems, and so on). The radiative forcing associated with CFC gases is equal to 0.32 W/m^2. The radiative forcing contribution of other fluorinated gases [hydrofluorocarbons (HFCs), perfluorocarbons (PFCs), sulfur hexafluoride (SF6)] is small (0.017 W/m^2) but unfortunately is rapidly increasing.

While anthropic radiative forcing is by far the most important contribution to radiative forcing, natural contributions also occur. Natural radiative forcing comes from solar and explosive volcanic eruptions. Since the start of the industrial era, solar changes have contributed a small positive radiative forcing ($\approx$0.12 W/m^2). This is in addition to the cyclic changes of solar radiation. Volcanic eruptions may lead to a negative radiative forcing over a short period of time ($\approx$1–3 years) because of an increase in sulfate aerosols in the stratosphere. The last important explosive volcanic eruption was Mt. Pinatubo in 1991.

4.6. WATER AND AEROSOLS

The emission of water vapor by human activities has a negligible contribution to radiative forcing for the reasons explained in a previous section. Estimates give a very small positive radiative forcing ($0.07\,W/m^2$). However, water vapor has an influence on the average temperature because it has a positive feedback on greenhouse effect. The quantity of water vapor emitted in fossil fuel burning or by nuclear plants is by far smaller than the quantity of water vapor emitted in agricultural activities. There are still uncertainties regarding the contributions of water vapor emitted in the stratosphere, in particular by the oxidation of CH_4 or by planes.

The atmosphere contains fine suspended particles constituting aerosols. These nongaseous particles can also contribute to the greenhouse effect. The clouds, made up of droplets of water or ice, are one example of aerosols. Smoke and dust emitted by vehicles and by the fires of forest or savannah are others. Volcanoes are also an important source of aerosols emitted in a sporadic way. Sometimes their dust reaches the stratosphere and remains there during two or three years, as was the case at the time of the eruption of Mt. Pinatubo in Indonesia. Aerosols can travel over very long distances. During drought periods when rain does not remove aerosols from the atmosphere, gigantic clouds of dust and smoke can be observed in winter on the northern half of the Indian Ocean.

Aerosols have negative radiative forcing because they prevent part of solar energy from reaching Earth's surface. Transportation, deforestation, construction, and intensive agriculture are human activities that generate dust. Industry is also an important source of aerosols. Even in nonindustrial countries which consume little energy, aerosols are produced by the combustion of wood. Some aerosols give a positive radiative forcing, but most of them give a negative contribution. In total one gets about $-0.5\,W/m^2$.

4.7. GLOBAL WARMING POTENTIALS

In order to compare the impact of radiative forcing contributions from different gases on the greenhouse effect the global warming potential (GWP) has been introduced. The GWP is defined as the cumulative radiative forcing from the emission of a unit mass of gas relative to a reference gas, which has been chosen to be CO_2. This is estimated for a specified period of time (e.g., 100 years). Of course, such comparisons are approximate, but they are useful to get order-of-magnitude estimates. Global warming potentials for some gases are displayed in Table 4.1. The lifetimes in the atmosphere are also indicated. This corresponds to the mean time during which a molecule stays in the atmosphere. The lifetime depends on the chemical reactivity of the gas. The weaker it is, the more it stays in the atmosphere.

These estimates indicate that, after 100 years, releasing 1 liter of CH_4 in the atmosphere has the same effect on radiative forcing as releasing 23 liters of

TABLE 4.1. Mean Lifetime in Atmosphere and GWP for Different Time Horizons for Different Gases

Gas	Lifetime (years)	GWP		
		20 years	100 years	500 years
CO_2	>100	1	1	1
CH_4	12	72	25	7.6
N_2O	114	289	298	153
HFC-134a (CH_2FCF_3)	14	3,830	1,430	435
PFC-14 (CF_4)	50,000	5,210	7,390	11,200
SF_6	3,200	16,300	22,800	32,600

Note: CH_2FCF_3 is used as a fluid in automotive air-conditioning applications; CF_4 and SF_6 are used in the semiconductor industry for etching processes and cleaning of chemical vapor deposition systems.
Source: Data from www.ipcc.ch, Fourth IPCC assessment report, 2007.

CO_2. Looking at the evolution of the GWP as a function of time shows that it decreases for CH_4 while it increases for SF_6. The reason for that is related to the mean lifetime of the gas in the atmosphere. If this lifetime is shorter than for CO_2, the GWP will decrease. If the lifetime is larger than for CO_2 the GWP will increase.

4.8. INCREASE OF AVERAGE TEMPERATURE

As can be seen in Figure 4.9, there has been a well-documented increase of the average global temperature since the start of the industrial era. Further, the present rate of change is far above the average of the past 1000 years (Figure 4.10). Note that this value is just an average and the increase of the mean temperature can vary from place to place. It is higher at higher latitudes. For example, in Alaska or in the north of Siberia the increase of the mean temperature is between 2 and 4 °C. The rise of the average temperature in the north regions has important implications. An area of about $1 \times 10^6 km^2$ of permafrost, equal to the combined area of France and Germany, is presently thawing. A continued increase in the average global temperature will have major implications for the global climate and could dramatically change local environments.

Whether the observed temperature increase reflects radiative forcing from anthropogenic emissions or whether other effects might be responsible is a major question for mankind. Although releasing greenhouse gases in the atmosphere should increase the radiative forcing and the average temperature, the $1.6 W/m^2$ of radiative forcing contributed by human activities is small compared to the incoming radiation ($342 W/m^2$). Indeed, it represents an increase

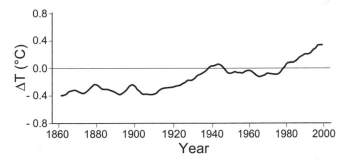

Figure 4.9. Evolution of ΔT over past 140 years, where ΔT is average global tempera-ture minus average global temperature observed during 1961–1990. *Source*: Data from www.ipcc.ch.

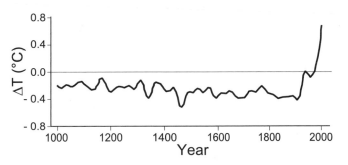

Figure 4.10. Evolution of ΔT over past 1000 years in northern hemisphere, where ΔT is average global temperature minus average global temperature observed during 1961–1990. *Source*: Data from www.ipcc.ch.

of only about 0.5% compared to the preindustrial era. Nevertheless, the current scientific evidence indicates that this is probably the leading contribu-tion to the ~0.75 °C increase of the average global temperature which has been observed.

Today, about 20% of Earth's land surface is covered by *permafrost* or glaciers. Permafrost is a soil having a temperature below the freezing point of water for at least two years. The permafrost belongs to the *cryosphere*, which consists of all snow, ice, and permafrost on or beneath the surface of the continents and oceans. One of the consequences of thawing permafrost is that many trees are falling because their roots are no longer fixed in solid ground.

Methane hydrate is a crystalline combination of natural gas and water which exists in large quantities both trapped in the permafrost and under the oceans. It is a molecular structure in which water molecules form cages inside which CH_4 molecules are trapped (a clathrate). The thaw of the permafrost may release important quantities of methane into the atmosphere. Methane is a strong greenhouse gas. Large additional quantities would produce a large positive-feedback mechanism, further increasing the anthropogenic greenhouse effect. The increase of the average temperature also has an impact on the temperature of the water in the oceans. The mean temperature of that water down to a depth of 300 m has increased by about 0.3 °C since the pre-industrial era. Increasing the temperature of the ocean decreases its capacity to dissolve gases. A decrease in the O_2 content can affect aquatic populations. Release of dissolved CO_2 into the atmosphere contributes to the greenhouse effect. If the water temperature were to increase a few degrees, methane hydrates trapped in the bottom of the oceans might also be released in some locations.

4.9. MODEL PREDICTIONS

Predicting the influence of human activities on the evolution of the climate is a major concern. This can only be done with well-constructed models which incorporate the best current scientific information and take advantage of the most sophisticated current computational methods. Several such models have been developed. While they differ in their quantitative predictions, all of them predict anthropogenic greenhouse gas emissions will lead to an additional increase of the average temperature within this century.

In interpreting these results it should be recognized that the complexity of the problem necessitates approximations. Even if these models are very successful at reproducing past climate trends, there remain real uncertainties in the application to the prediction of future climate changes. Further, the results depend upon the assumed intensity of greenhouse gas emissions. If we continue "business as usual," the average temperature predicted for the end of the century is high. A reduced rate of fossil fuel consumption can reduce the temperature increase. Keeping these points in mind, we note that the results of model calculations indicate a temperature increase between 1.8 and 4 °C by the year 2100. The higher value supposes that there is no catastrophic amplifying phenomenon of the greenhouse effect. We do not know if there is a point at which unforeseen positive-feedback effects might amplify the warming effect.

It might be felt that a few degrees increase is not much. However, it should be noted that this is an average value. The local increase in northern countries may be much larger. It should also be recalled that the difference of average temperature between the glacial period and the present corresponds to a change in the average temperature of only 4 °C. During the glacial period some

European countries were covered by several kilometers of ice. An additional 4 °C change is also expected to have a major impact on Earth's climate.

4.10. ENERGY AND GREENHOUSE GAS EMISSIONS

As we have seen in Figure 4.4 the energy sector is responsible for over 80% of anthropogenic greenhouse gas emissions. Of this, 95% is CO_2 (Figure 4.5). It is estimated that the concentration of CO_2 in the atmosphere will reach 540–970 ppm depending on the assumed evolution of the world's fossil fuel consumption. This is far above the 280 ppm of the preindustrial era and reflects the rapid increase of CO_2 emissions over the last 150 years (Figure 4.11). These CO_2 emissions have increased from almost zero in 1750 to 26.6 Gt of CO_2 (7.25 GtC) in 2004. Furthermore, CO_2 emissions from fossil fuel combustion have almost doubled in the last 30 years.

For the year 2004, the share of CO_2 emissions between the different types of fossil fuels is displayed in Table 4.2. They mainly come from solid and liquid fuels. There is only a small contribution from cement production and gas flaring.

In recent years, the combustion of coal in developing countries has contributed strongly to the rise of CO_2 emissions. Nevertheless, the emission of CO_2

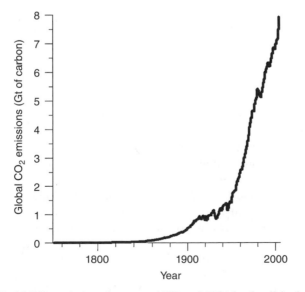

Figure 4.11. World CO_2 emissions between 1750 and 2004 for fossil fuels. This includes emissions from liquid, gas, and solid fossil fuels and those due to cement production and gas flaring. *Source*: Data from the Dioxide Information Analysis Centre, Oak Ridge, TN, www.cdiac.ornl.gov.

TABLE 4.2. Share of CO$_2$ Emissions from Fossil Fuels by Origin, 2004

Origin of Emissions (2004)	CO$_2$ emissions (Gt of Carbon)	Percentage
Liquid fuel	3.289	41.6
Gas fuel	1.434	18.1
Solid fuel	2.838	35.9
Cement production	0.298	3.8
Gas flaring	0.051	0.6
Total	7.910	100

Source: Data from the Dioxide Information Analysis Centre, Oak Ridge, TN, www.cdiac.ornl. gov.

per capita has leveled off since the 1970s, as can be seen in Figure 4.12. This is true only because the world population has risen. It went from about 3 billion inhabitants in 1960 to almost 6.7 billion by the end of 2007.

Fully 80% of the earth's primary energy production is consumed by only 20% of the inhabitants of planet. The average life expectancy for this 20% is more than 75 years. The next 60% of the population consumes 19% of the world's energy. The average life expectancy of this group is for more than 50 years. The remaining 20% of the world population consumes only 1% of the world's energy. Their life expectancy is less than 40 years. Not surprisingly, greenhouse gas emissions are closely related to energy consumption. Three quarters of the greenhouse gases are emitted by 25% of the earth's population.

For the year 2004 Figure 4.13 depicts the level of emission of CO$_2$ from fossil fuels for the top 10 emitting countries. Collectively they emitted a total of 17.2 GtC, 65% of the 26.6 GtC released at the world level. Estimates for the year 2007 indicate that, if cement production is included, China has probably overtaken the United States and has become the leading emitter of CO$_2$ on the planet. The large increase in Asian countries like China and Korea comes partly from the fact that a lot of the world's manufactured goods are now produced there. The energy is consumed there and consequently the CO$_2$ emission also occurs there.

The emission per capita for different countries varies over a wide range. A few examples are given in Table 4.3 for the year 2004. The United States emits more than 80 times as much CO$_2$ per inhabitant than does Bangladesh and more than 250 times than Ethiopia. However, Qatar emits 2.5 times more CO$_2$ per inhabitant than the United States. Fortunately for the planet the population of Qatar is small.

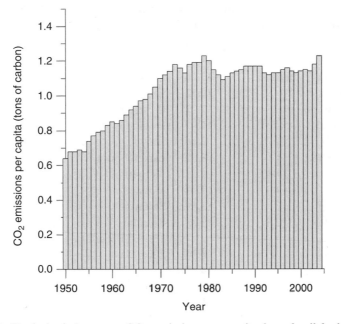

Figure 4.12. Evolution of average CO_2 emissions per capita from fossil fuels between 1950 and 2004. *Source*: Data from the Dioxide Information Analysis Centre, Oak Ridge, TN, www.cdiac.ornl.gov.

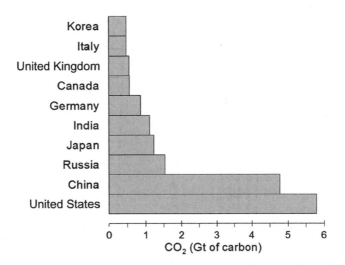

Figure 4.13. Carbon dioxide emission from fuels of top 10 countries, 2004. *Source*: Data from *CO$_2$ Emissions from Fuel Combustion 1971–2004*, IEA, April.

TABLE 4.3. Total CO$_2$ Emissions from Fuel Combustion per Capita for Selected Areas or Countries in Tons of CO$_2$, 2004

World	**4.19**	Austria	9.19
		Denmark	9.42
OECD North America	**15.64**	Finland	13.18
OECD Pacific	**10.32**	France	6.22
OECD Europe	**7.72**	Germany	10.25
OECD Total	**11.09**	Iceland	7.72
Non-OECD	**2.46**	Italy	7.95
Former USSR	**8.09**	Luxembourg	24.94
		Netherland	11.41
Canada	17.24	Russia	10.63
Mexico	3.59	Spain	7.72
United States	19.73	Sweden	5.80
		United Kingdom	8.98
Bangladesh	0.24	Ukraine	6.42
Brunei	14.18		
China	3.66	Israel	9.15
India	1.02	Kuwait	26.36
Japan	9.52	Quatar	49.64
Korea	9.61	Saudi Arabia	13.57
Vietnam	0.96	United Arab Emirates	23.86
Algeria	2.41	Argentina	3.54
Congo	0.18	Brazil	1.76
Ethiopia	0.07	Chile	3.63
Morocco	1.19	Colombia	1.28
South Africa	7.55	Venezuela	4.91
Africa	0.93	Latin America	2.05

Source: Data from *CO$_2$ Emissions from Fuel Combustion, OCDE 2006*, IEA, Paris.

It is interesting to compare the emissions of France, Germany, and Sweden, three developed countries. Germany emits, on the average, 1.65 times more CO$_2$ per inhabitant than France. This reflects the fact that 80% of the French electricity is produced using hydropower and nuclear power, which do not emit CO$_2$. Sweden, which produces more than 90% of its electricity without CO$_2$ emission, has similar results, 1.8 times less emission per inhabitant than Germany.

Oil, gas, and coal all produce CO$_2$ during combustion. However, the quantity which is emitted per 1 kWh of electricity production is not the same. This depends upon the carbon content per unit mass of fuel and upon the technology employed. A better energy yield produces less CO$_2$ per kilowatt-hour. It is also important in the evaluation of CO_2 emissions to take into account the entire production life cycle from "cradle to grave."

There are two types of CO_2 emissions:

• Direct emissions which are released during the production of electricity.
• Indirect emissions resulting from the energy used in all steps associated with making the energy production possible. This includes in particular the plant construction stage (use of concrete, transportation of materials, production of any manufactured components necessary for the operation), the operation stage (production of and transportation of the fuel, disposition of the waste), and, if necessary, a dismantling stage as well as other activities related to these.

Evaluations for different energy-producing technologies have been made by the International Atomic Energy Agency (IAEA) and some of them are displayed in Figure 4.14. We can see that renewable and nuclear energies make only an indirect contribution to CO_2 emissions. Fossil fuels have both direct and indirect contributions. It is worth noting that wind and photovoltaic energies have larger emissions than hydropower. With the present technology, the manufacture of photovoltaic cells demands a lot of energy. Wind turbines require a lot of concrete (360 t/GW) to be anchored to the ground, surprisingly much more for the same power output than that required to build a nuclear plant.

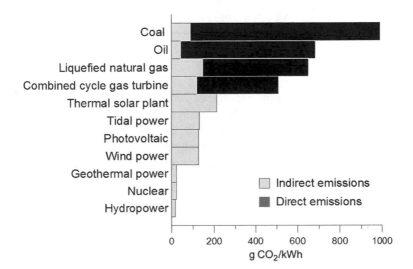

Figure 4.14. Magnitudes of direct and indirect CO_2 emissions of Electricity for Different Technologies. Depending on the technology used there can be differences from the average figures given here. *Source*: Data from www.iaea.org.

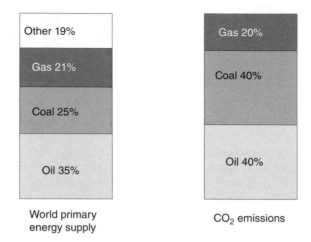

Figure 4.15. Share by fuel of world primary energy supply and CO_2 emissions, 2004. "Other" includes other sources of energy. *Source*: Data from *CO_2 Emissions from Fuel Combustion 2004*, IEA, Paris.

Among fossil fuels coal, which is almost hydrogen free (apart from some CH_4 trapped in the ore), produces the largest total emissions per kilowatt-hour. It is followed by oil (which has an empirical formula which is roughly CH_2). Finally natural gas (CH_4) is the cleanest fuel as far as CO_2 emissions are concerned. It is also much cleaner than coal in other respects (e.g., ash, sulfur dioxide, radioactivity). Furthermore new technologies such as combined-cycle gas turbines lead to much higher energy production efficiencies.

In Figure 4.15 the share of total primary energy supply supplied by fossil fuels is displayed on the left while the right-hand side shows the share of associated CO_2 emissions for the year 2004. Although coal represents 25% of the total primary energy supply, it contributes 40% of the total CO_2 emissions due to its high carbon content per unit of energy produced.

Carbon dioxide emissions also depend on the energy use sector considered. This is displayed in Figure 4.16, which compares the data for 1971 and 2004. Considering the year 2004 we see that heat generation and transportation account for almost two-thirds of CO_2 emissions associated with fossil fuels. In 1971 this was only 46%.

4.11. CONSEQUENCES

That we are in a global warming period is indisputable. The evidence suggests that this warming trend is very likely due to the rise in greenhouse gas concentrations which induces a positive radiative forcing. We have already discussed the increase of the global temperature in a previous section. Since the

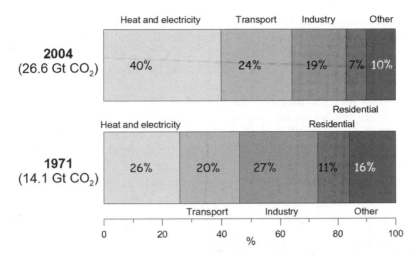

Figure 4.16. World CO$_2$ emission by sector. "Other" includes services, agriculture, energy industries other than electricity, and heat generation. *Source*: Data from *CO$_2$ Emissions from Fuel Combustion, OCDE 2006*, IEA, Paris.

1950s, a decrease in the number of very cold days and nights has also been observed, while the number of very hot days and warm nights has increased. Additional signals of a global warning can be found in the rising sea levels, melting glaciers, sea ice retreating in the Artic, and decrease of snow cover in the northern hemisphere.

This global warming trend also leads to increased evaporation of the oceans as well as the biomass, rivers, and lakes. Due to global warming, this leads to an increase of the water vapor concentration in the atmosphere. An increase of about 5% in the atmosphere over the oceans is estimated. This leads to more and heavier precipitation in certain localities, but the climate may vary differently from place to place. For example, wetter regions are observed in the northern part of Europe while drier ones are seen in the northern part of Africa. The eastern regions of North and South America have become wetter. In the long term several years of drought separated by years of heavy rains may happen more frequently. There are already some indications of that in the southwestern part of the United States where a wet winter in 2004–2005 followed a six-year drought. There has also been an increase in intensity and duration of tropical storms and hurricanes. A greater number of stronger storms are also observed out of the tropical regions. However, one has still to be cautious because there is not yet a definite proof that these observed facts are directly related to global warming.

Observations already show a decline of snow on a global scale. The snow cover has decreased by about 8% compared to the 1960s. Alpine glaciers are getting smaller and in the long run could even disappear if the increase in the global temperature were to reach 5 °C. The sea ice area in the Artic is decreasing by about 8% each year and could disappear by 2050, as indicated by model calculations. The melting of glaciers, the ice cap, and the ice sheet during the period 1993–2003 has resulted in an increase of about 1.2 cm in the sea level.

Estimates indicate that the melting of ice has caused the sea level to rise about 17 cm during the twentieth century. This is more than occurred in the 2000 previous years. Depending on the model calculation, the sea level is predicted to rise another 22–44 cm by the end of the century. The thermal expansion of the ocean due to the increase of the average global temperature is expected to contribute to more than half of this rise in the sea level.

4.12. OTHER IMPACTS ON OCEAN

In the long run there are concerns about other possible sizable impacts of global warming on the oceans. These concerns may be overestimated due to the thermal inertia of the ocean. Any major changes will occur over a long timescale.

Oceans play a key role in Earth's climate management because they redistribute heat around the globe. The ocean is a large thermal reservoir that can exchange energy with the atmosphere but is out of equilibrium with it. The average depth of the ocean is equal to 3800 m and the heat capacity of the ocean is extremely high compared to that of the atmosphere. The heat capacity of the atmosphere is only equivalent to a part of the ocean corresponding to a 3.2-m depth. Interactions between the ocean and the atmosphere take place at the surface. The active layer of the ocean that is involved in energy exchange depends on the region and on the season. It can be less than 50 m in spring or early summer and over 100 m in autumn or winter.

The ocean has both surface and subsurface currents. The Gulf Stream is the fastest ocean current in the world with peak velocities reaching 2 m/s. It extends along the North Atlantic Drift. The flow is about $30 \times 10^6 \, m^3$ in Florida and can reach up to $150 \times 10^6 \, m^3$ near 65 °W. It starts in the Gulf of Mexico and exits through the strait of Florida. Driven by powerful winds from the southwest it then follows the eastern coast of the United States and Newfoundland before it crosses the Atlantic Ocean and splits in two parts—one going to northern Europe and a second one to western Africa—before it sinks to feed the thermohaline circulation. Some scientists think that global warming may slow down the Gulf Stream and even change its path.

The density of ocean water varies from place to place because the temperature and the salinity are different. There are special currents driven by these variations in temperature and salinity. In some high-latitude regions of the North Atlantic cool saline water sinks in the deep ocean. This leads to a cold deepwater flow moving to the south. The low-salinity water wells up in tropical or subtropical latitudes. This constitutes the thermohaline circulation. The thermohaline circulation is a closed loop (a so-called ocean conveyor belt). It takes about 1000 years for a water molecule to travel the loop. It was previously thought that global warming might slow down the thermohaline circulation and even stop it. This would have severe consequences on Earth's climate. However, scientists are now more cautious in their conclusions.

Global warming may also have a significant effect on other ocean current phenomena. El Niño is an ocean–atmosphere phenomenon originating in the Pacific Ocean which leads to major Pacific Basin climate fluctuations every 3–8 years. It generally lasts about 18 months. The name El Niño comes from the fact that it develops near Christmas time. It is also referred to as the El Niño Southern Oscillation (ENSO). The first sign of appearance of the phenomenon is a considerable increase of wind toward the west. This wind drives to an accumulation of warm water in the Western Pacific and the level of water increases. But, as soon as the power of the wind decreases, the warm water of the Western Pacific flows to the Eastern Pacific as a large warm sea current of an area about 1.5 times the area of the United States. This warm water leads to unusual weather conditions ranging from torrential rains to drought depending upon location. Little can be said about the impact of global warming on a phenomenon like El Niño. The only thing we know is that, since the 1970s, this phenomenon has occurred more frequently, lasts longer, and is more intense. Whether these changes are caused by global warming is not yet understood.

4.13. FACTOR 4

The "factor 4" concept, first introduced in a report to the Club of Rome in 1997, initially advanced the idea that the world could double wealth by halving the amount of natural resources consumed on a worldwide basis. Later this concept was restricted to greenhouse gases and then even further to CO_2 emissions.

Basically, to stabilize CO_2 emissions at the world level, we should emit the same quantity of CO_2 as can be removed from the atmosphere by natural processes. The anthropogenic emission of CO_2, around 7 GtC/year, is currently twice that number. Halving that to ≈ 3.5 GtC/year for a population of 6.5 billion inhabitants would mean about 500 kgC/year per inhabitant. This is a small quantity. It corresponds to the carbon emission of a midsize car driven over 10,000 km. Currently emissions average about 1 ton of carbon per year per inhabitant. In the United States the emissions are eight times the average. In France they are four times the average.

In 2003, the French government decided to embark on a program to reduce CO_2 emissions (which account for 70% of the French greenhouse gas emissions) by a factor 4 by the year 2050. While this is a very desirable goal, a

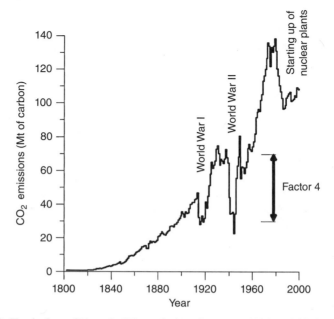

Figure 4.17. Evolution of French CO_2 emissions between 1800 and 2000. *Source*: Data from the Dioxide Information Analysis Centre, Oak Ridge, TN, www.cdiac.ornl.gov.

simple evaluation done on the back of an envelope shows that decreasing the French CO_2 emissions by a factor of 4 is really a very difficult job and would require constraints on life-style that the population is probably not ready to accept. To put the problem in historical perspective, consider Figure 4.17, showing the evolution of the annual French CO_2 emissions. Over a period of 200 years there was a regular increase in emissions associated with an increase in economic activity and living standards. One observes three significant dips. The first two correspond to the two world wars, during which the economic activity slowed significantly. These periods saw important declines in the living conditions of the people. The third dip corresponds to the commissioning of the nuclear reactors which replaced fossil fuel–driven facilities. The simple historical observation that a factor-of-4 decrease in emissions occurred only during the catastrophe of World War II suggests that achieving such a reduction is no easy task.

On the other hand, decreasing French CO_2 emissions by of a factor of 2 by the middle of the century, though very difficult, is probably possible given the evolution in energy sources and in the way energy is used.

The energy consumed annually in France is apportioned very roughly in the following way: 500 TWh in the form of electricity, 500 TWh for transportation, and 1000 TWh for the remainder, mainly heating and cooling. Since 90% of France's electricity is already produced without CO_2 emission (hydropower and nuclear energies), one can gain relatively little in this sector by expanding the facilities of electricity production based on renewable energies and nuclear power, 10% at best. However, significant reductions are possible in the other two areas:

- Innovations addressing the energy needs of transportation are necessary to address this area of consumption. Plug-in vehicles might meet 30% of transportation needs and second-generation biofuels could be employed to meet 20%. However, both of these technologies are still far from the industrial stage.
- Significant progress is possible in the heating and cooling sector, in particular by using heat pumps at a large scale, by insulating the buildings, with better monitoring and control of energy flow, and so on. In this field a factor of 4 can be achieved with existing technologies.

In sum, all of these possibilities could reduce emissions by a factor of 2.

These simple estimates show that the objective of a-factor-of-4 decrease of CO_2 emissions will be very difficult for France to reach by the 2050 horizon.

This illustration for a particular developed country shows the difficulty of emitting only as much CO_2 as nature can absorb. The problem is even more difficult to solve for the United States, where long traveling distances are common and where air conditioning is often necessary. The question is therefore to what reasonable extent can we realistically reduce CO_2 emissions in order to mitigate climate change?

4.14. KYOTO PROTOCOL

In the late 1950s a few American scientists warned the public that greenhouse gas emissions might lead to future problems. This idea later received the support of a small community of scientists, and measurements to monitor the CO_2 concentration in the atmosphere were begun in Hawaii and Antarctica. Between 1960 and 1980 more extensive and precise measurements were made to study the role of CO_2 in climate change.

Since, the ratio of ^{14}C to ^{12}C in fossil fuels is smaller than for living systems, measurements of radioactive ^{14}C have allowed the tracking of carbon coming from fossil fuels. More recently, several analyses of ice cores taken in the Antarctic have allowed deductions of the atmospheric CO_2 concentration and the temperature during the last 800,000 years. These data show a strong correlation between average global temperature and CO_2 concentration. According to this record, Earth has gone through a number of warming and cooling periods. For example, about 180 million years ago, in the early Jurassic Era, an important warming of 5 °C was associated with a rapid increase in the CO_2 concentration, perhaps arising from volcanic activity and/or gas hydrate decomposition.

The first initiative to reduce anthropogenic greenhouse gas emissions was proposed in 1992 at the United Nations Conference on Environment and Development in Rio de Janeiro. The idea was to maintain the anthropogenic concentration of greenhouse gases at a level that would not induce a dangerous modification of the climate. An agreement, the United Nations Framework Convention on Climate Change (UNFCCC), was achieved which would lead later to the Kyoto Protocol in December 1997. The Kyoto Protocol is an international treaty involving 174 countries and governmental entities belonging to two categories: annex I parties who have accepted greenhouse gas emission reduction obligations and non-annex I parties who have not accepted reduction obligations but may participate in the clean development mechanism.

The objective of the Kyoto Protocol is to achieve a "stabilization of greenhouse gas concentrations in the atmosphere at a level that will prevent dangerous anthropogenic interference with the climate system." It requires that, by 2012, industrialized countries reduce their greenhouse gas emissions by an average of 5.2% relative to 1990 levels. The actual objectives vary with country. For example, the European Union should reduce emissions by 8%, Canada and Japan by 6%, Russia should remain constant (0%), while Australia and Iceland may increase their emissions by 8 and 10%, respectively. The United States is supposed to reduce emissions by 7%. China, India, and other developing countries are non-annex I parties and have no commitment to reduce their emissions. Most of the participating parties have officially ratified the protocol. However, a few, including the United States, have not. The agreement came into force on February 2005 after being ratified by at least 55 countries accounting for at least 55% of the global greenhouse gas emissions. The 5.2% reduction may appear to be small. Without these constraints, however, the predicted normal rate of increase of greenhouse gas emissions would lead to 30% more emissions in 2012 compared to the 1990 level.

Emissions of six greenhouse gases are covered by the Kyoto Protocol: CO_2, CH_4, N_2O, SF_6, HFCs, and PFCs. All the members within an annex group do not have the same constraints. Within the European Union, for example, Germany must reduce its emissions by 21%, Portugal can increase them by 27%, and France, which had already strongly reduced its emissions due to nuclear energy, must stabilize its emissions.

Flexible mechanisms have been introduced to help reach the overall goal of the Kyoto Protocol:

- An emission trading market has been created for exchange of carbon credits. An industrial enterprise which anticipates exceeding its emission quota can buy credits from others producing less.

- The Clean Development Mechanism allows annex I parties to invest in projects that reduce emissions in developing countries. This allows creation of new carbon credits and is an alternative to carrying out more expensive reductions in their own country.
- The Joint Implementation Mechanism is similar to the Clean Development Mechanism but employs carbon credits associated with a specific project in an annex I country.

Unfortunately there are potentially undesirable effects of these mechanisms, one being externalizing greenhouse gas emissions but also externalizing the associated jobs.

The Kyoto Protocol has proved very important in educating people about the importance of greenhouse gas emissions and preparing them for future stronger emission regulations and changes in life-styles. From the scientific point of view the impact of the Kyoto Protocol on climate change is small. It is estimated that if the Kyoto commitments are fulfilled, the global average temperature would be reduced by only about 0.15 °C, and this would delay global warming effects by about four or five years. From the political point of view, the commitments of the Kyoto Protocol have important economic and social costs that some countries are not ready to accept.

It is already proving difficult to meet the Kyoto Protocol commitments and post-Kyoto discussions are meeting with difficulties. At the end of 2007, the Bali conference reached no agreement on quantitative objectives for anthropogenic greenhouse gas reduction. The real issue is a cost–benefit one. It has to be decided whether it is better to invest now to prevent global warming and its consequences or to wait to address the problems it will generate. Many economists think that prevention has the more favorable cost–benefit ratio. Others argue that the funds required to address this problem would be better used to address more pressing problems.

The Nicholas Stern report, published in October 2006, says that the economic cost of climate change would amount to about $5500 billions if nothing is done to prevent it. This represents something like 5% of the world's GDP (gross domestic product) per year. This report estimates that the cost of greenhouse gas emission reductions would be 1% of the GDP per year. Of course, these evaluations must be taken with great care. However, they indicate that it is economically advantageous to prevent the problem rather than to manage it when it occurs.

4.15. CONCLUSION

Anthropogenic greenhouse gas emissions are likely to cause an important climatic change on planet Earth. Greenhouse gases are invisible and distributed throughout the planet's atmosphere. A solution to the emission problem cannot occur without the participation of most of the inhabitants of Earth. This is difficult to obtain because so many diverse economic and political interests are involved that the drastic measures required will be difficult to accept.

Returning to the composition of the atmosphere that was present before the industrial revolution is not feasible. However, stabilization of the current concentrations in the atmosphere appears to be an achievable goal within a sustainable economic development. There is however some urgency if this is to be accomplished. Lowering emissions today will have an effect only in the long run. Nevertheless, it is very desirable that we act now to decrease our greenhouse gas emissions. This will be useful to mitigate global warming. In the longer term, it is necessary and possible to employ more carbon-free energy sources and decrease our demand for fossil fuels.

■■■■■■ **CHAPTER 5**

Energy from Water

Humans have used water as an energy source for a long time. The flow of liquid water can be exploited and used for power generation. In a country where it is available, hydropower is one of the first resources developed to produce electricity. This chapter is devoted to means of extracting energy from water by either using hydropower generation or harnessing the energy from the ocean.

5.1. HYDROPOWER

Moving water contains energy which can be used. It has been employed for centuries to do mechanical work, for example, milling grain. The Roman Empire used water to power equipment for the sawing of timber and stone. Water was used for a long time for hushing techniques in the extraction of metal ore. Water produced a large part of the mechanical energy used in the preindustrial era. For that reason, growing industries were located close to rivers. During the Middle Ages and until the nineteenth century, paddle wheel–operated water mills were common. The power delivered by these water mills was small, below about 10 kW. The efficiency was also small, of the order of 20% or so. Nevertheless these were very important installations.

The turbine appeared in the nineteenth century. In 1827, a French engineer, Fourneyron, built the first completely submerged turbine. The system was installed in a waterfall of 1.4 m height at Pont-sur-l'Ognon, France. The power delivered was small, 4.5 kW, but the efficiency was 83%, which is more than three times larger than that of paddle wheels. Later, in 1837, Fourneyron used an 108-m head in the Black Forest and used penstocks to feed the turbines. After these first attempts, the technology of turbines was improved in France, England, and finally the United States by Pelton and Francis, who gave their names to turbines which are still used today.

Our Energy Future: Resources, Alternatives, and the Environment
By Christian Ngô and Joseph B. Natowitz
Copyright © 2009 John Wiley & Sons, Inc.

..1.1. Hydropower: Important Source of Electricity

Hydropower is a renewable energy source which is governed by the solar energy reaching the earth. About 22% of the solar power incident on the earth is used in the evaporation of water. Water evaporated mainly from the sea but also from rivers, lakes, vegetation, and so on, forms clouds which are blown by the wind and travel in the sky. Under appropriate conditions the water vapor contained in the clouds condenses into rain, snow, or hail. Although precipitation is an intermittent phenomenon and corresponds to a small amount of energy (36 liters of water falling from a height of 10 m carries only 1 Wh), a natural concentration of potential energy occurs as the water flows on the ground, reaching streams, rivers, and then the sea. This flow may be harnessed to do useful work and produce electricity. Natural or artificial reservoirs may be used to store large amounts of water which can then be used on demand to produce electricity.

Hydropower is the major renewable energy source employed to produce electricity. Globally, hydropower production of electricity has increased dramatically over the last 60 years. For example, global hydropower electricity production in 1950 was 340 TWh. This met a little more than one-third of the global electricity needs. It reached 680 TWh in 1960, 1150 TWh in 1960, 1500 TWh in 1975, and 2994 TWh in 2005. The latter may be compared to the global consumption of 15,000 TWh and global production of 18,306 TWh. The presently installed hydropower capacity in the world is about 780 GW. Although absolute production has increased, hydropower's share of the global electricity production has actually decreased. For example, if we focus attention on the United States, the installed capacity increased from 56 GW in 1970 to 95 GW in 2008 (an almost 70% increase). Nevertheless the contribution of hydroelectricity has fallen to 10% whereas this contribution was 14% twenty years ago. This is because the demand has increased faster than has hydropower capacity. In many developed countries most of the possible areas where hydropower facilities can be installed have already been used.

> By the end of the eleventh century there was, in England, about one watermill per 400 inhabitants (≈5600 watermills for the whole country). In France, in 1848, there were 22,500 watermills. At that time there were only 5200 steam engines.

Hydropower produces 20% of all the electricity in the world. In emerging countries, hydropower produces one-third of the electricity used and there are many run-of-river hydrostations of about 5 kW. It is estimated that streaming water on the earth contains an energy of around 40,000 TWh/year. About 15,000 TWh/year is technically exploitable but only part of it is economically profitable and is acceptable in terms of environmental impact. Most of the new possibilities for development are located in the former Soviet Union, South Asia, and South America.

In 2007, in France, the installed capacity of hydropower was 20.4 GW and the electricity production amounted to 42.6 TWh. This represented 8.8% of the total electricity production of France. This was provided by 447 hydropower plants and 220 dams. The power of the hydropower plants ranged from a few tens of kilowatts to 1800 MW.

The distribution of hydroelectricity production between OECD and non-OECD countries is displayed in Figure 5.1. A little more hydroelectricity is produced in non-OECD countries where this renewable energy source contributes a larger share of the total energy production.

ITAIPU: LARGEST HYDROPOWER PLANT IN WORLD

The largest operational hydropower plant in the world is Itaipu, located on the Paraná River. Developed jointly by Brazil and Paraguay, it produces 25% of the electric power consumed by Brazil and 90% of that consumed by Paraguay. The output of the power station was originally 12.6 GW (18,700-MW units). The first generator entered into service in May 1984, the last on April 9, 1991. The station was extended in 2006 with two new generators, increasing the output to 14 GW. The annual production of Itaipu is greater than 100 TWh. The construction of this station required a quantity of iron and steel equivalent to that of the construction of 380 Eiffel Towers.

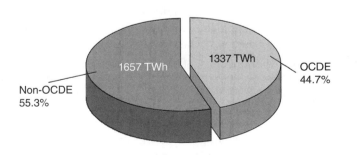

Total = 2994 TWh

Figure 5.1. Distribution of hydroelectricity production between OECD and non-OECD countries, 2005. *Source*: Data from www.iea.org.

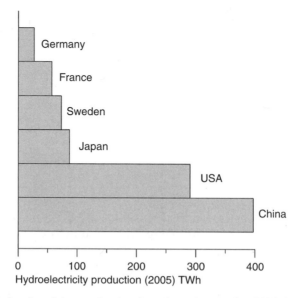

Figure 5.2. Hydroelectricity production for selected countries, 2005. *Source:* Data from www.iea.org.

The hydroelectricity production for selected countries in 2005 is displayed in Figure 5.2. We see that China and the United States are large producers of hydroelectricity.

THREE GORGES PROJECT

The Three Gorges Dam is located on the Yangtze River in China. The decision to build it was made in 1992 and the work started in 1994. (A dam project started in 1919 and another one in 1944 were both abandoned.) It will be completed in 2009 and will be the largest area reservoir in the world, but second in terms of production of electricity. The leader in that will remain the dam of Itaipu referred to above. The Three Gorges Dam is 2.3 km long and 185 m high. It required $27 \times 10^6 \, m^3$ of concrete to build it. The reservoir has an area of 1084 km^2 and flood storage capacity is 22 km^3. It stretches about 600 km upstream. Filling of the reservoir began in 2003. The installed capacity (18.2 GW) is expected to represent 10% of the capacity installed in China, but, due to the fast increase of demand for electric power, it may represent only 3% of the consumption of China by the time the hydropower station is completed. The power station started commercial operation in 2003, with a working installed capacity of 5.5 GW. The production of electricity by this power station will be approximately 84.7 TWh/year.

Figure 5.3 shows the top-10 producers of hydroelectricity in 2005. China's production of about 400 TWh is equivalent to the production of about fifty 1-GW$_e$ nuclear plants.

Even flat countries can produce electricity with hydropower. This is, for example, the case of the Netherlands, which has an installed capacity of 38 MW.

The distribution of hydroelectricity production between the different regions of the world, as defined in the BP annual review, is shown in Figure 5.4.

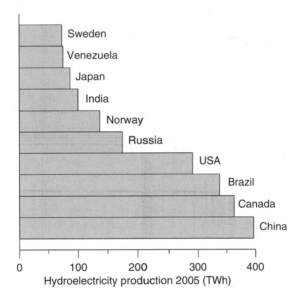

Figure 5.3. Top-10 producers of hydroelectricity, 2005. *Source:* Data from www.iea. org.

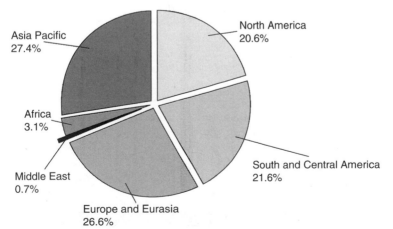

Figure 5.4. Distribution of hydroelectricity production between different regions of world, 2007. *Source:* Data from BP annual statistical review.

5.1.2. Dams and Diversions

A dam is used to store a large amount of water which is used later to produce electricity. The reservoir created by the dam helps to concentrate the water flowing from a river, for example. The dam is used to raise the level of the river and to create falling water. There are several types of dam. Their structure varies according to ground conditions. Some, like gravity dams, rest on the ground. Others lean on the banks of the river if the banks are stable enough. In Figure 5.5 some types of dams used in hydropower plants are schematically displayed. Raising the water level of the river creates a height difference between the upstream part of the dam and the downstream part. This difference is called the hydraulic head or simply the head.

> It is also possible to generate electricity without dams. This process is known as diversion or run of the river. The principle is to divert part of the water of fast-flowing rivers through a penstock to a turbine. Diversion is used, for example, in Niagara Falls. Another run-of-the-river technology would be to use a water wheel put on a floating platform. It is a low-cost technology which can be easily implemented. However, the amount of energy which can be harnessed is small, of the order of what our ancestors could harness. For example, covering the entire Amazon River with such platforms would correspond to a power of only 650 MW.

Dams must be watched closely because there is a risk of accidental failure. Large dams can enhance seismicity. Dams are presently monitored by routine surveillance and maintenance. Nevertheless, by the end of the twentieth century, about 0.6% of all dams have failed.

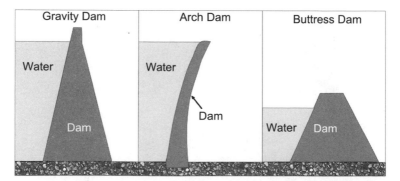

Figure 5.5. Examples of dams.

There are a number of examples of dam failure and some of them have resulted in many casualties. For example, at Vaiont, in Italy, $270 \times 10^6 \, \text{m}^3$ of rocks slipped in 1963. About $169 \times 10^6 \, \text{m}^3$ of water was released, killing 2118 people in the river valley. On August 11, 1979, a dam located about 6 km upstream from the city of Morvi collapsed. Around 5000 people died and the urban and rural areas downstream were destroyed. Between 1959 and 1987 there were 30 serious dam failures, causing about 18,000 casualties.

5.1.3. Head and Flow

The relevant parameters of a hydropower site are the effective head H, which is the height through which the water falls, and the flow rate of water, Φ. The power which can be recovered is just the potential energy per unit of time due to the flow of the water falling from height H:

$$P(\text{kW}) = H\Phi g \approx 9.8 \times H\Phi$$

where g is the acceleration due to gravity ($9.8 \, \text{m/s}^2$). This is of course the maximum theoretical power because the system has less than 100% efficiency for harnessing this mechanical energy and converting it into electricity. The mechanical yield (transformation of the kinetic energy of the fluid at the blades of the turbine to the shaft of the generator) can be larger than 90%. The rotational energy obtained at the shaft is then converted into electricity with a generator which can have a yield larger than 95%. Therefore, the overall efficiency for producing electricity from the water flow is of the order of 85%. Since there are also flow losses of the water before reaching the turbine blades, the yield factor between the reservoir potential energy and the electricity output is of the order of 75–80% depending upon the site and the technology. This is much larger than any fossil fuel electricity plant.

The same output power can be provided by different kinds of hydroelectric plants. It can be provided by a high head reservoir with a small flow, as is the case in the mountains, for example, or by a low head with a large flow, as in rivers. A high head corresponds typically to a difference of height larger than 100 m between the low-altitude part and the high-altitude part. A low head corresponds to a difference smaller than about 10 m. A medium head is in between. Actually, the limits between the different classes are not so sharp. With a high head, the penstock, which is the pipe carrying the flow, has to resist high pressures. Indeed, because of the high density of water, the pressure increases by 1 atm every 10 m. A 1000-m head means an additional pressure of 100 atm at the outflow level. A schematic showing hydropower for different classes of heads is displayed in Figure 5.6.

Dams are also used in pumped storage systems. These are treated in Chapter 12 on energy storage.

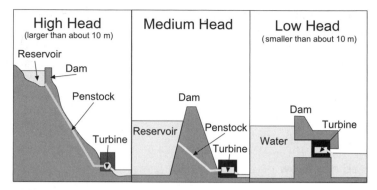

Figure 5.6. Schematic of different kinds of heads depending upon situation.

5.1.4. Turbines

Turbines are machines which harness the kinetic energy of water and transform it into rotational energy. There have been many different types of turbine technology developed for this purpose. Turbines can be basically classified in three main categories: reaction turbines, impulse turbines, and axial flow turbines. Reaction and impulse turbines use not only the kinetic energy of water but also its pressure. The choice of turbine technology to be employed depends on the nature of the hydropower site.

A typical reaction turbine is the Francis turbine. A Francis turbine has a runner (the turning part of the turbine) with nine or more vanes. Powers up to 800 MW can be reached. It is notable that the same turbine technology can be used over a large dynamical range of powers, from small ones to huge ones. It works completely submerged and can run with its axis horizontal or vertical. It is well suited for a head of water in the range of about 5–400 m. This technology is widely used in medium- and large-scale hydropower plants. The turbine is put inside a housing allowing a continuous pressure against the blades. Francis turbines are the most efficient when the speed of the blades is a little smaller than the speed of water flowing by them. In reaction turbines, water enters radially but leaves the turbine axially in parallel to the shaft, which rotates the generator and produces electricity. Efficiencies of the order of 95% can be achieved with Francis turbines run in optimum conditions. The efficiency can be smaller if the direction of the incoming water relative to the runner blades is not correctly adjusted. This may happen if it is necessary to reduce electricity production by reducing the input flow.

Impulse turbines (Figure 5.7) are better suited for high heads, typically higher than 250–400 m. In impulse turbines, one or several jets of pressurized water impinge on vanes or cups located on the perimeter of a wheel. Almost all the incident kinetic energy of water is transferred to the wheel. The best known turbine of this type is the Pelton wheel. Efficiencies of the order of 90% are obtained in large installations, but the efficiency can drop to about 50% in small ones.

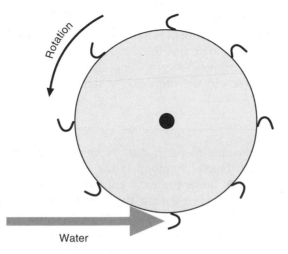

Figure 5.7. Principle of impulse turbine.

The potential energy of a mass M of water with an effective head H is MgH. The kinetic energy is $\frac{1}{2}Mv^2$, where v is the velocity of the mass M. If all of the potential energy is converted into kinetic energy, we have $\frac{1}{2}Mv^2 = MgH$. Therefore $v = (2gH)^{1/2}$. This is the maximum velocity of the water. For an effective head of 1000 m the maximum velocity of the water arriving at the turbine blades is 140 m/s = 504 km/h. The flow rate (volume flowing out in each second) is equal to the product of the area S times the velocity v, or $\Phi = Sv = S(2gH)^{1/2}$.

Since the power is P (kW) $= H\Phi g$, we have P (kW) $= HS(2gH)^{1/2}g \approx 43 \times S$ (m²) $[H$ (m)$]^{3/2}$.

If there are many jets, as on a Pelton wheel, which has several buckets in the perimeter, the output power is multiplied by the number of jets.

The output power of a Pelton wheel can be varied by changing the flow rate. In contrast to reaction turbines, Pelton wheels operate in air. There are variants of the Pelton wheel, but the principle is the same.

Axial flow turbines are also called propellers. They look similar to boat propellers with a runner and blades. They are of interest for small heads of water, typically between 3 and 30 m, with a large water flow. A propeller usually has a runner with three to six blades in which the water flows. Water is always in contact with the blades. The optimum speed of the blades is about twice that of water, allowing a fast rotation even for small water velocities. There is an axial flow along the axis of the turbine. The efficiency can be varied by changing the angle of the blades with respect to the incident water flow.

Propellers are particularly well suited for large volume flows and are used in low-head situations. Many propellers have blades with a fixed pitch. They have a high efficiency at full load, but this efficiency can drop to about 50% at partial loads. Kaplan turbines are propellers using variable-pitch blades. They provide a high efficiency in any situation (of the order of 90%).

The number of revolutions per minute (rpm) of a turbine depends on the site and the turbine technology. If the turbine is directly coupled to the generator, only limited rates of rotation are allowed. The reason is that alternating current used on the grid has a definite frequency (50 Hz in Europe, 60 Hz in the United States). For 50 Hz frequency, the revolution with a two-pole generator (magnet and a pair of coils) has to be 3000 rpm. More generally, the rpm has to be a submultiple of 3000 rpm. A 20-pole generator would require only 300 rpm. For 60 Hz frequency the number of rpm should be submultiples of 3600.

To summarize, impulse turbines are usually used for high heads (typically larger than a few hundred meters) and small flow rate (typically smaller than $1 \, m^3/s$). Reaction turbines are used mainly for intermediate heads (typically between 10 and 100 m) and medium flow rates (typically between 1 and $100 \, m^3/s$). Finally propellers are used for small heads (typically smaller than 10 m) and high flow rate (typically larger than $10–100 \, m^3/s$).

The rate of rotation of the turbine is also a relevant parameter which is characterized by a so-called specific speed N_s given by the expression

$$N_s = \omega \sqrt{\frac{P}{H^{3/2}}}$$

where ω is the rate of rotation (in rpm), H the effective head (in meters) and P the power (in kilowatts). Each family of turbines has a range of N_s at which they operate the best: 10–80 for propeller turbines, 100–350 for reaction turbines, and 70–500 for impulse turbines.

5.1.5. Small-Scale Hydropower

There is no global agreement for defining small-scale hydroelectricity plants. This depends on the country. For example, in France, small-scale hydroelectricity corresponds to installations with a power smaller than 10 MW. But this threshold is larger in the United States (30 MW) and China (50 MW). Depending on the output capacity of the facility, small-scale hydroelectricity

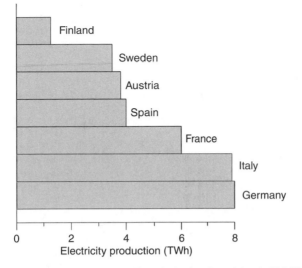

Figure 5.8. Top-seven producers of small-scale hydroelectricity (<10 MW) in European Union, 2006.

facilities are sometimes called mini-, micro-, and pico-hydroelectricity. In developed countries there are often incentives to develop small-scale hydroelectricity, but there are also several barriers preventing quick development due to regulatory environmental constraints. However, small-scale hydroelectricity is worth developing because it is a renewable resource which does not emit CO_2 during operation.

Figure 5.8 shows the top seven small-scale hydroelectricity-producing countries in the European Union. Germany and Italy produce about 8 TWh/year and France about 6 TWh/year. This is about what is produced by a 1000-MW$_e$ nuclear plant over a year. In Figure 5.9 the installed small-scale capacity is shown for the top seven hydroelectricity-producing countries in the European Union. This gives an idea of the overall efficiency of such installations. As we can see from the figure, Germany has a pretty good efficiency since its produces more electricity with a smaller installed capacity.

5.1.6. Environmental Concerns

As for any other source of energy, extensive use of hydropower has an impact on the environment. Some impacts are positive, others are negative. Small-scale hydropower has a relatively low impact because the civil engineering requirements are smaller. Environmental concerns are more visible for larger scale hydroelectricity installations because large quantities of energy are produced with these sources.

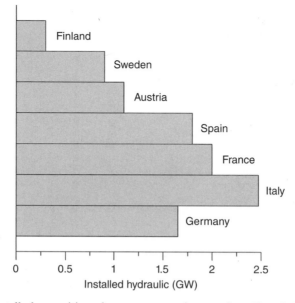

Figure 5.9. Installed capacities of top seven producers of small-scale hydroelectricity (<10 MW) in European Union, 2006.

No CO_2 emission is produced during the production of electricity using hydropower. Some is emitted during the construction of the hydroelectric facility but in the end the amount of CO_2 emissions per kilowatt-hour produced is very small compared with fossil fuel sources even if the whole life cycle is taken into account. It is estimated that large-scale hydropower emits between 3.6 and 11.6 g CO_2/kWh. The amount of NO_x emissions is negligibly small (3–6 mg NO_x/kWh) and it is the same for SO_2 (9–24 mg SO_2/kWh). Small-scale hydropower emits about 9 g CO_2/kWh and other emissions are similar to large-scale hydropower.

Very large dams have an environmental impact during construction, as is the case of any large civil engineering work: emissions, dust, noise, accidents, and so on. There is sediment disturbance and a possible impact on water quality. When a dam is built, the river habitat is replaced by a lake habitat. Relocation of populations who were living there can be a critical issue. For the Three Gorges Dam close to 2 million people have been relocated. Furthermore, building a dam submerges upstream land. Often the land which is submerged was fertile land or had a rich wildlife habitat. If the reservoir is flooded without taking care of the existing plant and animal life, problems may occur. This happened in Brazil with the Tucurui Dam. The trees and plants were flooded and rotted in the water. The consequence was a decrease of the oxygen concentration in the water, which killed the fish and the plants living in it. During flooding metals naturally contained in the rocks may be leached and pollute the reservoir. This happened, for example, Canada with mercury pollution. Problems may also occur because the flow o. the river is slowed: Stratification of water temperatures or accumulation of sediments may occur.

Another negative effect of hydropower is that fish cannot travel the river if no accommodations for this are made. This is in particular a major issue for salmon. For this reason solutions such as the building of fish ladders or transporting fish in barges have been developed. Nevertheless the number of salmon going upstream generally decreases as these projects come into operation.

5.1.7. Costs

Hydropower demands high capital expenditures but low operational expenditures. Investment costs range from 1000€/kW for a good site to 6000€/kW for the more difficult small-scale hydropower installations. In Figure 5.10 these investment costs are compared with the investments for other 1-kW sources of power. The overall cost of hydropower depends on the site and on the distance to where the electricity is consumed. After the initial investment has been paid for, which may demand between 15 and 30 years depending upon the facility, the cost of electricity is very low, typically around 2c€/kWh. The great advantage of hydropower is that when a dam is built it can last for one century or so. Large hydropower plants produce electricity at a lower cost than small-scale hydropower plants, typically by a factor of about 2. Hydropower

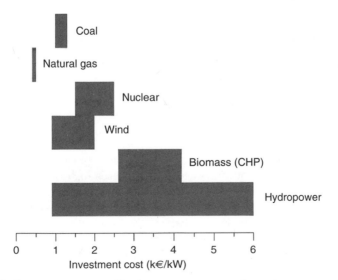

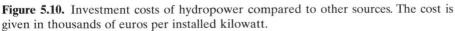

Figure 5.10. Investment costs of hydropower compared to other sources. The cost is given in thousands of euros per installed kilowatt.

is a very competitive source to produce electricity. Actually, it is normally the first energy source exploited to produce electricity when it is available in a country. In France, for example, hydroelectricity was the major energy source used to produce electricity for a long time. Actually, 56% of the total French electricity was produced by hydropower in 1960.

5.2. ENERGY FROM THE OCEAN

With a combined area of $360 \times 10^6 \, \text{km}^2$, oceans cover 71% of the surface of the earth and contain 97% of the total water available on the earth. The average depth of the ocean is 3800 m, much larger than the average height of the land above sea level, 850 m. The total volume of seawater is $1320 \times 10^6 \, \text{km}^3$. The water of the ocean weighs 300 times more than the earth's atmosphere and has a heat capacity 1200 times larger. Annually the ocean absorbs about 37,000 Gtep of energy from the sun. This corresponds to $4.3 \times 10^8 \, \text{TWh}$. These statistics provide a striking illustration of the huge amount of energy which we receive from the sun, energy which is the base source of most renewable energies as well as of fossil fuels. The energy arriving at the surface of the earth is not uniformly distributed: More energy arrives at the equator than at the Poles. About 10% of the energy absorbed by the ocean contributes to heat transfer from tropical areas to the polar regions via ocean currents, such as the Gulf Stream. This heat transfer is of utmost importance for our present climate. It is worth noting that the total amount of ocean energy presently exploited by mankind represents only 1% of the heat energy carried by the Gulf Stream alone.

There are plenty of metals contained in the oceans. The IFREMER (www.ifremer.fr) estimates that $1 \, \text{km}^3$ of ocean contains about 10 tons of zinc; 3 tons of uranium, copper, and tin; 2 tons of nickel; 1 ton of titanium; 500 kg of cobalt; 250 kg of silver; and 5 kg of gold.

Derivation of energy from the ocean is an interesting option because about half of the world's population lives a distance smaller than 200 km from the coastline and close to 40% live less than 100 km from a coastline. The United Convention on the Law of the Sea currently allows countries to exploit sea energy up to a distance of 200 miles from their coastlines. Metropolitan France with a land area of $551,000 \, \text{km}^2$, has access to $350,000 \, \text{km}^2$ of sea area when the Island of Corsica is included. This area extends to $11 \times 10^6 \, \text{km}^2$ if France's overseas territories are included. This is slightly less than the United States, which ranks first with $11.35 \times 10^6 \, \text{km}^2$, and more than Australia, which has $8.2 \times 10^6 \, \text{km}^2$.

If we exclude crude oil, more than a quarter of which has an offshore origin (treated extensively in Chapter 2), we can identify six different energy resources available from the sea:

- Offshore wind energy
- Wave power
- Marine current energy
- Ocean thermal energy conversion
- Tidal power
- Osmotic power

The first four reflect directly the sun's action on the earth. One of these, wind power, is also available on land and the basics of utilization of this source are discussed in detail in Chapter 9. Tidal power originates from the gravitational interaction of the moon. Osmotic power is derived from the thermodynamic drive to equilibrium for solutions of different concentration.

Although there is a large potential for sea energy, the harnessing of this energy, apart from the use of tidal and offshore wind energies, has been relatively small up to the present. Most of these other technologies are at the demonstration or pilot project stage. While so far they could only be competitive in niche markets, they could prove to be very competitive on a large scale in the future as the price of oil becomes very high.

5.2.1. Offshore Wind Energy

The total installed wind power capacity in the world is larger than 75 GW. Of this, 49 GW is found in Europe. Since it becomes more and more difficult in some places to install wind turbines inland, a number of studies and projects have been carried out to put wind turbines in the sea. There are several advantages to that. The most important is that the average wind velocity at sea is larger and more stable than that on land. This allows an increase of load factors up to 40% above those of the same wind turbines installed on the land. Sometimes, load factors may even be doubled. In addition, more space is generally available and few people live close to sites of sea wind turbines. They are often erected in places where the depth of the water is larger than 20 m. Even though there are additional engineering considerations for construction at sea, bringing materials to the sites can prove easier because ship transport has fewer constraints than for inland transportation. Ships have been built, and existing ones have been adapted specifically for transporting and installing offshore wind turbines.

The first sea farm consisting of offshore wind turbines was commissioned in 1991 in Denmark in a region of the Baltic Sea where the sea bed was only a few meters deep. The wind turbines used were similar to those used inland but with reinforced foundations.

We refer the reader to Chapter 9 where offshore wind energy is presented in more detail.

5.2.2. Wave Energy

Wave energy is derived from the wind and may be viewed as a concentrated form of wind energy. In the Atlantic Ocean, for example, the top 20 m of water is swept up to an energy density of about $2.5 \, kW/m^2$. For the European Atlantic coast the mean power contained in sea waves is about 45 kW/m of coastline. This value is obtained at about 50 m depth. Values of the order of 25 kW/m are available, for example, in the southwest coast of Wales. The theoretical annual amount of energy available from wave energy is about 1400 TWh, which represents about 10% of the present world electricity consumption. Installations need to be built to withstand the effects of waves under exceptional circumstances, like gales. In this case waves can exert forces 10 times stronger than normal ones.

How do we arrive at estimates of available power such as those given above? The energy carried by a wave is the sum of its potential and kinetic energies. In deep water, the dispersion relation tells us that the wave frequency is proportional to the square of the wave period T. Consequently, the power of a wave is proportional to the square of the height of the wave h (crest to trough) and proportional to the wave period. The theoretical maximum available power is then $P \approx 1.92h^2T$ and the maximum *extractable* power is $P \approx 0.96h^2T$. This corresponds to half the power of a single deep-water wave. Actually, only 15–20% of P_{th} can be extracted in practice. The power P is therefore between $\approx 0.15h^2T$ and $\approx 0.2h^2T$ kilowatts per meter, where h is expressed in meters and T in seconds. Given transmission losses, the available power represents at best about 10% of the theoretical power of a single deep-water wave. For typical values of $T \sim 10 \, s$, wave lengths of 156 m and crest-to-trough values of $h = 2 \, m$ to $h = 4 \, m$, one gets P values of ≈ 15–20 kW/m. This is the actual power extractable from wave motion which is lower than the theoretical one. The most promising technologies known so far require about 65 km of coastline to get a power of 1000 MW. Projects of that size for the capture and conversion of wave energy will require about 10 million tons of structural material. In terms of surface area, it is estimated that one can harness about $30 \, MW/km^2$.

The available power varies from country to country. Taking the example of France, the largest resources are found in the Gulf of Gascogne with power reaching values of the order of 40 kW/m. The estimated power which can be delivered is about 28 GW. The power delivered along the Atlantic coastline is greater than along the Mediterranean coastline where it is about 10 times smaller (≈ 4–5 kW/m).

There are several techniques currently used to extract the energy from waves and produce electricity. The basic idea is to use elements that are displaced due to motion of the water. Energy may be extracted from the waves because the mechanical forces applied to the different elements of the structure are different. Devices enabling recovery of wave energy have been in development since the 1970s.

Among the current devices for wave energy conversion, the "Salter duck" is the best known (Figure 5.11). It is an asymmetric cam which extracts energy using a semirotational motion. The geometry of the cam is such that its front surface moves as the incident wave touches it but the back surface is designed so that it does not disturb the water behind. For optimal operation, the size of the cam should be large, typically having a radius of about 15 m, which is quite expensive to make.

Another promising technology is the oscillating water column (OWC). An OWC consists of a large column open to the sea at the bottom and having a turbine at the top (see Figure 5.12). When a wave hits the OWC, the water inside the column rises and the air inside is compressed and forced upward. When the wave falls, the air is sucked back down. A specially designed turbine spins when the air is forced upward by a wave and continues to spin in the same direction when the wave drops and the air is drawn back down again. The air turbine at the top of the column is connected to an alternator, which produces electricity.

The first successful OWC device was produced in Japan and was used to power lights for navigation. Other OWCs have been built or are under construction in Japan, Norway, India, China, Scotland, and Portugal. As an example, the island of Islay OWC, located off the west coast of Scotland, has been operating for 10 years at a power of 75 kW. Another interesting example is the one developed in Norway. It will have an electric power of 500 kW and consists of a 19.6-m steel chimney pipe plunging 7 m down into the ocean. Most OWCs are still at the experimental stage.

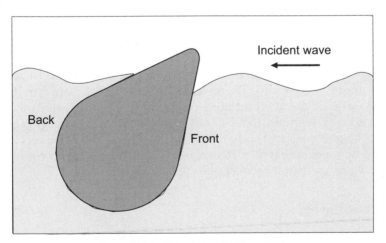

Figure 5.11. Principle of Salter duck.

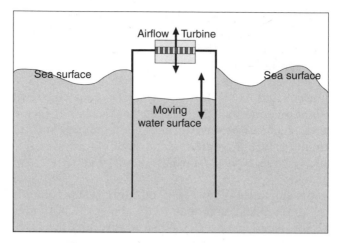

Figure 5.12. Principle of offshore OWC.

A third technology is that of the tapered channel station (TAPCHAN) in which the waves funnelled into a narrowing channel push water up 10–15 m into a dammed reservoir. The water from the reservoir then flows back down to the sea through a turbine which produces electricity.

Techniques employed so far can be categorized as first-generation systems. They have only been used to harness wave energy close to the coastline where the contained energy is smaller because a part has already been dissipated in reaching the shore. Such energy systems, located close to the shore, are only effective in particularly favorable sites. They also have a visual impact which can cause resistance to their installation. Second-generation, deep-water systems are being explored. They would allow the harnessing of wave energy farther from the shore before that energy has been dissipated. These second-generation systems will be real offshore devices. They must be designed to resist strong storms and will have to be associated with energy storage systems to level rates of energy production.

The typical power that can be obtained with such systems organized in sea farms is about 30 MW/km^2. The investment cost varies between 1000 and 3000 €/kW depending on the technology and the area. Anticipated operational times for such systems are about 4000 h/year because wave energy is intermittent (the heights of waves vary from time to time). The price of electricity produced by wave energy is currently between 50 and 100 €/MWh but could decrease when industrial production stages are reached.

5.2.3. Tidal Energy

Harnessing tidal energy is not a new idea and many tide mills, water mills designed particularly to take advantage of the ebb and flow of the tides, have been used in the past. More recently, larger scale harnessing of tidal power has taken place in specific areas. The basic techniques are to block estuaries with a barrier, forcing the water to flow through turbines in order to generate electricity. An alternative is to couple a dam and a storage basin for water. When the tide rises, the dam is open and seawater fills the basin. The dam is closed when the tide goes down and the water which is contained in the basin can be released to flow through a turbine and produce electricity. The principle of tidal power generation is shown in Figure 5.13.

Consider the *useful volume* of the storage basin, V, which is equal to the product of A, the basin's area, and the tidal range R, which is the difference between the heights of the high and low tides. For a country like France, for example, where there are four tides every 24.8 h, the maximum power available is $P_{max} = 0.22AR^2$ megawatts if R is expressed in meters and A in square kilometers. In practical applications, the power available is only one-fourth of this value, or $P = 0.056AR^2$ megawatts. The tidal range varies significantly from place to place and is sometimes negligible. Saint Malo, France, is a place where large-amplitude tides are measured. For example, on March 1, 2006, the difference between high and low tide was close to 13 m. It was only 3 m during the period at which the tidal range is the lowest. Three meters is the average tidal range along the Atlantic coast of North America.

The most expensive part of a tidal power plant is the dam. Its price is proportional to its length L. In order to be economically competitive, the rule of thumb is that the length-to-area ratio of the dam, L/A, should be smaller than 80.

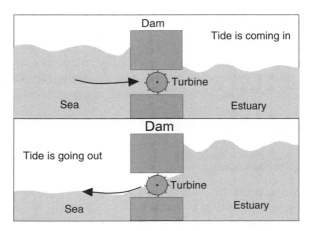

Figure 5.13. Principle of tidal power generation where the sea is separated from an estuary by a dam.

The largest recorded tidal range was in the Bay of Fundy, Canada (17 m). In France it was in the Bay of Saint Michel (14 m). The highest recorded sea wave ever recorded was seen on February 7, 1993, in the North Pacific by the *USS Ramapo*. It was 34 m high. Tsunamis, which are generated by undersea earthquakes, propagate at large velocities in the open sea (velocities can reach 800 km/h, the speed of a plane). Initially, the height of a tsunami wave is usually smaller than 1 m. As tsunamis approach the coast, they slow down and the height of the wave increases. A 1-m wave in the deep sea leads to heights reaching 16 m at the shore. They carry a huge amount of energy and water, which makes them extremely devastating for coastal populations.

The largest plant harnessing tidal power is located in La Rance, France. It has a power of 240 MW and produces about 500 GWh of electricity annually. The electricity production is not continuous and takes place for 4–5 h during each tidal cycle.

Two new projects based on this technology are in progress in China (≈300 MW with an artificial sea wall) and in South Korea (≈250–260 MW). The latter should be finished in 2009 for a total cost of $250 million and is expected to produce about 550 GWh/year. The major limitation of tidal energy as an energy source is that good locations are not so frequent. Furthermore, there is a nonnegligible environmental impact on the coastline. The barriers can prevent both fish and boats from traveling through. The habitat of some birds can be destroyed and sediment, trapped behind the barriers, can accumulate, which has a consequence of reducing the volume of the reservoir.

Initially, there was a strong interest in developing tidal power plants in estuaries where large volumes of water flow through a narrow channel leading to high current velocities. The La Rance power plant in France is such an example. The trend now is to develop tidal fences which are of smaller size and can be installed between small islands or between the mainland and an island. The environmental impact turns out to be smaller. In such systems, the electrical equipment is put above the water, which simplifies maintenance operations. Projects based on this technology are under way in southeast Asia, the first of these in the Philippines.

5.2.4. Marine Current Energy

The availability of marine currents as an energy resource is due essentially to tides and, to a lesser extent, to thermal and density differences in the water. The strength of the current depends on the distance and relative positions of the moon and the sun with respect to the earth. Since this strength varies as a function of the time period, the power of the marine current varies also. The relative importance of the moon and the sun depends on their masses and distances to the earth. The magnitude of the tide-generating force is about 68% for the moon and 32% for the sun.

The strength of marine currents driven by tides depends on the site, the shape of the coastline, and the sea bed. Along straight coastlines and in the middle of deep oceans, marine currents are slow. High marine currents are found in narrow straits, between islands, around peninsulas, and sometimes at the entrances of lakes, bays, or harbors. High marine currents, in terms of water flow, generally exist where the water depth is shallow and the tidal range large. The best sites are found at distances smaller than 1 km from the shore and in water depths of 20–30 m. In the best sites, it is estimated that about 10 MW/km² can be harnessed. More than a hundred promising sites have been identified in the European Union, 40% of them located in the United Kingdom.

The strength of such currents is periodic and follows a sinusoidal curve. The maximum is obtained at midtide. The marine current is almost reversed in direction at ebb tide. It is often slightly larger during ebb tide than at flow tide.

Energy from water currents can be harnessed with water turbines which work on the same principle as wind turbines. The kinetic energy of the moving fluid is transformed into mechanical energy in a turbine and subsequently into electricity. The ratio between the seawater density (1025 kg/m³) and the air density (1.225 kg/m³ at 15 °C) is equal to 837. Therefore, despite the fact that current velocities of water are smaller than those of air, sea turbines are smaller in size than wind turbines for the same mechanical energy output.

As for wind turbines, the power of a sea turbine is given as

$$P = \tfrac{1}{2} K \rho S v^3$$

where ρ is the density of seawater in kilograms per cubic meter, S is the area of the rotor blades in square meters, v is the current velocity in meters per second, and K is a coefficient measuring the efficiency of the turbine. Sea turbines may have either a horizontal or a vertical axis.

Optimum efficiencies for tidal turbines are obtained for coastal current velocities around 2–2.5 m/s. Smaller current velocities are not interesting economically and higher ones put too much stress on the equipment. These velocities are quite small compared to the velocities of winds exploited by wind turbines, which is closer to 15 m/s. Nevertheless, due to the much higher density of water compared to that of air, sea turbines are more efficient and produce an energy density about four times that obtained from wind. Therefore, for a given nominal power, tidal turbines are of smaller size than wind turbines. In general, tidal turbine facilities have a low visual and environmental impact.

5.2.5. Ocean Thermal Energy Conversion

As indicated earlier in this chapter, the oceans covering 71% of the surface of the earth, collect the bulk of the solar energy incident on the earth. This energy is stored in the form of heat. This is a considerable amount of energy. Indeed, it is estimated that, during an average day, tropical seas ($\approx 60 \times 10^6 \, km^2$) absorb an amount of energy equivalent to about 35 Gt of crude oil. This is more than three times the total annual world energy consumption. The exploitable resource is estimated to be about 80,000 TWh/year, corresponding to an installed power of 10,000 GW$_e$. This energy resource is stable in time (there being not much change between summer and winter seasons) and it is available 24 h a day. There are obviously some compelling advantages to recovering even a small part of this energy.

In tropical or subtropical areas, the temperature difference between the surface of the sea and 1000 m below it can be larger than 20 °C. The technology used to harness part of the thermal energy contained in the sea, called *ocean thermal energy conversion* (OTEC), is based on the principle of a heat engine which works between two sources at different temperatures: in this case, the hot one located at the sea surface and the cold one at large depths. The OTEC plants can be floating or land based. The mooring of the floating plants can take advantage of techniques learned from deep offshore technology. Floating plants have the advantage of requiring shorter pipes to reach down to the cold source. However, they have the disadvantage of needing additional systems to transport produced electricity to the shore. This transport is most efficiently done using high direct-current (DC) voltages. Land-based plants do not need power transmission to the shore and there are no mooring costs. However, longer pipes are needed to reach the cold-water source.

The OTEC capital costs are still large ($\approx$5000–10,000 $/kW, that is, about 10 times the investment cost of conventional power systems), and this has prevented its development for the moment. The operating cost is expected to vary from 0.07 to 0.22 $/kWh. In the future ocean thermal energy might be a viable option. There is also the possibility that cold water extracted from the deep sea could be used for air conditioning and nutrients could be extracted from the cold water for use in aquaculture.

5.2.6. Osmotic Energy

Consider two vessels separated by a semipermeable membrane. This membrane is chosen in such a way that it lets water molecules pass through but not salts and other materials (e.g., sand, silt) contained in a water sample. If the first vessel contains freshwater and the second one seawater, there is a difference in *osmotic* pressure at the membrane level. The reason is that the total system is out of equilibrium as far as the chemical potential is concerned. This is analogous to the case of two vessels at different temperature separated by a metal plate: Heat will flow through the plate from the high-temperature vessel to the low-temperature vessel until temperatures in the two vessels are equal. Similarly, water flows from the low-concentration (low-chemical-potential) vessel to the high-concentration (high-chemical-potential) one. Osmotic pressure phenomena can be used to produce energy. If a typical seawater salt concentration is initially 3.5 g/L, the osmotic pressure difference is 28 bars. Actually, a difference of 3% in the salinity is equivalent to a 250-m waterfall.

When the water flows through the membrane, it increases the pressure on one side and decreases the pressure on the other side. Actual plants operate with an overpressure of 10 bars in the salted compartment. A rate of flow through the membrane of 1 m³/s gives a power of 1 MW. With the present technology, a very large membrane area is necessary: about 200,000–250,000 m²/MW. This corresponds to about 4–5 W/m². A schematic of an osmotic device is shown in Figure 5.14.

> Seawater consists of 96.5% water. It contains 3.5% other matter (salt, dissolved gases, organic materials, and solid particles). Eighty-five percent of the dissolved matter is sodium chloride. On the average there is 35 g of salt per liter of seawater. In total there is about 50 trillion tons of salt in the ocean. This amount of salt put on the land surface of the earth would cover it with approximately 320 tons/m².
>
> The salinity of the seawater varies with location. In the Red Sea it is 42 g/L but only 8 g/L in the Baltic Sea. It depends also on temperature and pressure. It increases by 1 g/L when the water is cooled by 5 °C or at 200 m depth.

In the investigations done so far, it is not the direct osmotic pressure which is used to produce electric power but the resulting flow. In the so-called SHEOPP converter, submarine hydroelectric power plants are anchored to the sea floor. Freshwater taken, for example, from a river is brought down from the surface. The best efficiency of this type of technology is at a depth of about

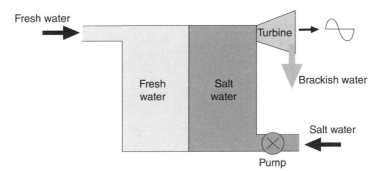

Figure 5.14. Principle of osmotic power generation.

110 m. In the undersea pressure retarded osmosis (PRO) plants, an intake structure at the sea surface brings water to a depth of 90 m (the pressure is 9 bars there). Here the osmotic pressure difference between two saline solutions is exploited.

The main use of osmotic pressure is in reverse-osmosis applications to produce freshwater from salt water. Energy is used to create high pressure on the salt water vessel in order that the freshwater migrates through the semipermeable membrane into the freshwater vessel.

Currently osmotic energy is not economically competitive with other energy-producing methods. There are still significant technological problems to solve. The cost, maintenance, and cleaning of the membranes are among these. If the pores are not clogged earlier, the useful operational lifetime of a membrane is only about six months. However, in the long term, it is an energy resource which may prove to be truly useful.

CHAPTER 6

Biomass

Biomass is living or dead biologic material which can be used as a source of energy. It is a source of carbon which will be of utmost importance in the future. When fossil fuels become less readily available, synthetic fuels can be made from biomass. The biomass inventory is dominated by plant matter, which accounts for about 90% the total but also includes animal matter and biodegradable wastes. Combustible wastes are included in the biomass inventory provided they are derived from living or dead biologic materials. This is the case for wood and crop wastes, animal materials and wastes, and even black liquor (the alkaline spent liquor from the digesters used in paper manufacturing). Also included are municipal wastes coming from residential, commercial, and public service sources. The latter encompasses hospital wastes, but these require special treatment in order to be sure that the process has no negative impact on human health. As commonly used, the term *biomass* does not include organic fossil fuels, coal, oil, and gas, which were formed hundreds of millions of years ago.

Wood is the biomass fuel which has been used the longest as an energy source. Archaeological evidence for the early controlled use of fire using wood as a fuel dates as far back as 1.4 million years ago. However, reliable fire-making techniques appear to have been developed only within the past 10,000 years. By the end of the seventeenth and eighteenth centuries, wood was becoming a scarce resource in some locations because it had so many different uses (e.g., energy, buildings, boats). Yet the energy consumption was much less than today (14 times less for a Frenchman of 220 years ago compared to now). The price of wood was increasing. Coal replaced wood as an energy source. This allowed industry to develop quickly. Astonishingly, the forest area in France today is twice as much as it was before the industrial revolution.

Biomass is a renewable source because the timescale for harnessing that which has been planted is within the timescale of a human life. This is not the case for fossil fuels, which require millions of years to be created. Biodegradable

Our Energy Future: Resources, Alternatives, and the Environment
By Christian Ngô and Joseph B. Natowitz
Copyright © 2009 John Wiley & Sons, Inc.

wastes are also a renewable resource since we can anticipate that humans will continue to produce wastes as long as they exist on the earth.

Biomass is not only useful for producing energy; it is also valuable as a raw feedstock for building materials, paper, biodegradable plastics, and so on. It can be used as well to produce basic chemical compounds for organic chemistry applications and development of new materials.

In addition to being a renewable energy source biomass is also a sustainable one. As long as that biomass which is harvested and burned is replaced by replanting, the impact on the anthropic greenhouse effect (the additional greenhouse gases produced by human activity) is actually much smaller than that of fossil fuels. That is to say that a significant fraction of the CO_2 emitted during burning will be absorbed during the growth of the new plants.

In this chapter, following some general discussion of biomass, we concentrate on its use for production of electricity and heat. We then turn to a discussion of biofuels, which are of growing importance in transportation.

6.1. PRODUCING BIOMASS

Biomass is directly or indirectly produced from photosynthesis. The basic chemical transformation can be written in a simplified form as

$$\text{Sunlight} + CO_2 + H_2O \rightarrow CH_2O \text{(carbohydrate)} + O_2$$

In this process, carbohydrates are produced. Oxygen, required for breathing by all living animals, is also produced in the photosynthetic process. For each mole of CH_2O produced (30 g) this reaction absorbs 500 kJ of energy. This energy is provided by the sunlight. We see that CO_2 and H_2O are needed. Carbon dioxide is plentiful everywhere in the air but there are regions, like the deserts, where little water is available.

Oxygen (O_2) was not initially abundant in the earth's atmosphere. It was initially released about 3.8 billion years ago by blue algae (which could be considered as polluters since they definitely changed the nature of the early environment). At that time the UV rays coming from the sun were not stopped by the earth's atmosphere and life could only develop in the sea or in protected places. Living beings adapted themselves, with the breathing mechanism, to survive in this new medium and obtain their energy needs from this gas. The oxygen concentration in the earth's atmosphere progressively increased, reaching a concentration of 20.95% by volume in modern-day air. With the development of an ozone (O_3) layer all around our planet, the most dangerous UV rays were stopped and it became possible for living beings to leave the sea and settle on land.

Plants and other organisms need each other to maintain an equilibrium concentration of O_2 in the atmosphere. Our planet contains a large variety of chemically reducing species capable of reacting with O_2 molecules and removing them from the air. In the absence of photosynthesis or breathing, the oxygen concentration in the air would gradually decrease. The oxygen would disappear from the atmosphere over a period of approximately 4 million years.

The total exposure to the sun varies with region on the earth, as does the mean energy received. For example, the average power received in the United States is $185 W/m^2$, while it is only $130 W/m^2$ in France. In Brazil or Australia it is $200 W/m^2$. Close to the Red Sea an average power of $300 W/m^2$ is realized. If the efficiency of photosynthesis were 100%, it would be possible, with an average solar radiation of $160 W/m^2$, to produce about 3000 t/year of dry biomass per earth inhabitant. This corresponds to about 1225 t/year of carbon per inhabitant. In fact, the efficiency is far from 100%. This results from the absorption spectrum which limits the range of wavelengths useful in photosynthesis and on the quantum efficiency of photosynthesis. Taking these into account, the theoretical yield is 9%. There are additional losses due to photobreathing (photobreathing is a wasteful mechanism in which about 50% of the quantity of CO_2 absorbed by photosynthesis is released back to the atmosphere; it does not occur in all crops) and the fact that real cultivation and harvesting operations also have some limitations. Consequently, in usual situations the realizable transformation of solar energy into biomass has an efficiency of between 0.5 and 2%. The relative apportionment of different types of land in the world is shown in Figure 6.1. At the world level, one can estimate the energy obtainable from forest and meadows to be about 1.3 toe/year/inhabitant and the energy which could be obtained from arable land to be about 0.7 toe/year/inhabitant. Since land is also used for other purposes like

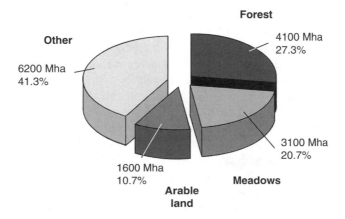

Figure 6.1. Apportionment of different types of land in the world [15,000 million hectares (Mha)]. *Source:* From J. Bonal and P. Rossetti, *Energies alternatives*, Omniscience.

food production, we could not get enough energy to satisfy our needs using biomass only.

It seems difficult to significantly improve the yield from photosynthesis. Therefore, different strategies must be employed to increase the amount of biomass produced per unit of cultivated surface. One way to improve the yield would be to increase the number of cultivation cycles each year. Interestingly, some prognosticators point to this possibility as a potential benefit which would result from moderate global warming. The full ramifications of global warming remain under active discussion.

Another approach to increase the benefits realized from biomass is to use the entire plant to produce energy or for other applications. Actually, the cultivation of sugarcane for ethanol production in Brazil is an example of a major success for this type of approach. Not only does this crop have a high photosynthetic yield but the bagasse, which is the waste obtained after sugarcane crushing, is used to produce heat and even electricity. So a large part of the sugarcane is used.

6.2. AN OLD ENERGY RESOURCE

Although biomass is the oldest renewable energy resource used by humans, its contribution to the world energy production has gradually decreased. This has occurred because fossil fuels have higher energy densities and have been relatively cheap. This is especially true in industrialized countries. It is nevertheless worthy of note that biomass is, in fact, presently used by a larger number of people than in the past. In 1800, the entire world used biomass for energy production, but the total population was around 1 billion inhabitants. Today, biomass is still the main source of energy for about 3 billion of the earth's estimated 6.6 billion inhabitants.

Several convergent reasons have combined to create a greatly renewed interest in using biomass to produce energy. Among these are concerns over fossil fuel availability and costs, concerns about greenhouse gas emissions, and the availability of new biomass conversion technologies. A great advantage of biomass compared to other renewable sources is that it also plays the role of an energy storage system. It can be viewed as a stock of solar energy. The energy density of dry biomass is about one-third of that oil. It can be used on demand and, except for some special biomass materials, which may evolve significantly as a function of time (e.g., some wastes), it can be stored easily until it is needed.

Although burning biomass produces carbon dioxide, the creation of biomass by the mechanism of photosynthesis also absorbs this gas. Even if the global balance is not at equilibrium, biomass can produce energy with a low net contribution to CO_2 emissions. One could even achieve negative greenhouse gas contributions in centralized energy plants where CO_2 is captured and stored rather than released during the energy production process.

The main problem of employing biomass derived from crops as an energy source is that it must compete with the use of these crops for other purposes, for example, food and building construction. Since the area available to produce biomass is too small to meet all energy needs, strong competition between the different uses could lead to strong price increase in the raw material and hence in the goods and services for which it is used. For example, in the United States promoting ethanol derived from corn as a fuel has led to an increase in the cost of corn and foodstuffs using corn. On the other hand, a larger demand for the crops has helped create local jobs.

6.3. ELECTRICITY PRODUCTION

Different technologies can be used to produce electricity from biomass: direct combustion, cofiring, gasification, pyrolysis, and anaerobic digestion. The biomass which is used as a fuel can come from plants, animals, or waste. It can be either directly burned in a thermal plant or intermediate fuels can be produced. The fuel can be in different physical states: solid, liquid, or gas. Since the energy density of biomass is three times smaller than that of oil, fuel storage, transport, and pretreatment are usually an important part of the cost of electricity production.

Several different sources of biomass are available. Among these are short-rotation forestry crops (e.g., eucalyptus, poplar, willow), perennial crops like miscanthus (elephant grass), or crops of sugarcane or rapeseed. Residues from primary biomass production are also interesting as sources: for example, wood residues from forestry operations, straw residues from food and industrial crops such as cereals, sugarcane, oil and coconut palms, rubber trees, coffee, and tea. By-products and waste from different processes can also be used, among these sawmill waste, black liquor from paper mills, manure, sewage sludge, organic waste, and used vegetable cooking oil. In early 2000, the United States had an installed capacity of 11 GW for electricity production from biomass. This capacity was shared between forest product and agricultural industry residues (7.5 GW), municipal solid waste (3 GW), and other sources, such as landfill gas, for example (0.5 GW). In 1850, wood provided about 91% of the total energy supply of the United States. Today 50% more wood is used, but this contributes only a very small part of the total energy consumed (about 1% of the total electricity-generating capacity).

Figure 6.2 shows the contributions of different biomass energy sources in the United States for the year 2004. This includes all energy applications of biomass, electricity, heat, and production of fuel for transportation. When compared to statistics for the year 2000, these figures reveal that in four years there was a 12% decrease in the contribution of wood-derived energy, a bit less than the 10% increase for energy derived from waste and a very strong 113% increase in the use of alcohol fuels.

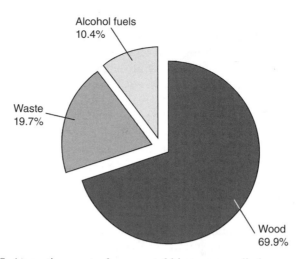

Figure 6.2. U.S. Apportionment of sources of biomass-supplied energy consumption, 2004. This includes electricity, heat, and transportation fuel production. *Source*: Data from the U.S. Department of Energy (DOE), www.energy.gov.

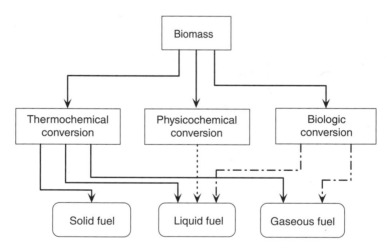

Figure 6.3. From various types of biomass, it is possible, depending on the transformation method, to produce solid, liquid, or gaseous fuels which can be used to produce heat or electricity or in some cases as fuel for transportation.

Biomass can be converted into energy using thermochemical, physicochemical, or biologic transformation techniques, as summarized in Figure 6.3:

- In thermochemical conversion, biomass can be used in direct combustion to produce heat, transformed into solid fuel, that is, charcoal, converted by pyrolysis into liquid fuel, or converted into gas in a gasification process.

- In physicochemical transformations, a liquid fuel can be obtained directly by pressing and extraction or esterification.
- Biologic conversion can be done either by fermentation/hydrolysis, which produces a liquid fuel, or by anaerobic digestion, which leads to a gaseous fuel (biogas).

With all of these fuels, solid, liquid, or gas, it is possible to produce electricity using the appropriate system: engines, steam turbines, gas turbines, or fuel cells. The fact that the feedstock is often dispersed and the energy density is low compared to fossil fuels leads to a high cost for transporting biomass to power plants. Consequently, smaller scale, widely distributed, bioelectricity plants which usually have a smaller efficiencies than those of large-scale fossil fuel power plants are often used.

In the United States, there is a renewal of interest in the generation of electricity using biomass. While the installed power capacity was smaller than 200 MW in 1979, it has now reached about 7000 MW. This still represents only about 1% of the total electricity demand in the country. Biomass power is particularly well suited to rural areas where much of the biomass supply is generated. Here small power plants may be a better solution than large power plants. For farmers, using crop residues as a fuel can be very cost effective.

6.4. TECHNOLOGIES

There exist several technologies to produce electricity from a wide variety of biomass.

6.4.1. Direct-Combustion Technologies

Most bioelectricity plants used today are based on direct combustion. In this technology, suitably prepared solid biomass is burned and used to heat a boiler and generate steam. This is used in a steam turbine which drives an alternator to produce electricity. Direct combustion can be performed in a fixed bed, in a fluidized bed, or by dust combustion techniques.

In a fixed-bed technology, the biomass is burned in a layer on a grate which moves through the furnace. Ashes are removed at the exit. The advantage of this method is the low investment cost compared to other technologies, but it can only be used with certain biomass fuel types.

Fluidized-bed technology allows the burning of different biomass fuel types provided the size of the fuel particles is roughly the same. In these systems, the fuel burns in a mixed suspension with a hot and inert granular bed material (silica sand or dolomite). Air enters below the furnace to feed the combustion. This technology requires higher investment and operating costs.

In dust combustion, small-size fuel particles (e.g., sawdust) are injected together with air, making a suspension which easily burns.

Direct-combustion biomass plants usually have a power lower than $100\,MW_e$. Despite their higher cost, fluidized-bed systems are preferred for larger installations (>10–$20\,MW_e$). Because of their higher efficiencies, dust combustion systems are reserved for smaller power capacities.

6.4.2. Cofiring Technologies

The cofiring technology consists of burning a mixture of a fossil fuel, such as coal, with biomass fuels. In practice 3–5% in mass of biomass is included, but this could go up to 40% in principle. It is an interesting approach since the extra investment compared to the fossil fuel plant is small while one takes advantage of the biomass to give a better net balance for CO_2 emissions and a reduction of NO_x and SO_x emissions compared to fossil fuels. Most cofiring plants use a pulverized coal system and have powers in the range 50–$700\,MW_{th}$.

The cofiring technologies use a direct, indirect, or parallel method for fuel injection. In the direct method a blend of biomass and coal fuels is injected directly into the furnace. It works best when the percentage of biomass fuel is low. In indirect cofiring, there is first a gasification of the biomass. The gas obtained is then burned with coal. In parallel cofiring, biomass is burned in a separate boiler to produce steam which is injected into the coal-fired power station. There, the temperature and pressure increase lead to a better yield of electricity production from the biomass. Since a duplicate installation is needed, cost is higher than for the other cofiring methods.

6.4.3. Biomass Gasification

Biomass gasification consists of producing a gaseous fuel by oxidization of the biomass with air, oxygen, or steam at high temperature. The gas produced is a mixture of hydrogen (H_2), carbon monoxide (CO), methane (CH_4), carbon dioxide (CO_2), and water vapor (H_2O). It also contains some quantities of hydrocarbons. In the process, both inorganic and oil-tar residues are also produced. The heating value of the "synthesis gas" produced is small, between one-tenth and half that of natural gas, depending on the nature of the biomass and the gasification process. The produced gas can be burned directly in boilers. It can be cleaned up by removal of tars and other species before it is used as a fuel in engines or gas turbines. As we shall see, it can also be used to synthesize other fuels like methanol or hydrogen. There exists a number of operating gasification systems with dedicated use of the gas produced.

There are basically two families of biomass gasification technologies—fixed bed or fluidized bed—working along the same principles as described for direct combustion.

In fixed-bed reactors, the biomass fuel is injected into the top of the unit. There are two methods to provide air for the gasification process:

- Air can be fed from the bottom through the grate and exit the gasifier at the top. This situation corresponds to an updraft gasifier because biomass fuel and air move in countercurrents. The gas exits at the top of the gasifier. Large amounts of tars are produced in this process.

- In the other system, called a downdraft gasifier, air and biomass move in the same direction. The synthesis gas exits at the bottom of the reactor. The amount of tar produced is smaller than in the previous method because the temperature in the hot zone, around 1000 °C, causes some cracking of the tars.

In fluidized-bed gasifiers, a mixture of air, biomass particles, and some inert bed material undergo drying, pyrolysis, and gasification at high temperature. The gasification process is more efficient due to the better heat transfer between the gas and the solid phases. Since this system operates at high temperature, a part of the tar undergoes cracking. There are two main methods to perform fluidized gasification of biomass. The first one corresponds to a bubbling bed, the second to a circulating bed. The advantage of fluidized-bed gasifiers compared to fixed ones are their higher capacities, the possibilities of using a larger variety of biomass fuels, and the fact that moist biomass can be used.

6.4.4. Anaerobic Digestion

Anaerobic digestion is a biological process in which organic materials are decomposed by bacteria in the absence of oxygen. The bacteria may be present in the original biomass material or may be introduced externally. The process is carried out in an airtight container called a digester. Any organic solid or liquid waste can be transformed by this process. It is particularly effective for treating moist organic waste. Feedstocks that can be used are agricultural, industrial and household organic wastes, sewage sludges, animal by-products, and solid organic municipal waste.

The biogas produced from waste by anaerobic digestion is a mixture composed mainly of methane (60–70%) and carbon dioxide (30–40%). It has a low heating value ranging from 5 to 8 kWh/Nm^3. That of pure methane is 10 kWh/Nm^3. The biogas contains between 20 and 40% of the heating value of the feedstock used to produce it.

In order to produce electricity, biogas is used in small-capacity installations ranging from a few tens of kW$_e$ to several MW$_e$. This is done using internal combustion engines and heat can be recovered simultaneously. Biogas can also be burned in gas turbines or, for large-scale equipment, in combined-cycle systems.

6.4.5. Pyrolysis

Pyrolysis is the thermal decomposition of biomass without oxygen. The result of this process is solid char, bio-oil (a liquid also called pyrolysis oil), and a mixture of combustible gases. The relative proportions of solid, liquid, and gases can be tuned with the temperature of the process and the time of residence. Typical shares between liquid, char, and gas are shown for different operational conditions in Figure 6.4. Pyrolysis with a primary goal of favoring bio-oil production appears to be an interesting technology for the future. In flash pyrolysis, where higher temperatures and shorter residence times are used than in fast pyrolysis, the product is 75–80% bio-oil. Bio-oil has a lower heating value of about 4.5 kWh/kg. After transformation, this bio-oil can be used in diesel engines or gas turbines. Bio-oil has a larger energy density than the solid biomass from which it is derived and is easier to store. Therefore the costs of transport and handling are lower. Like oil, bio-oil is a means to store energy and can be used as needed.

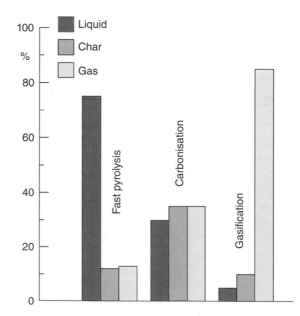

Figure 6.4. Different phases produced in biomass pyrolysis under different operational conditions. Fast pyrolysis involves moderate temperatures and short residence times. Carbonization uses low temperatures and very long residence times. Gasification needs high temperatures and long residence times. *Source*: Data from www.eusustel.be.

6.5. HEAT PRODUCTION

In the production of electricity from biomass one also gets heat. Both energies can be exploited simultaneously increasing, in this way, the total efficiency of energy extraction.

At present, the most important use of biomass is for heat production. In the European Union wood is the largest renewable energy resource used for this purpose. One can produce between 3 and 15 tons of dry biomass per hectare. Wood has a low water content compared with the other vegetable materials: 40–60% for fresh wood, 20–25% for air-dried wood. Its calorific value is 3.9–5 kWh/kg if it is dried in an oven and 2.8–3.6 kWh/kg if it is air dried. Taking into account the efficiencies of the boilers, 1 ton of air-dried wood delivers 0.25 Mtoe.

Wood does not burn. When warmed up, it first absorbs heat to eliminate its moisture before emitting, near 200 °C, gases. In the presence of air, the gases combust and heat is released. This increases the temperature up to about 800 °C. The wood is transformed into charcoal, a fuel whose calorific value is 9.1 kWh/kg. Temperatures larger than 1000 °C can be obtained with charcoal.

In 2005 the energy production in Europe amounted to 58.7 Mtoe and represented an increase of 5–6%/year. In that year, France was the largest producer, with 9.6 Mtoe, followed by Sweden, Germany, and Finland. However, Finland led in proportion of energy requirements met. It covered about half of its needs with wood.

In France more than 40% of the individual houses use wood for space heating (sometimes as a complementary source). Significant progress has been made in the design of fireplaces used to burn wood. Old fireplaces with open hearths have efficiencies less than 10%. Newer open-hearth fireplaces can reach 30% efficiencies. Old fireplaces with closed hearths have efficiencies between 30% and 50% while new ones can reach 60–85%. For a single loading of wood, open hearths have operating times of a few hours. This increases to more than 10 h for closed-hearth fireplaces.

Charcoal is usually obtained by heating wood in the absence of oxygen. It is 85–98% carbon and has a larger energy density than wood. In addition it burns hotter and cleaner. It was extensively used in the past before the advent of fossil fuels, mainly as a fuel but also in metallurgy as a reducing agent, for manufacturing pencils, and as one of the constituents of gun powder. Charcoal was critical bound to the beginning of metallurgy, about 5000 years ago. In earlier times the demand for charcoal was one of the major causes of deforestation in Europe.

A wood-fired boiler coupled to a heat network can be used to provide space heating and hot water to buildings. In central heating systems, wood-fired boilers have efficiencies from 55–60% for the less efficient ones to 75–85% for the best ones.

Care must be exercised when burning wood. Particles, volatile organic compounds, carbon monoxide, and nitrogen oxides are emitted in quantities which depend on the technology used and the filters installed. These emissions are dangerous for health and bad for the environment.

6.6. BIOMASS FOR COOKING

In developing and emerging countries, about 2.5 billion people use biomass to meet more than 90% of their household energy needs. They burn wood, charcoal, animal dung, and agricultural residues. This occurs mostly in rural areas but is common also in some urban areas. Biomass is used mostly for cooking, which is the primary energy demand in these countries. Obtaining fuelwood is a difficult job in some countries. For example, in sub-Saharan Africa, people (women and children) must often carry an average load of 20 kg of wood over distances of a several kilometers.

The problem with the use of wood, agricultural residues, and animal dung as fuel is the high levels of carbon monoxide, hydrocarbons, and particulate matter produced. Animal dungs produce a lot of hydrocarbon emissions while agricultural residues generate large amounts of particulate emissions. As a consequence, there are more than 3400 premature deaths per day in the world traceable to the use of biomass. Women and children are the most affected by indoor pollution caused by biomass combustion because they use very primitive means to burn biomass. More than half of the victims are children under five years of age. When indoor use of coal is included, the number of deaths per day attributable to indoor air pollution associated with energy production increases to more than 4000 (~1.5 million per year). This figure is of the same order of magnitude as the number of deaths per year due to malaria (1.2 million). Figure 6.5 shows the distribution of premature deaths among different regions of the world.

As a rule, the higher the energy density, the lower the pollution. In this respect, charcoal is better than fuelwood or animal dung. Improving the way biomass is used in the developing countries and helping those countries to switch to more modern energy sources (liquefied petroleum gas, biogas, ethanol gel, electricity) should be a high priority for the international aid community.

6.7. ENVIRONMENTAL IMPACT

Using biomass on a large scale has a noticeable impact on the environment on both a local and global scale. Some of this impact is negative, some positive.

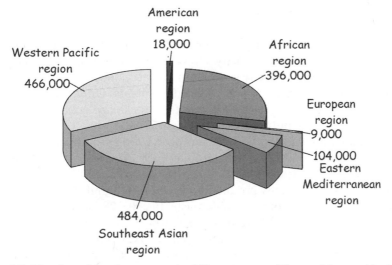

Figure 6.5. Number of deaths per year in different areas of the world caused by indoor air pollution (biomass and coal). *Source:* Data from the World Health Organisation, www.who.int (2004).

The impact is related to both the needs associated with cultivation of the biomass crops and the use of the biomass to produce energy. The great advantage of vegetal biomass is that CO_2 is captured during biomass growth. However, it is then released in the atmosphere when the biomass is burned. While the fact that fossil fuels may be needed for cultivation, harvesting, transport, and processing of biomass must also be taken into account, the net CO_2 emission balance is positive and less CO_2 is emitted than would result if only fossil fuels were used instead of biomass.

Cultivation of perennial crops is preferable to that of annual crops because perennials reduce erosion of the land and improve soil organic matter. Short-rotation crops such as coppice and miscanthus are interesting in terms of yield but also because they need much less fertilizer than conventional agricultural crops. However, transforming grassland and removing forests in order to cultivate new crops can have a negative impact on greenhouse gas emissions since this can release significant amounts of the carbon dioxide trapped in the soil. Lignocellulosic materials, like wood and its by-products, are particularly interesting because one does not use fertilizers to grow forests. Fertilizer as well as other agricultural runoffs may create pollution of soil and water. For any of these crops large quantities of water may be required. Meeting this need may reduce the amount of water flowing into rivers and have a negative impact on groundwater quality as well as on ecosystems. (On the other hand, in regions which are prone to flooding, reduction of flows might be an advantage.) Finally we note that, in the particular case of organic wastes, their use to produce energy has the positive effect of reducing the amount of end wastes. Producing energy provides a bonus.

6.8. MARKET SHARE

Biomass is an expanding energy resource. Mostly devoted to heat production, it is also used to produce electricity. However, without incentives to promote it, biomass-produced electricity is not normally economically competitive with electricity produced by large-scale fossil fuel plants. Only some small-scale installations using waste and by-product feedstocks are economically viable because their primary purpose is waste destruction.

Biomass accounted for 2.2% of the European Union's total electricity production in 2005 and nonrenewable waste produced 1%. If we restrict ourselves to the renewable part of electricity production, biomass provided 15.1%. The average annual growth for electricity production between 1995 and 2005 was 12.5% for biomass and 13.5% for nonrenewable waste. In 2005, 28.1 TWh were produced from solid biomass, 13.7 TWh from biogas, and 6.6 TWh from renewable waste.

The capital cost for power generation depends very much on the technology chosen. It may vary between 250 €/kW$_e$ for existing coal cofired plants to something like 2500 €/kW$_e$ for fluidized-bed boilers or gasification coupled to a diesel engine or a gas turbine. It may even be very expensive and reach 5000–6000 €/kW$_e$ for gasification combined-cycle plants. The cost of the produced electricity also varies over a large range, from 0.02 to 0.14 €/kWh depending upon the technology. Efficiencies run between 25 and 55% according to the technology. However, price is not the only point to be considered. Biomass encourages rural development and crop cultivation also creates local jobs.

In Figure 6.6 we show the eight leading producers of wood energy for the European Union (EU) in 2004. The ninth one, Latvia, produces 1.3 Mtoe. In

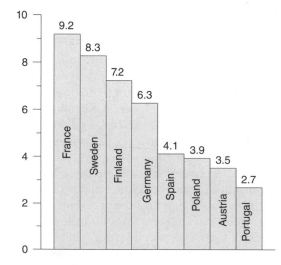

Figure 6.6. Primary energy from wood energy in eight leading EU countries, 2004. Units are in Mtoe. *Source*: Data from EurObserv'ER, www.energies-renouvelables.org.

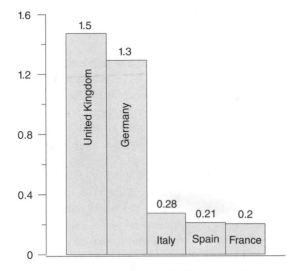

Figure 6.7. Crude biogas production for five EU first producers, 2004. *Source*: Data from EurObserv'ER, www.energies-renouvelables.fr.

that year, a total of 55.4 Mtoe was produced in the EU. The production of electricity was 34.6 TWh, often in combined heat and power plants.

Figure 6.7 shows the five leading producers of biogas in the EU in 2004. The sixth one, Sweden, produced 119 ktoe. A total of 4.12 Mtoe was produced in the EU. In contrast to the two EU leaders that generate a lot of electricity from biogas, France uses it principally to produce heat (56 ktoe for heat and 42 ktoe for electricity).

As discussed in the next section, the use of biomass fuels for transportation is also an area under fast development.

6.9. BIOFUELS

Biofuels are fuels derived from biomass. Fossil fuels are also derived from dead biomass, but the production timescale is not the same: a year or years for biofuels, hundreds of millions of years for fossil fuels. The increase in the price of oil, the need to decrease greenhouse emissions, and the desire to develop local agricultural resources all favor the development of biofuels as an alternative to fossil fuels for powering vehicles or for other purposes, cooking, for example. It should be kept in mind that while biofuels can decrease our demand for oil they cannot totally replace oil for powering vehicles. They may be able to provide 10–20% of the total fuel needed for transportation.

A major issue confronting biofuels is the competition existing with food biomass, which can lead to an increase in the price of food. This has already been observed in Mexico, for example, with the rising price of tortillas resulting from the decision to develop corn-based biofuels in the United States.

The area of cultivated land in the world is decreasing. It went from 0.5 ha/ inhabitant in 1950 to 0.3 ha/inhabitant in 1990. This area is expected to continue to decrease to about 0.1–0.2 ha/inhabitant by 2050. This is due to the fact that, even though the population is growing, the growth is compensated for by an increase of the production yield per unit of cultivated area. The yields of crops dedicated to biofuels have also increased. For example, in the United States the production yield of ethanol per unit of area has increased by 2.7%/year over the last 20 years and in Brazil by 3.8%/year during the last 30 years. One factor in determining the need for cultivatable land is that meat consumption continues to increase. If we take a vegetable diet as a reference, sustaining a white meat diet requires five times more cultivated area and sustaining a red meat diet requires nine times more cultivated area. The additional pressure on land use caused by cultivating energy crops could have a major influence on food prices.

Burning biofuels or using them to power a vehicle leads to CO_2 emissions, as in fossil fuel burning. However, the plants used to produce biofuels absorb CO_2 during the growing process while for fossil fuels the CO_2 absorption occurred hundreds of millions of years ago. The absorption of CO_2 in modern biomass production partly compensates for the amount of CO_2 emitted in the final use. In evaluating the balance of the CO_2 emission, all stages of the biofuel production (biomass farming, fertilizer production, transportation and extraction) should also be taken into account.

Although there is a new emphasis on biofuel development, the use of biofuels is not a totally new concept. By the end of the nineteenth century France was already considering the use of denatured alcohol as a fuel for engines. The main concern at that time was to reduce the dependence of the country on foreign oil supplies and ensure energy security. Agricultural surpluses existed and provided another good reason to use crops to produce biofuels. The same situation exists today.

In France in 1902, there was a car race on the so-called *circuit du nord*. The total distance to be covered was 900 km on roads which were in very bad shape. Leon Serpolet, who set the speed record with a car in 1902 with 120.7 km/h, had four cars in this race. He was financially supported by Franck Gardner, an American citizen. These cars were powered with alcohol and finished at the third, fourth, fifth, and sixth positions with an average speed of 71 km/h. The winner of the race was Maurice Farman, driving a Panhard & Levassor, 40 HP.

Before World War I, Paris buses were powered by a mixture of gasoline and denatured alcohol (30%). Since the problem of alcohol is its low calorific value compared to gasoline, the denatured alcohol was previously mixed with coal benzine (up to 50%), a benzene-containing mixture of aromatic hydrocarbons obtained from coal.

Between 1920 and 1950 ethanol was widely used as a fuel for vehicles. The production was of the order of 2 Mhl ($1\,hl = 0.1\,m^3$) per year. That is about twice the amount produced today to power cars in France. This contributed a significant part to the fuel dedicated for cars at that time. For example, France consumed $1.4\,Mm^3$ of gasoline in 1925 and $3.4\,Mm^3$ in 1950. In 1950, gazole was also used ($1.1\,Mm^3$). In the 1950s it was possible to fill the tank of a car with a mixture called tri-super-azur sold by the Antar Company. This mixture consisted of 75% gasoline, 15% ethanol, and 10% benzole (a mixture of benzene, toluene, and xylene).

In the 1960s ethanol was no longer being used as a fuel for vehicles for two reasons. The first is that there was a plentiful supply of cheap oil. The second was that there were no longer sugar beet surpluses. The sugar was being used in the food-processing industry and in the chemical industry. It was only after the oil shocks of 1973 and 1979 that there was a renewed interest in biofuels. This was supported by the reappearance of agricultural surpluses.

6.9.1. First-Generation Biofuels

First-generation biofuels are those produced today. There are two ways to produce biofuels. The first is to use crops which contain a high sugar concentration such as sugarcane, sugar beet, or sweet sorghum or starch such as corn, wheat, and barley. Yeast fermentation is then used to produce ethanol. The second way is to grow plants such as oil palms, soybeans, rapeseeds, and sunflowers that contain a high concentration of vegetable oil. Lignocellulosic biomass such as wood and wood residues, for example, will be used as second-generation biofuels. These are discussed in another section of this chapter.

6.9.1.1. Biofuels for Gasoline Engines

Biofuels for gasoline engines are ethyl alcohol (ethanol) or a derivative such as the ETBE (ethyl-*tert*-butyl ether). Methanol is another alcohol which can be used in principle. It is cheaper than ethanol. It can be produced from wood (wood alcohol) or from natural gas. (In the latter case it cannot be classified as biofuel.) However, methanol is a toxic product and is particularly dangerous because it can be easily mixed with water. MTBE (methyl-*tert*-butyl ether) is a chemical derivative from methanol analogous to ETBE, but it is a serious risk to the environment.

The sugar discussed in this chapter is called sucrose by chemists. It is a disaccharide with the formula $C_{12}H_{22}O_{11}$. It is made by photosynthesis according to the reaction

$$12CO_2 + 11H_2O \rightarrow C_{12}H_{22}O_{11} + 12O_2$$

Sugar is produced by some plants, like sugarcane or sugar beet, to store extra energy. Most of the commercial sugar is produced from sugarcane ($\approx70\%$) and the remaining part from sugar beets ($\approx30\%$). Sugarcane is mostly grown in tropical countries and sugar beets in temperate zones of the north. More than 120 million tons of sugar are produced each year in the world.

Ethanol is the primary biofuel used in the world. Ethanol has a greater octane number than gasoline (of the order of 110 compared to about 95 for gasoline). It can be blended in any ratio with gasoline. It is often blended in concentrations up to 5%. This does not require major changes in an automobile engine. Higher concentrations, between 5 and 10%, are used in the United States (gasohol E10) and between 20 and 24% (or even 100%) in Brazil. The product called E85 contains 85% ethanol. In the EU, a target of 5.75% of ethanol blending by 2010 has been adopted.

The presence of oxygen in the chemical formula has a positive impact on the combustion in the engine, reducing the emissions of incompletely burned hydrocarbons and partially oxidized products. Carbon dioxide emissions are reduced but those of nitrogen oxides increase in most cases except for some types of vehicles for reasons which are not well understood. However, the energy content of ethanol is about 25–30% smaller than that of gasoline. Consequently, if, for example, the E85 blend is used, an increase of fuel consumption of 25–30% is observed.

Special cars called "flex fuel vehicles" have been developed by manufacturers. They can use any combination of ethanol and gasoline. They adjust the engine control automatically according to the fuel blend used by measuring the exhaust oxygen. Ethanol is corrosive for many parts of the vehicle—rubber hoses, aluminum, combustion chambers—and this requires special treatment in the construction of the vehicle.

About 80% of the ethanol produced is now used as biofuel. Brazil and the United States account for more than two-thirds of the global ethanol production (see Figure 6.8). Brazil employs sugarcane—a very efficient crop as far as sugar production is concerned—and the United States produces ethanol from corn. In 2005 the global production of ethanol was equal to 34 Mm³. In Brazil, ethanol accounts for 40% of the fuel consumption. An area of 5 Mha is devoted to sugarcane cultivation.

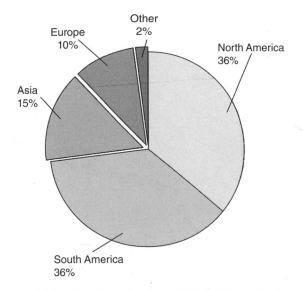

Figure 6.8. Share of bioethanol production, 2005 (total production 37 Mt). *Source*: Data from www.ifp.fr.

Brazil is able to produce ethanol at a very competitive price: about 0.15 €/L. This is about 40% less expensive than ethanol obtained from corn in the United States and 70% less than ethanol obtained from sugar beets or cereal crops in Europe. Developing crops to produce biofuels requires not only land but also adequate water. In that respect Brazil is in a favorable position since it has 18% of the world's freshwater resources. Furthermore, 90% of sugarcane grown in the south of the country does not need irrigation.

There has been a swift increase in bioethanol production, as can be seen in Figure 6.9. The steep rise since 2000 reflects the great increase in interest in the production of bioethanol and its use as fuel for vehicles. The replacement of 10% of the fuels used for transportation by biofuels would require the use of about 20% of the cultivable lands in Europe and 25% in the United States.

Table 6.1 shows the potential energy available from 1 ton of sugar cane grown in Brazil. In total 1 ton of sugar cane (0.154 toe) is the energy equivalent to about 1 barrel of oil. If we use only the sugar in sugarcane, the average amount of ethanol produced per hectares is about 6 tons, which is equivalent to about 3.8 toe/ha. If the bagasse is used, the amount of energy which can be harnessed is larger. If all the available energy were used, one could obtain the equivalent of 85 barrels of oil per hectare. So far, among the residues, only the bagasse is used to produce energy (heat or electricity). The other lignocellulosic residues are not used. Sugar beet processing also has by-products: There is typically 0.8 kg of flesh for each kilogram of produced ethanol.

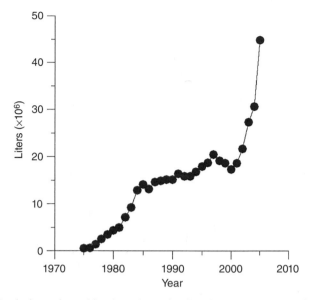

Figure 6.9. Evolution of world ethanol production between 1975 and 2005. *Source*: Data from www.earth-policy.org.

TABLE 6.1. Potential Energy Available From 1 Ton of Sugarcane in Brazil

Products Obtained	Energy Available
150 kg (sugar) = 85 liters (67.3 kg) ethanol	0.042 toe
280 kg of bagasse (50% of dried material)	0.056 toe
280 kg of lignocellulosic residues (50% of dried material)	0.056 toe

The yields of some crops are shown in Figures 6.10–6.12. Among sugar plants, sugarcane has the best yield. One of its advantages is that it also produces lignocellulosic residues. The bagasse in particular is used to produce heat, which decreases the cost of ethanol production. With sugar plants, between 2 and 4 toe can be produced per hectare of crop. Cereals are rich in starch. For this type of plant, corn has the best yield. Potatoes can also be transformed into ethanol. It is possible to produce about $4 m^3$/ha from potatoes. Manioc can also produce alcohol ($\approx 370 l/t$). Barley contains sugars with five atoms of carbon in their formulas (pentoses). At present these C_5 sugar are not fermentable into ethanol. Finding out how to biologically transform C_5 sugars into ethanol would open a huge number of possibilities for transformation of some types of biomass into alcohol.

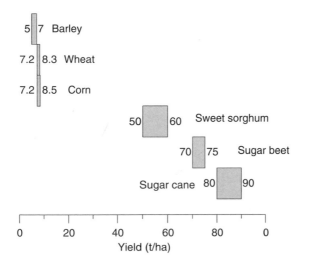

Figure 6.10. Yields of selected crops. *Source*: Data from D. Ballerini, *Les biocarburants*, Technip, 2006.

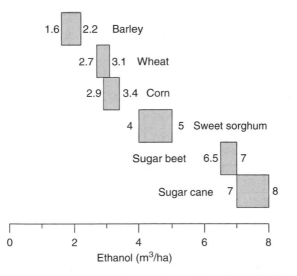

Figure 6.11. Yields of selected crops. *Source*: Data from D. Ballerini, *Les biocarburants*, Technip, 2006.

In 1815, the French chemist Louis Joseph Gay-Lussac determined that the theoretical yield for producing ethanol from glucose is 51.1%. Forty years later Louis Pasteur showed that a more realistic yield is 48.4%. Thus, under the best conditions one can get 48.4 kg of ethanol out of 100 kg of glucose.

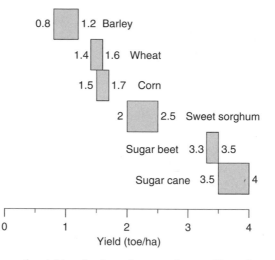

Figure 6.12. Energetic yields of selected crops. *Source*: Data from D. Ballerini, *Les biocarburants*, Technip, 2006.

To summarize the situation for production of first-generation biofuels:

- One hectare of land produces about 75 tons of sugarbeet. Since 1 ton of sugarbeet gives 70 kg of ethanol, 5.25 t/ha of ethanol (or 6.6 m³/ha since the ethanol density is 0.792) can be produced. The energy content of ethanol is 26.8 MJ/kg, which makes 140.7 GJ/ha. Taking for 1 toe the average value of 42.5 GJ gives 3.3 toe/ha.
- One ton of corn gives 400 liters of ethanol (317 kg). There are 800 kg of residues for 1 m³ of ethanol produced. Taking a yield for corn of 8 t/ha gives 3.2 m³ of ethanol (2.5 tons) and a yield of 1.6 toe/ha.
- Three hundred and seventy liters or 295 kg of ethanol can be produced from 1 ton of wheat. One ton of residue is obtained for each cubic meter of ethanol produced. A yield of 8 t/ha gives about 3 m³ of ethanol or 2.4 t/ha. The energetic yield is 1.5 t/ha.

Bioethanol has some disadvantages which can be removed using a derivative compound: ETBE. Ethyl-*tert*-butyl ether is synthesized from ethanol and isobutene. In contrast to ethanol, ETBE does not induce evaporation of gasoline and does not absorb humidity from the atmosphere. Except for Sweden ETBE is used in European countries. The octane number of ETBE is similar to that of ethanol and better than that of gasoline. It reduces CO emissions by about 5–10% and incompletely burned hydrocarbons by about 5%. However, there is an increased emission of nitrogen oxides and aldehydes. Because supplies of isobutene derived from oil refining are limited, ETBE cannot be extensively used.

Methyl-*tert*-butyl ether is a chemical compound, $(CH3)_3C(CH3)$, that has also been used blended with gasoline. It increases the octane number (its octane number is 118) and reduces the pollutant emissions. It was widely used in the 1980s, especially in the United States. This chemical compound is very soluble in water and is an environmental hazard. It can pollute both surface and groundwater. It is difficult to clean up the polluted water. The use of MTBE is being phased out in the United States.

6.9.1.2. Biofuels for Diesel Engines

Biodiesel is a biofuel used in diesel engines. It represents only a small part ($\approx$10%) of biofuels. Its use has developed rapidly in Europe where the share of diesel cars is large (about two-thirds of the brand new cars have a diesel engine). Biodiesel is produced from oils or fats by transesterification. Most common derivatives are RME (rapeseed methyl ester) and FAME (fatty acid methyl ester). A fatty methyl ester is obtained from the reaction between fats or fatty acids and methanol. In biodiesel the FAMEs are obtained from vegetable oils by transesterification. Frying oil or animal fat can be used. The RME can be used in diesel engines at various blend concentrations.

The yield of biodiesel production is shown in Figure 6.13 for different crops. Oil palm trees provide by far the most efficient way to produce oil, but this can only be done in some regions of the world.

In 2005, Germany produced more than half of the European biodiesel (1.7 million tons out of a total of 3.2 million). This was spurred by tax incentives (see Figure 6.14). The United States and Brazil have a low production of biodiesel, but this is increasing. Other types of biodiesel could be synthesized by hydrothermal catalytic treatment using vegetable oil, animal fats, and used oil.

With older diesel engines it is sometimes possible to use vegetable oil directly. However, with modern engines this is not recommended because it may age the engine and lead to problems in the long run. Indeed, vegetable oils have too small a cetane number and cracking of molecules occurs, leading to formation of deposits in the engine. Using vegetable oil in a diesel engine generally increases emissions of CO, incompletely burned hydrocarbons, and particulates but reduces NO_x emission. Synthetic diesel [biomass to liguid (BTL)] is a very pure fuel. Compared to diesel oil, the CO and incompletely burned hydrocarbons are reduced by 85 and 78%, respectively.

The transesterification of vegetable oil leads to products which can be blended with diesel in any proportion. They have a cetane number and a viscosity comparable to diesel oil derived from petroleum. Compared to diesel, FAMEs have a value about 10% smaller than diesel, which means a

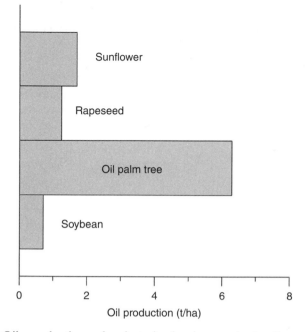

Figure 6.13. Oil production of selected oleaginous plants. *Source*: Data from D. Ballerini, *Les biocarburants*, Technip, 2006.

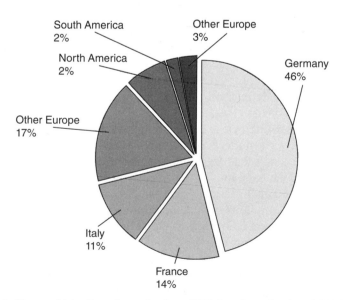

Figure 6.14. Share of bioethanol production, 2005 (total production 3.21 Mt). *Source*: Data from www.ifp.fr.

corresponding increase of fuel consumption if the pure product is used in the engine. Instabilities occurring during storage can be cured with specific additives.

About 1 ton of oil and 100 kg of methanol is needed for transesterification. This process gives about 1 ton of biodiesel and 100 kg of glycerine. The yield of the reaction is close to 99%, the remaining part corresponding to soap formation.

About 900 MJ of heat is necessary to produce the amount of steam necessary to treat 1 ton of vegetable oil and 100 MJ of electricity is needed. Instead of using methanol esters, ethanol esters can be used. Pure ethanol cannot be blended with diesel because it is insoluble while ethanol esters are soluble. This could be another use for bioethanol. Research on ethyl esters are being carried out in Brazil because this country produces large quantities of ethanol but has to import methanol.

6.9.2. Second-Generation Biofuels

The problem with first-generation biofuels is that the biomass used to produce them is grown in competition with food production. Furthermore, some crops cannot be grown at the same place every year and they should be grown at a different place for several years before they can use the initial place. This puts an additional constraint. It will not be possible to completely replace petroleum products by biofuels and producing too much of them may increase the price of crops dedicated to food. This is why second-generation biofuels are interesting to develop. The idea is to use lignocellulosic biomass (wood, forestry or farming residues, waste biomass, and eventually dedicated nonfood crops). Dedicated crops might be miscanthus, for example, or short-rotation coppices or undergrowths.

Woody and fibrous biomass contains cellulose, hemicelluloses, and lignin. Using enzymes, steam, or other treatments, it is possible to free the sugar molecules from the cellulose and leave lignin, which can be burned as a fuel. The liberated sugars can, for example, be fermented and transformed into alcohol, but this does not work for those obtained from hemicelluloses.

Lignocellulosic biomass is made of cellulose (between 35 and 50% of the dried material), hemicelluloses (between 16 and 34% depending upon the plant), and lignin (between 11 and 29%). Glucose can be obtained from cellulose and fermented into ethanol. Hemicelluloses contain pentoses, but they are difficult to convert into ethanol. Conversion of pentoses into alcohol is an important area of current research.

Gaseous fuels can also be obtained from lignocellulosic biomass but, as noted in the Chapter 13, liquid fuels are more convenient. What seems to be the most interesting for the future is to produce liquid fuels using the Fischer–Tropsch process after gasification of the biomass. This is the BTL route. As discussed in Chapter 13, if external energy and hydrogen produced elsewhere in a clean way are used to garner all of the carbon available, we can get two or three times more synthetic oil with BTL than with crops used for the first-generation biofuels.

6.9.3. Third-Generation Biofuels

In the future biofuels could also be produced in the sea. Some micro-seaweeds (with a size between 2 and 40 µm) contain a large proportion (up to 80%) of lipids, fats which could be used to make biofuels. The yield is much higher, up to 30 times more than from oleaginous plants like rape or sunflowers. With 1 ha of crop, it is possible to produce several tons of seaweed.

Currently cultivation of seaweed for biofuels is at the research project stage, but it might be an interesting possibility in the future, especially because of its small impact on the environment.

Algae fuel, or oilagae, can be produced with yields which can be an order-of-magnitude larger than what can be obtained with crops in land. The DOE has even calculated that an area in the sea the size of the state of Maryland could provide enough biofuel to replace the petroleum used in the United States.

6.10. FROM WELL TO WHEELS

A few years ago, there was a strong cry for development of biofuels. The argument was that it would significantly reduce our dependence on petroleum. Now that biofuels are being produced in significant quantities, some environmentalists are opposed to the further development of first-generation biofuels. One of the reasons is that the process is in competition with food production, which leads to an increase in food prices. It also needs water, which is scarce in some regions. To add to the confusion different well-to-wheel life-cycle studies attempting to carry out a net energy and CO_2 emission accounting give results which are quantitatively different. These studies indicate that using biofuels decreases CO_2 emission but they differ on the amount of reduction obtained. This is because they are based on different assumptions.

Biofuel production leads to significant quantities of by-products. On the average 2 tons of by-products are produced per ton of biofuel. These by-products are oil cake, which can be used as foodstuff for animals, straw, glycerine, and so on. Some of the studies do not take into account these by-products and charge the energy consumed in biofuel production only to the biofuel. A second method takes the by-products into account in so-called avoided impacts. This can be simply explained using glycerine production, for example, taking into account the fact that producing glycerine by a conventional method would consume energy and produce pollution. Since the glycerine produced during biofuel production can be substituted for the glycerine that would be produced by other means there is a net cost saving. The same accounting argument can be made for CO_2 emission.

There are other sources of discrepancies. Nitrous oxide emission from agriculture associated to biomass growth depends very much on the quantity of fertilizers used, on the nature of the soil, and so on. Because N_2O is almost 300 times more efficient than CO_2 as far as the greenhouse effect is concerned, any difference in the evaluation assumptions leads to an important final difference in the conclusions. All well-to-wheel evaluations indicate that biofuels can reduce greenhouse gas emissions. Estimates range from 30 to 94% reductions compared to petroleum fuels. The largest reduction of greenhouse gas emissions is obtained with second-generation biofuels. The estimated energy gain varies between 22 and 90%.

If wood were used to produce ethanol, using lignin as a heat source, almost no external energy would be required in the process. Using this ethanol would mean that between 75 and 80% of CO_2 emissions would then be saved.

If we consider a rapeseed crop producing, for example, 1.4 t/ha of biodiesel or 1.3 toe/ha, the total yield after subtracting the energy necessary for producing the fertilizers (e.g., for transportation) is 0.85 toe/ha.

6.11. CONCLUSION

Biomass has been an important resource in the past, and it will also be an important one in the future. While it is useful in producing heat or electricity, this is generally not the best use of biomass unless it is done using residues which cannot be used in better applications. Organic waste is also a useful biomass for energy production. Using bacteria or enzymes it could even be employed to produce high-valued products such as biogas or hydrogen.

Synthesizing liquid fuels for transportation (one expects to have hybrid vehicles consuming average quantities of the order of 1 L/100 km) will become increasingly important because a concentrated form of energy is needed to power transportation. Second-generation biofuels will be particularly interesting in this regard because they should be produced in higher yields for the same cultivated area. Using micro-seaweeds is a very appealing possibility because the yields are an-order-of-magnitude larger than that which can be obtained from land crops. Farming this resource from the sea could greatly extend the amount of biomass which can be recovered from the earth.

One of the critical issues in the future will be the availability of organic molecule carbon. This type of carbon exists in fossil fuels (coal, oil, and gas) but also in biomass. This makes biomass an important source of carbon in the future.

The main problem is that biomass has many possible uses. Food is the first priority. If fossil fuels become scarce, developing new "green-chemistry" techniques, in which one finds other sources from which to synthesize the important organic compounds that are presently being obtained from crude oil, will also be a very high priority.

CHAPTER 7

Solar Energy

Annually, the earth intercepts about $342\,\text{W/m}^2$ of energy from the sun in the form of solar radiation. The atmosphere itself reflects $77\,\text{W/m}^2$ and absorbs about $68\,\text{W/m}^2$. Thus, on average, the radiation reaching the earth's surface is $197\,\text{W/m}^2$. Although this is a very small proportion, less than a billionth of the energy emitted by the sun (0.46×10^{-9}), the amount received annually is more than 10,000 times as much energy as that consumed annually by humankind. This solar energy can be exploited to get heat or produce electricity. Presently, the solar energy incident on only about 5% of the desert area of the earth would be sufficient to produce the entire amount of electricity needed in the world. Of course, it would be a major effort to store this electricity and transport it from these remote areas to the populated areas where it is needed, but solar energy has a very important advantage over other energy sources: It is available in inhabited regions everywhere on the earth (although the extent to which solar power is available and its intensity vary with location and time).

Mankind was early to recognize that concentrated sunlight contains a lot of energy. In the seventh century B.C. magnifying glasses were used to focus the sun's rays to make fire. Later, in the third century B.C., the Greeks and Romans used mirrors to concentrate the sun's rays and light torches. Legend has it that during the siege of Syracuse in 211 B.C. the Greek Archimedes set fire to the wooden ships of the Roman Empire, besieging Syracuse by focusing sunlight using bronze shields as mirrors. Archimedes was killed by a Roman soldier despite orders that he should not be harmed.

Our Energy Future: Resources, Alternatives, and the Environment
By Christian Ngô and Joseph B. Natowitz
Copyright © 2009 John Wiley & Sons, Inc.

7.1. SOLAR ENERGY: A HUGE POTENTIAL

Solar energy is only directly available during daylight hours. During a day the intensity at a given location may vary significantly depending upon weather conditions. Solar intensity also varies from season to season. The power of solar radiation reaching the earth in the upper atmosphere, perpendicular to the axis joining the center of the sun and that of the earth, is equal to $1366\,W/m^2$. At the equator, about 6% of the incoming solar radiation is reflected and 16% absorbed by the atmosphere. This means that a power of about $1\,kW/m^2$ reaches the ground at the equator. This value decreases to $0.5\,kW/m^2$ at higher latitudes because the thickness of the atmosphere is larger. Variations in atmospheric conditions (e.g., clouds, pollution, aerosols) can further reduce the solar energy reaching the earth's surface by about 20%. In North America, the average energy available varies between 3 and $9\,kWh/m^2/day$. On the average, northern Europe receives between 2 and $3\,kWh/m^2/day$ whereas it is about $6\,kWh/m^2/day$ in the tropics. The received solar radiation varies by about 20% between the tropics. In France it varies by a factor of 2.5 between the southern and the northern regions.

Figure 7.1 shows the average solar radiation for different cities in the United States. There is more than a factor 2 between Chicago and Albuquerque, for example.

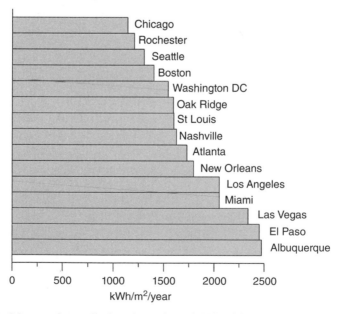

Figure 7.1. Mean solar radiation for selected U.S. cities. *Source*: Data from www.solar4power.com.

Many technologies are used to exploit solar radiation. Some of them use solar energy directly to produce heat. Others produce electricity from solar power. The use of solar energy is particularly efficient in countries located between the tropics. Many of them need energy to ensure their economical development. The main problem with solar energy is that transforming it into electricity is very expensive.

7.2. THERMAL SOLAR ENERGY

The simplest form of energy in which to harness solar radiation is as thermal energy. Before humankind mastered fire, energy from the sun was the only major energy source available. This is the reason why humankind first developed in warm regions of the world.

7.2.1. Producing Hot Water for Domestic Purposes

Using solar energy is a good way to heat in a temperature range of 30–150°C. For domestic purposes (bath, shower, heating buildings, etc.) the simplest and most economical way to harness solar energy is to heat water. About a quarter of the total energy consumed in the United States or European Union is devoted to heating or cooling. A large part of it is used for space heating of buildings or to heat water for domestic purposes.

In the United States, heating water accounts for about one-third of an average total household energy consumption and also about one-third of its greenhouse gas emissions. About 14% of the total energy consumed in the country is dedicated to heating water. Solar water heaters are particularly appropriate devices to heat water in a cost-effective way. Depending on the climate and the technology used to harness solar energy, a solar water heater may meet between 50 and 90% of the total demand of hot water for a household.

The total world installed capacity of solar hot water systems is around 88 GW$_{th}$ (thermal gigawatts). It increases at a rate of 14% per year. Solar water heaters typically employ solar collectors or panels to absorb the thermal energy coming from the sun (see Figure 7.2). Water in these collectors is

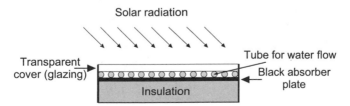

Figure 7.2. Example of a solar collector. Water tubes located behind a glazing and in front of a black absorber harness thermal solar radiation. The temperature of water in the tubes can increase up to 50°C.

heated and flows into a thermally insulated tank where it may be stored for later use.

In passive systems water flows naturally between the collectors and the tank. Active systems employ pumps to effect this transfer. The storage tank is usually complemented by a supplementary heating source (powered by electricity or fossil fuel) to heat water when there is no sunlight and meet peak hot-water demands as required. This extra energy source often also has another purpose—to destroy microorganisms such as *Legionella* bacteria which might develop at lower temperatures.

The higher the temperature desired, the more sophisticated the solar collector must be. A collector with a black painted surface is adequate for systems used to temperatures of 30–35 °C. A glass located in front of the black collector will produce a greenhouse effect and allow temperatures to reach between 50 and 60 °C (Figure 7.2). If the black paint is replaced by a selective absorptive material which prevents reemission of infrared radiation, it is possible to reach temperatures around 70–90 °C. With collectors under vacuum (usually glass cylinders) temperatures higher than 100 °C can be obtained.

Some countries use solar hot-water systems extensively. In Israel 90% of the homes are equipped with this technology. In a temperate country like France, about $4\,m^2$ of collectors is sufficient to provide hot water for a family of four persons. The cost of such a system lies in the range of 2000–4000 €. Solar water heaters can also be used to heat the interior space of a house (Figure 7.3). This is done using an in-floor radiant system where hot water flows into pipes below the floor or inside the ceiling.

Usually, solar collectors can harness between 200 and $800\,kWh/m^2$ depending on the technology, the region, and the needs. In France, between 10 and

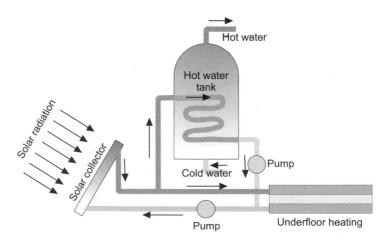

Figure 7.3. Schematic of system using solar collector to provide hot water and in-floor heating. The system presented here is not a passive system and requires pumps to circulate water. This is an additional source of energy consumption.

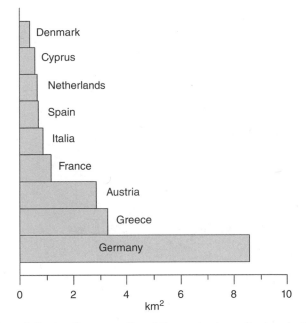

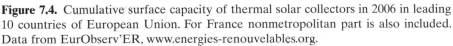

Figure 7.4. Cumulative surface capacity of thermal solar collectors in 2006 in leading 10 countries of European Union. For France nonmetropolitan part is also included. Data from EurObserv'ER, www.energies-renouvelables.org.

$20\,m^2$ of collectors are needed to meet between 50 and 60% of the energy necessary to heat the interior space of a house for a family of four persons. Solar hot-water systems are also useful to heat swimming pools.

Increasing use of solar water heaters should be a major energy priority. It is a technology which is cheap and simple compared to photovoltaic cells but could significantly decrease reliance on wood and fossil fuels. In some countries, incentives are given to boost the development of thermal solar energy and solar energy in general.

Figure 7.4 shows the cumulative surface capacity in 2006 of thermal solar collectors in the 10 countries of the European Union which lead in capacity. In Figure 7.5 the cumulative power capacity of thermal solar collectors is shown for the same 10 countries of the European Union. The total installed capacity in the European Union in 2006 was $20.4\,km^2$.

7.2.2. Heating, Cooling, and Ventilation Using Solar Energy

In the United States and the European Union, conventional systems for heating, cooling, and ventilating buildings consume about 40% of the energy used. Solar energy could meet much of this energy demand. Using solar energy for this purpose would be particularly advantageous in locations where the energy currently used is derived from fossil fuels. There are many clever active or passive techniques available to utilize solar energy. In passive solutions,

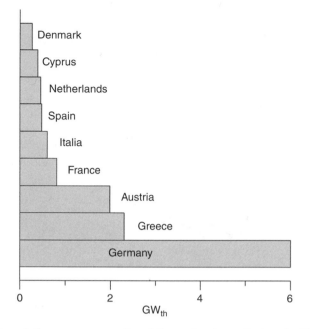

Figure 7.5. Cumulative power capacity of thermal solar collectors in 2006 in leading 10 countries of European Union. For France nonmetropolitan part is also included. Data from EurObserv'ER, www.energies-renouvelables.org.

natural energy flows allow the transfer of thermal energy into or out of a building without the consumption of conventional energy.

In passive solutions, materials such as stone, concrete, and water may be used to store solar energy during the day and release it during cooler periods. The techniques involved are particularly simple. The specific architectural arrangement which is most effective varies with region and a solution which works well in one country can be inefficient in another country. For example, houses with thick walls made out of stone are efficient in the South of France because the night temperature is rarely high. Heat stored during the day diffuses through the wall, reaching the interior of the house at night. The same type of construction would be quite inefficient in regions where temperatures remain high during the night.

Solar chimneys are passive systems which use a hollow channel constructed from thermally conductive material to connect the inside and outside of a building. During the day solar energy heats the chimney and the air inside. This creates an updraft of air, producing an airflow which is used to ventilate and cool the building. Chimneys of this type have been used for ventilation since the time of the Romans and are still common in the Middle East.

A Trombe wall is a passive solar system which is basically an air channel confined between a sealed insulated glass and a wall built with a material which has a high specific heat, that is, capable of storing large amounts of thermal energy (e.g., stone, concrete, tank of water). Sunlight passes through the insulated glazing and solar energy is stored in the thermal mass. The air channel is warmed, producing natural ventilation. A number of variations on this technique exist. In some cases vents are added. The vents have one-way flaps which control the direction of the air (and heat) flow. In the absence of sunlight the wall radiates the stored heat. This technique will also be discussed in Chapter 14.

A solar roof pond is also a passive system for heating or cooling buildings. It consists of a water tank in a transparent plastic tank located behind a movable insulating cover which can be opened or closed. For heating application (winter months), the tank is uncovered during the day and sunlight warms the water of the tank. Heat is stored in the tank, and this energy is used during the night while the tank is covered again. The house is heated by radiation inward from the roof. If cooling is required (summer months), the cover screens the tank during the day and heat from the building is used to warm the tank. During the night the system is uncovered and the heat is radiated away.

7.2.3. The Solar Cooker

Using solar energy to cook food is also relatively simple. Solar cookers have been developed. They can have a power of several hundred watts. A solar cooker can reach temperatures of the order of 150 °C (302 °F) (Figure 7.6). Different systems have been developed; solar panel cookers, hotpots, and solar kettles are seeing increasing use. Larger scale solar cookers allowing preparation of larger quantities of food are also being developed. Hybrid systems, combining solar power and the use of an electric oven when the weather conditions are not good enough, also exist. The great advantage of a solar

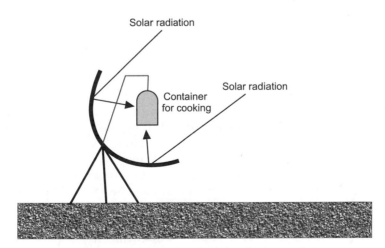

Figure 7.6. Schematic principle of solar cooker.

cooker is that no fuel is needed. Extensive use of solar cookers in developing countries could slow down deforestation and desertification because they reduce the demand for firewood. Furthermore, they are better for the health. Indeed, using firewood under conditions of insufficient ventilation, as is done in many countries, has health risks equivalent to smoking two packs of cigarettes a day.

The first known western solar cooker was built in 1767 by a Swiss aristocrat, Horace-Benedict de Saussure. A physicist, he is considered the father of modern alpinism. He made many scientific measurements in the mountains of the Alps (pressure, temperature, hygrometry, wind speed, etc.).

7.3. CONCENTRATED SOLAR POWER PLANTS

Solar power plants concentrate sunlight in order to reach very high temperatures and produce electricity. This is done through an associated thermal power plant in which steam generated by the solar power plant drives an electrical generator. The basic idea of a solar thermal plant is to use some device, a parabolic reflector, for example, to focus solar radiation on a vessel containing a fluid, typically oil or water. The fluid can be heated to high temperatures, $300-100\,°C$. This technology is effective in regions where sun is available at least $2500\,h/year$. In a country like France, for example, the amount of energy which can be harnessed is about $70\,kWh/m^2$. Most of the present projects are at the demonstration stage, although electricity is already produced in some regions, California, for example.

7.3.1. Parabolic Troughs

A parabolic trough combines a parabolic mirror with a Dewar tube containing a heat transfer fluid (usually oil) positioned along the focal line of the mirror (see Figure 7.7). Sunlight illuminating the mirror is reflected and concentrated (to between 30 and 80 times its normal intensity) on the Dewar tube and heats the fluid. The trough is usually oriented along a north–south axis and can be rotated to track the sun. The fluid inside the tube reaches a sufficiently high temperature, up to $390\,°C$, or $735\,°F$, to produce steam and drive a turbine generator. The efficiency of the system (the ratio between the incoming energy from the sun and the amount of electrical energy produced) is about 15%. A solar plant consists of many troughs operated simultaneously in parallel.

In commercial plants hybrid systems employing fossil fuels (generally natural gas) may be used to supplement the energy production during the

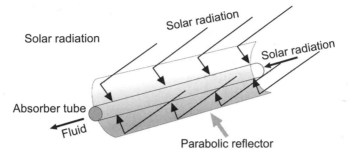

Figure 7.7. Schematic of parabolic trough module.

night or in periods of insufficient sunlight. Nevertheless, to remain qualified as a renewable energy source, the maximum portion of electricity produced by fossil fuels is limited to around 25%. Another possibility to deal with intermittency of sunlight is to store heat energy in a tank filled with a heat transfer fluid, a molten nitrate salt, for example.

The largest hybrid plant (using natural gas to supplement sunlight) of this type is located in the Mojave Desert in California. It consists of five 33-MW solar thermal electric-generating facilities corresponding to 165 MW of total installed power. In this area, which is one of the sunniest in the United-States, a total of nine different solar power plants were commissioned between 1984 and 1991. This group of solar power plants is the largest in the world. They utilize more than a million mirrors covering an area over 6.4 km². The installed power is 354 MW and can meet the daily energy needs of about 500,000 people. A smaller 64-MW power plant of this type exists in nearby Nevada.

Solar trough systems are the most economically competitive plants using sunlight. When the sun shines, the availability of this type of solar plant is high: near 93%. The price of the generated electricity is about 10¢–12¢/kWh. This is about half the price it was 20 years ago. While this remains above the cost of electricity generated by natural gas, for example, future cost increases in fossil fuels may well change that comparison.

In California, the 80-MW Luz solar power station consists of 850 troughs (each containing several mirrors). The total active area of the mirrors equals 465,000 m² and the array occupies a ground area of 1.5 km². The temperature of the heat transfer fluid reaches 390 °C. The amount of solar energy received on the mirrors is 2700 kWh/year/m². The electricity production is equal to 175 GWh/year or 380 kWh/year per square meter of mirror. This corresponds to a net conversion efficiency of 14%. The Luz Company went out of business in 1992, but the plants still continue to produce electricity.

In Europe, sunny Spain is making a major commitment to solar power. The Andasol 1 solar power station near Guadix is Europe's first parabolic trough commercial power plant. It has an installed power of 50 MW and is coupled to a large heat storage reservoir sufficient to provide electricity for 7 h in the absence of sunlight. Close to 180 GWh is expected to be produced each year, meeting the demand of about 200,000 people. Parabolic trough power plants are also planned in Crete, Egypt, and India. The price of electricity is expected to be about 8 ¢/kWh. Usually, the solar field installation represents about half of the total investment.

7.3.2. Power Towers

Power towers use a set of flat mirrors, each of them being able to move in two directions to track the motion of the sun, and focus sunlight on a collector target located in a tower. These solar plants are also sometimes called "central tower" or "heliostat power plants." [Heliostat comes from the Greek word *helios* ("sun") to which is added *stat* for "stationary".]

Compared to parabolic trough installations, higher temperatures can be obtained leading to a better efficiency for production of electricity. This follows from the Carnot principle, which states that the higher the temperature difference between two heat sources, the higher is the yield of useful work. These are more sophisticated systems because each mirror must be able to move in two directions while, for parabolic troughs, all the mirrors positioned along the same line have to move but only along one degree of freedom. Large-area heliostats are more efficient and more economically competitive. Heat has to be stored in order to produce electricity at night or during overcast days. Molten salts can be used to do that. Power towers do not need to be built on flat surfaces.

Several power tower projects were started in the 1980s because industrialized countries feared shortages of low-cost oil. Most of them relied upon production of steam from water to produce electricity. The largest project was SOLAR 1 in California in 1982. It had 71,500 m^2 of mirrors. This was followed by the SPP-5 project with 40,000 m^2 of mirrors in the Ukraine.

Other water–steam power towers are SUNSHINE in Japan (1981, 12,900 m^2 of mirrors), CESA 1 in Spain (1983, 11,900 m^2 of mirrors), and EURELIOS in Sicily (1981, 6200 m^2 of mirrors). Two other projects using different heat transfer fluids were also constructed. One using molten salt was developed in France (THEMIS, 1982, 11,800 m^2 of mirrors) and the other in Spain using liquid sodium (CRS, 1981, 3700 m^2 of mirrors). SOLAR 1, in California, was then transformed into SOLAR 2 using molten salt instead of water–steam. It functioned for three years (1996–1999).

The concentration factor for the sunlight depends upon the project. It was around 700 for THEMIS in France but only 235 for SOLAR 1.

Since March 2006, the Spanish company Abengoa has provided electricity from the PS-10 solar thermal power plant near Seville. This 11-MW plant provides electricity for about 6000 households. It employs 624 heliostatic solar mirror panels that track the sun automatically, focusing sunlight on a tower where water is heated and converted to steam. This steam is then used to power turbines and generate electricity. By 2013 the capacity of this plant is expected to increase to 300 MW. This would supply ~180,000 households.

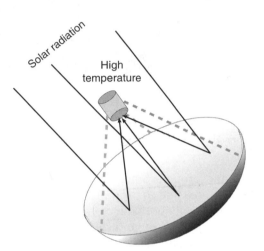

Figure 7.8. Principle of parabolic dish collector.

7.3.3. Parabolic Dish Collectors

It is possible to collect direct solar radiation using a parabolic dish which focuses sunlight onto a single point above the mirror (see Figure 7.8). The parabolic dish receiver is driven by a computer and uses a dual-axis system for tracking the sun. For some systems a Stirling engine is used to generate electricity. The concentration factor for the sunlight can in principle reach 10,000, but in practice values around 4000 are typical. Temperatures higher than 1000 °C can be reached at the receiver focal point, allowing a high efficiency for solar energy conversion.

7.4. SOLAR CHIMNEYS OR TOWERS

It is also possible to produce electricity using passive thermal solar systems without concentration of the sunlight. In solar updraft towers (Figure 7.9) a tall chimney stands in the middle of an extended circular greenhouse collector

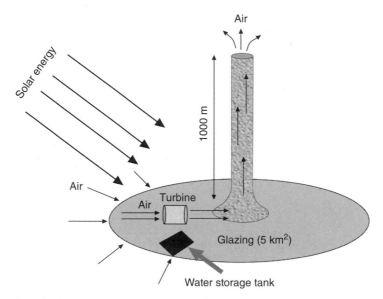

Figure 7.9. Principle of solar tower project in Australia. Just one turbine is shown (there are 32). There are also several water tanks for heat storage.

structure located on the ground. Since there is a temperature difference between the greenhouse area and the top of the chimney, natural convection causes air heated in the greenhouse collector to rise and escape through the chimney. The air current passes through turbines to produce electricity. If heat is stored inside the collector during the day, this system can also be used to produce electricity during the night.

The idea of using a solar chimney to produce electricity was first proposed in 1903 by the Spaniard Isodoro Cabanyes. The idea was elaborated upon in 1931 by the German science writer Hans Gunther. In 1975 the American Robert Lucier filed a patent request based upon a more complete design. This patent was granted in 1981. In 1982, the German architect Jörg Schlaich built a prototype powerplant. A full-scale plant under construction by Environmission in Australia will provide electricity for 200,000 homes. It incorporates a 1-km-high tower which, when completed, will be the tallest man-made structure on earth. For comparison, the height of the Eiffel Tower is 324 m, and that of the Empire State Building is 449 m. Presently, the world's tallest supported land structure is a 628.8-m television tower in North Dakota. In 2002 *Time Magazine* identified this project as one of the best inventions of the year. The operating principle is considered revolutionary but is based on very common knowledge: Warm air rises.

In 1982, a 50-MW prototype solar tower was built in Manzaneres, Spain. Its greenhouse collector is a circle 244 m in diameter covering 46,000 m², or 4.6 ha. The chimney is 195 m tall and 10 m in diameter.

Now under construction in New South Wales in Australia is a 200-MW solar tower. The greenhouse collector is a circle 7 km in diameter (38.5 km²) and the chimney is 1 km high. Typical ambient temperatures are expected to be around 30 °C at ground level and 20 °C at 1000 m. Inside the collector, due to the greenhouse effect, the temperature of the air will be increased to 70 °C. The airflow will be 15 m/s. Energy will be extracted from this flow using 32 turbines. Because heat can be stored during the sunlight hours, the solar tower is expected to operate 24 h/day, producing electricity continuously.

A solar tower requires a large initial investment cost but relatively low operating costs are expected. Nevertheless, since the yield of conversion is small compared to concentrating solar systems, this leads to higher electricity costs. For the Australian project the construction cost is estimated to be about $800 million. The extracted energy is 5 W/m², which corresponds to a yield of 0.5% of the impinging solar energy, much smaller than that achieved with concentrating solar systems. The cost of electricity is higher than for conventional means of production. Reasonable estimates give a cost between 20 ¢ and 35 ¢ per kilowatt hour. It could provide electricity for 200,000 homes and would reduce CO_2 emissions by 0.83 Mt/year.

7.5. PHOTOVOLTAIC SYSTEMS

Sunlight can be transformed directly into electricity using the photovoltaic effect, which was discovered in 1839 by Antoine Becquerel, a French physicist and the grandfather of Henri Becquerel, who discovered radioactivity 57 years later. Antoine Becquerel noticed that some semiconductor materials exposed to light (photons) could create electron–hole pairs which could be collected to produce an electrical current. The phenomenon occurs for photons having an energy above a certain threshold and thus wavelengths shorter than a certain value, characteristic of the material. In the case of silicon, the threshold for electron–hole production is located in the infrared region of the electromagnetic spectrum. An electron is negatively charged while a hole is a positively charged electron vacancy. In order to produce a current, it is necessary to separate the electrons and holes. This is done naturally if an electric field can be produced in the medium. This is the reason why composite *np* semiconductor materials in which electron-rich *n*-type material is in contact with hole-rich *p*-type material are employed. Such an architecture is called a *p–n* junction and provides the operational basis for solar, or photovoltaic, cells made out of semiconductor materials.

Most of the photovoltaic cells used today are made of silicon. They have areas ranging from 1 to 100 cm². Several cells are associated together to make a module which has a typical area between 0.5 and 2 m². Modules are themselves connected together in arrays which are used to produce electricity. A direct current is produced, but it can be converted to an alternating current (AC) with electronic devices. There is a difference in efficiency between a single cell and a photovoltaic module as well as between laboratory prototypes and industrial photovoltaic systems. For example, for monocrystalline silicon, the maximum recorded cell efficiency reaches about 25% in the laboratory while the corresponding module efficiency is about 23%. In a commercial application, the typical efficiency of a module is between 12 and 15%, well below that obtained in laboratory conditions.

The average solar radiation power received on the horizontal surface of the earth in the United States varies from a bit less than $150\,W/m^2$ in the north (e.g., New Hampshire) to about $250\,W/m^2$ in the south or southwest (e.g., New Mexico). A large part of the United States receives a yearly 24-h average of the order of $200\,W/m^2$. In December, the monthly average drops to between 50 and $150\,W/m^2$.

7.5.1. Market Dominated by Silicon

Silicon is the material most commonly used in commercial photovoltaic applications. Starting in the 1950s a number of technologies for using both monocrystalline and polycrystalline silicon in different forms (e.g., wafers, ribbons, sheets) were developed. This element is obtained by reducing natural silica (SiO_2), in an electric oven. The purity of the resultant metallurgical silicon is about 98%. For applications in the microelectronics industry silicon is transformed into gaseous products which are subsequently purified in the gas phase. Silicon used in photovoltaic cells is obtained by pyrolysis of the gaseous silicon components.

Monocrystalline silicon is used to produce close to 27% of commercial photovoltaic modules. In this technology, the starting point is a large silicon monocrystal (a cylinder of 30 cm diameter and more than 1 m height) made by the Czochralski method. From 1 kg of SiO_2, one obtains around 100 g of monocrystalline silicon. The process requires about 1 MWh of energy. The final purified ingot is then sliced into wafers of 200–250 µm thickness. During the sawing operation, around half of the material is lost.

Over the years there has been a large improvement in photovoltaic cell efficiency. Before 1980 the efficiencies of monocrystalline cells were between 8 and 10%. Today, in the laboratory much higher efficiencies are obtained, up to 24.7%. Typical efficiencies between 14 and 17% are observed in commercial applications. As noted above, the module efficiencies are smaller than those of individual cells. Furthermore, the efficiency depends upon the operating temperature of the module, decreasing with increasing temperature.

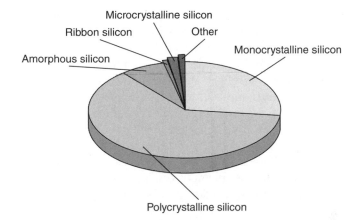

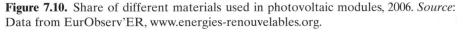

Figure 7.10. Share of different materials used in photovoltaic modules, 2006. *Source*: Data from EurObserv'ER, www.energies-renouvelables.org.

Until 2002, photovoltaic silicon needs were met with by-products of the microelectronic industry. Because of increasing demands, the price of the raw silicon material has strongly increased in recent years. For example, between 2000 and 2005, the price of metallurgical silicon went from $15/kg to $75–$100/ kg. In the future, shortages of low-cost silicon may become a limiting factor in the development of photovoltaic energy. For that reason, simplified techniques to produce silicon or other materials with a quality good enough for photovoltaic purposes and other solutions to transforming sunlight into electricity using the photovoltaic effect are being investigated.

Solar cells can also be made using polycrystalline silicon. In this technology, offcuts of silicon coming from the microelectronics industry are melted at 1430 °C. As the temperature slowly drops, polycrystalline silicon is formed. Ingots weighing several hundred kilograms are formed. These are later sliced into wafers similar to those derived from monocrystalline ingots. The cell efficiency for these polycrystalline cells is lower, between 12 and 14%, but the cost is also lower. New production technologies avoiding ingot slicing exist. In these processes, which are less energy consuming, silicon is then directly produced in the form of ribbon or sheets.

The photovoltaic module market is dominated by silicon, as can be seen in Figure 7.10.

7.5.2. Other Photovoltaic Technologies

A goal for the future is to produce thin-film photovoltaic cells needing less material and less energy for processing. Thin film techniques also offer the possibility of making very large areas and, compared to wafer-based silicon technology, the cost should be very low. The basic idea is to use low-cost substrates such as glass or plastic to support micrometer layers of materials having high light absorption and energy conversion efficiencies for the incident solar radiation. The manufacture should be much easier and can easily be automated.

Such thin-film technologies are already being applied to produce amorphous silicon films. Initially, hopes were very high for this technology as amorphous silicon has a very high light absorption allowing thicknesses of about 1 μm to be used in practical applications. Also the process needs five times less energy than required for processing monocrystalline silicon. However, the 6–8% efficiencies of amorphous films are much smaller than those of crystalline cells. Further, amorphous cells have a special feature in that the efficiency drops during the initial operation before leveling off to a constant value.

Thin films of polycrystalline silicon baked on a low-cost substrate (e.g., glass, metallurgical silicon) are also currently being investigated. The thickness of the silicon ranges between 10 and 40 μm. Cell efficiencies close to 10% have already been obtained in some special cases. Still, among thin-film technologies, amorphous silicon is the one which has the largest share in commercial applications (see Figure 7.10). However, its share of the market is continuously decreasing. In 1999 the market share was 10–12%. It dropped to about 4.5% in 2004. For the future, hybrid cells combining amorphous and crystalline silicon show particular promise. Efficiencies up to 21% have been obtained in the laboratory and 16% in industrial production.

Copper indium diselenide (CIS) has a high potential and is being commercialized. Yields of 18.8% have been obtained in the laboratory in the United States and more than 13% on field-operated prototypes. Nevertheless, indium abundances will limit this development. Estimates have been made that all the indium resources in the world could not supply more than 1% of the present world electricity consumption. Therefore, this technology must be reserved to some specific niches.

Until recently cadmium-based technology (CdS–CdTe) was quite appealing because of the high efficiency and the low cost of the manufacturing process. However, cadmium is a chemically dangerous element and a health hazard. The future seems to be closed for this technology. While some other materials also show promise for thin-film photovoltaic cell development a much greater effort and investment will be required to find an ideal one.

While the thermodynamic limit of sunlight conversion is 87%, single-junction solar cells have a theoretical upper efficiency limit of 33%. There is therefore still a need to deploy new technologies with higher efficiencies. One such technology, that of multijunction cells, allows capture of almost all the incident photons coming from the sun. For example, yields up to 35% can be obtained using a 3-μm-thick stack of gallium indium phosphide, gallium indium arsenide, and germanium. This is not possible with silicon alone. Functionalized active layers may be used to extend the range of the solar spectrum to be exploited (in particular in the UV region). Since such devices are expensive, they are often employed together with an optical system that concentrates sunlight on a smaller area and therefore minimizes the surface of the photovoltaic material required. These systems effectively allow the replacement of an expensive semiconductor material by less expensive optical and mechanical components. For instance, with such devices, solar radiation falling on a 10-cm^2 surface could be concentrated and focused into a few-square-millimeter area of active semiconductor. Yields around 40% are expected with this technology.

Cells constructed from organic molecules and polymeric materials have the potential advantages of low cost, ease of fabrication, and small energy requirements to produce them. However, this technology is still at the research stage and producing viable organic photovoltaic cells is a big challenge. The stability of these cells is one of the main issues. In 2005, organic cells were made in the laboratory with a yield of 5.7%, but their useful lifetimes were below 500 h. The operation mechanism of a dye-sensitized solar cell constructed from organic materials is similar to that of photosynthesis. Efficiencies up to 8.2% have been obtained for such cells. There still remain a number of problems to build large modules from such cells.

7.5.3. Applications

A great advantage of photovoltaic modules is that they are reliable and there are no moving parts. Little maintenance is required and the systems are noise free. There are also no CO_2 emissions during operation. Electricity can be produced on demand at the place of consumption, avoiding transport losses over large distances. However, the power delivered by photovoltaic cells is reduced by cloudy conditions or if the angle of incidence of sunlight shifts from normal irradiation.

Although no greenhouse gas is emitted during operation of a photovoltaic cell, this is not currently the case for its fabrication. Silicon cell technology requires a lot of energy. It is estimated that, for crystalline silicon, four to five years of exposure to the sun are needed to reimburse the energy necessary for their manufacturing. Fortunately, in the case of grid applications photovoltaic modules can last more than 30 years. However, if the consumer is not connected to the electric grid, a supplemental means of energy storage is required. In stand-alone applications batteries may be required to store electricity for use in periods without sunlight. Battery fabrication also requires energy so the net energy payback time is longer.

The installed power for a photovoltaic system is usually rated in "peak watts." One peak watt is the maximum power delivered by a photovoltaic device, a cell or a module, under standardized test conditions (typically 1000 W/m^2 of sunlight, a temperature of 25 °C, a pressure of 1 atm, and almost no wind). One peak watt (1 W$_p$) corresponds to the typical maximum output of a 10 cm × 10 cm square of monocrystalline silicon cells. Therefore 1 m^2 of this material has a peak power of about 100 W/m^2. The quantity of electricity produced by 1 W$_p$ is about 1 kWh/year. This corresponds to about 100 kWh/m^2.

Photovoltaic arrays haves peak powers ranging from a few milliwatts to several megawatts. The main problem for photovoltaic energy is that it is expensive, although much progress has been made in reducing costs. For consumers coupled to a grid, the cost per kilowatt-hour ranges between 0.3 and 0.45 €. This is more than 10 times more expensive than electricity produced with natural gas or nuclear fission.

For stand-alone systems, including storage batteries, the price per kilowatt-hour reaches about 1.5 €. However, in countries where no electric grid is installed, this is competitive with other electricity generators and much less expensive than batteries alone. A major problem is that, in developing countries, initial investments far beyond what ordinary people can afford may be required. Innovative financial arrangements such as advancing installation costs to consumers who then reimburse this amount by periodic payments over an extended time would be needed to spur adoption of this technology.

High efficiencies of solar radiation conversion are not always necessary. For a given power, the cost of the total system is sometimes more important than having the highest performance photovoltaic cells. Larger surfaces of lower efficiency modules may be more cost effective. This is often the case for individual homes where for esthetic reasons it may be better to have the whole roof covered by photovoltaic cells rather just a part of it. Since the usual roofing material may be replaced (or protected) by the photovoltaic modules, costs for the former are reduced, helping to decrease the total.

With the present technologies, mostly based on crystalline silicon, a lot of energy is needed to fabricate photovoltaic cells. Some toxic chemicals are also used during their manufacturing. These can often be recycled. Since most of the energy needed to manufacture photovoltaic cells comes from electricity, the amount of greenhouse gases emitted during manufacturing depends on the way in which the electricity is produced. Although there is some debate about the evaluations, values between 20 and 45 g of equivalent emissions, CO_2 per kilowatt-hour of silicon cell capacity are generally estimated. Studies requested by the European Commission give even higher values. The actual value depends strongly on the origin of the energy used to manufacture the cells. For example, if electricity is used, it differs significantly depending upon whether the electricity is produced by coal power plants or nuclear plants.

The use of photovoltaic energy is increasing rapidly. Since 2002 the world's photovoltaic electricity production has increased by about 40% each year. The increase has taken place mainly in three domestic markets: Germany, Japan,

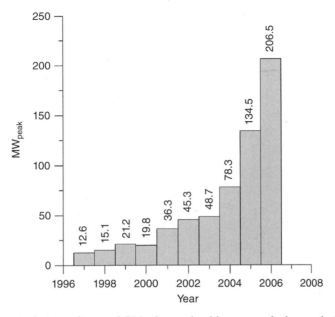

Figure 7.11. Evolution of annual U.S. domestic shipments of photovoltaics. *Source*: Data from www.eia.doe.gov/cneaf/solar.renewables/page/solarphotv/solarpv.html.

and the United States. For the United States, Figure 7.11 illustrates the strong increase of domestic shipments (export shipments have been extracted) since 1997. It went from $12.6\,MW_p$ in 1997 to $206.5\,MW_p$ in 2006. There has been a large increase of the global photovoltaic capacity during the last few years, as seen in Figure 7.12.

In 2005 the photovoltaic industry used about 15,000 t of silicon to manufacture 1.15 GW of crystalline photovoltaic cells. In that same year, the installation of $1\,GW_p$ of new photovoltaic power systems, most of this (85%) in Japan and Germany, brought the total world installed photovoltaic power to $3.7\,GW_p$. Of this, $644\,MW_p$ was installed in Europe, bringing Europe's total installed capacity to $1.79\,GW_p$. In Europe, Germany accounts for 86% of the total installed capacity. This capacity is overwhelmingly (94.4%) in grid-connected applications.

The United States is the world's largest producer of silicon, $\approx$16,000 t in 2005, of which 6300 t was for photovoltaic applications. Japan produces $\approx$8000 t, with $\approx$16–18% used for photovoltaic applications. Germany produces $\approx$5200–5500 t, with more than half dedicated to the photovoltaic industry (data from IEA, www.iea.org). Silicon demand has driven the price sharply higher in recent years.

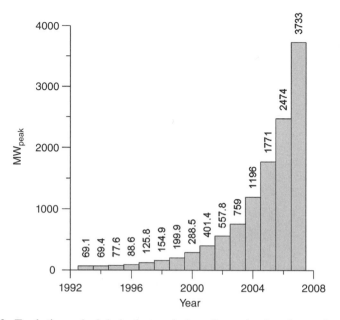

Figure 7.12. Evolution of global photovoltaic cell production in peak megawatts. *Source*: Data from EurObserv'ER 2008, www.energies-renouvelables.org.

In Figure 7.13, the geographic distribution of photovoltaic cell production is displayed. Seventy-five percent of the cell production occurs in three regions: Europe, Japan, and China.

Most of the new photovoltaic systems installed in Europe are on grid. By the end of 2007, on-grid applications represented 97.3% of the cumulative capacity installed in Europe. The installed on-grid applications during year 2007 represent 99.5%. Off-grid applications are of only minor importance for European countries except for France, which has a significant amount in its overseas territories. In Figure 7.14 the 10 countries of the European Union with the largest installed photovoltaic capacities at the end of year 2007 are shown.

The idea of harvesting solar energy in space using solar power satellites consisting of several kilometers of high-yield photovoltaic modules in geostationary orbits dates to 1968. At high altitudes the solar radiation is more readily accessible and photovoltaic arrays may be oriented to achieve maximum efficiencies. At 36,000 km of altitude an average of about eight times more energy per unit of cell surface may then be collected at the earth's surface. The collected energy can be beamed to the earth using microwaves. The beam intensity should not be too large, typically one-tenth of the sun's radiation power, in order not to be a health hazard. Transmission

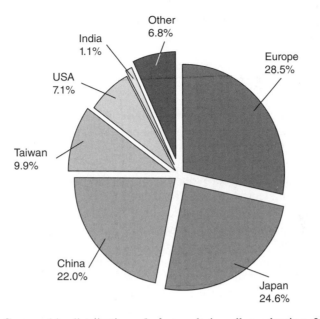

Figure 7.13. Geographic distribution of photovoltaic cell production, 2007. *Source*: Data from EurObserv'ER 2008, www.energies-renouvelables.org.

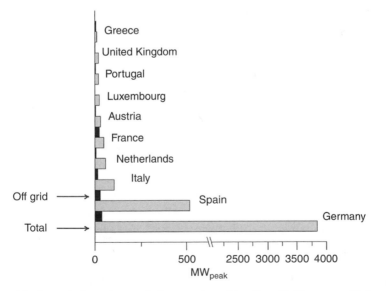

Figure 7.14. Cumulative photovoltaic capacity of leading 10 European Union countries, end of 2007. The grey bars correspond to the total capacity (on-grid and off-grid) and the black bars to the off-grid capacity. *Source*: Data from EurObserv'ER 2008, www.energies-renouvelables.org.

efficiencies of the order of 50% have been recorded in microwave energy transmissions.

7.6. ELECTRICITY STORAGE

Stand-alone photovoltaic applications cannot meet the demand during periods of no sunlight. A means of electrical storage is needed. In addition to providing electricity when the photovoltaic cells are inoperable, batteries can also be used to meet peak demands during the day. Electricity can then be drawn simultaneously from the photovoltaic module and from the battery. Batteries store energy in an electrochemical form with a one-way conversion efficiency of 85–90%. For photovoltaic applications rechargeable batteries are used. During the charging mode they convert electricity into chemical energy. This process is reversed in the discharge mode. An efficiency ranging between 70 and 80% is usually observed in the charge–discharge cycle. More details about batteries are contained in Chapter 12. The cost of lead–acid batteries, ≈$200–$500/kWh, is much less than for other technologies. For example, the cost for Ni–MH batteries is ≈$2500/kWh.

The low-cost lead–acid battery has a cell voltage of 2 V. This battery is widely used in photovoltaic applications despite the fact that lead batteries have smaller energy densities by weight and by volume than other technologies. For niche markets, other types of batteries can be used: Ni–Cd, Ni–MH, Li ion. Such batteries, while more costly, have better energy densities. For example, the energy density by weight of Li ion batteries is three times larger than that of lead–acid batteries.

In photovoltaic applications, deep-cycle lead–acid battery technologies that allow repeated charge and discharge cycles should be used. This technology is different from the one typically used for cars or trucks for which batteries are required to deliver short bursts of energy as needed and remain charged. The operating temperature has an influence on the efficiency of the charge–discharge cycle. A lead–acid battery can operate between −10 and 50 °C while a Li ion battery works between 10 and 45 °C. For a Ni–Cd battery operating at 20 °C, the charge efficiency is equal to 93% and the discharge efficiency to 100%. At −20 °C these figures become respectively 75 and 95%. The self-discharge of batteries is also temperature dependent. For Ni–Cd batteries the self-discharge rate per day is only 0.4% at 20 °C. For an operating temperature of 60 °C, this rate increases to 8%.

It should be noted that the number of possible cycles at a given temperature depends very much on the depth of the discharge. For example, the number of usable cycles can double by accepting a 25% discharge instead of a 50% discharge. This point is important for electric or hybrid vehicles, as we shall see in the discussion of the use of batteries in transportation applications in Chapter 13. Lead–acid batteries have between 500 and 1000 cycles in full-discharge cycles and more if the discharge is partial. These figures are similar for Ni–Cd and Ni–MH batteries but twice as few for Li ion and Li polymer batteries.

7.7. ECONOMY AND ENVIRONMENT

The price of an entire photovoltaic system depends very much on size of the system, location, intended application, and whether the system is connected to the grid or employed for stand-alone applications. Additional components needed for effective exploitation of the photovoltaic modules can represent a nonnegligible contribution to the total cost of a photovoltaic system. This contribution ranges from about 20% in the case of grid-connected applications to around 70% in the case of stand-alone systems. For stand-alone applications, installed capacity prices were ranging between $10 and $20/W in 2005. For grid applications, the average prices were $5.5–$6.5/W but lower prices could be obtained for large-scale applications.

Because operation and maintenance are inexpensive, the main economic problem with photovoltaic systems is the requirement for relatively large initial investments. Although a photovoltaic module can be used 30–40 years, storage batteries used in stand-alone applications have to be replaced regularly. This need increases the cost of photovoltaic energy significantly. For a typical stand-alone photovoltaic system, the photovoltaic module accounts for about 67% of the capital investment and the battery for about 14%. Good lead batteries (tubular ones) can last almost a decade, but normal ones have to be changed after about 5 years. After 20 years of operation, the additional costs for batteries cause the battery investment to increase to 48% of the capital investment.

The amount of pollution saved using electricity from photovoltaic modules instead of using electricity from the grid depends very much upon the way this latter has been produced. It is particularly effective to use photovoltaic modules in the case where most of the electricity coming from the grid is produced using fossil fuels, especially coal. To give a rough estimate, for the United States, where a large part of the electricity comes from coal-fired plants, each peak kilowatt of installed photovoltaic capability would do away with between 600 and 2300 kg of emitted CO_2. In addition, this would also lead to reductions of 16 kg of NO_x, 9 kg of SO_x, and 0.6 kg of particulates.

7.8. CONCLUSION

Among renewable energy sources, solar energy is probably the one which will see the largest development and exploitation. Our star, the sun, will be available as an energy source for about another 5 billion years before it runs out of hydrogen fuel and shifts to helium fuel. It will then expand into a red giant star and finally evolve into a white dwarf star. By that time we can expect that life on earth will have disappeared, ideally by relocating. Harnessing the energy of the sun at a competitive price is therefore a particularly effective means of meeting various energy needs.

Deriving thermal energy from solar radiation can be done rather cheaply and technologies exist to employ that energy to produce heat, cold, and electricity. These technologies will be improved and will be used more extensively in the future.

The production of electricity using photovoltaic devices is an important focus area for future development. Since sunlight is not available during the night, development of more efficient means of electricity and thermal storage are key requirements to provide energy continuously to the consumer.

Photovoltaic energy is more suited to small and local demands than for high-power requirements. Its high price is, for the moment, its largest drawback. Prices are improving slowly. If the price of photovoltaic modules becomes truly competitive with that of other energy sources, a rapid increase in the use of this technology can be expected. The timescale for this development may be shortened by the rise in cost of fossil fuels.

It is also possible, by a proper architecture of homes and buildings, to save a lot of energy either by better use of incident solar radiation (this was well known by our ancestors, who warmed their buildings by a proper orientation towards the sun) or by avoiding it as much as possible in hot countries. More recently, in the European Union about 10% of energy used in housing has been saved due to a better architecture pushed by European authorities. Sunlight can also be used to avoid using electric lights which consume electricity that is often produced with other energy sources. Optical fibers may be used to confine and transport light to the place where it is needed, increasing lighting efficiency.

Finally, in the future it would be worthwhile to develop DC appliances on a large scale because they are often more efficient in terms of power peak requirements. For example, less initial power is required to start a DC refrigerator than an AC refrigerator. An extensive development of photovoltaic electricity would make electric appliances even more economical.

CHAPTER 8

Geothermal Energy

The heat energy contained inside our earth, geothermal energy, has been used by mankind for several thousand years. The first applications were for the heating of water for baths, of homes, for cooking, and for agriculture. However, only relatively recently has it been used on a large scale.

Beneath the crust of our planet is a vast reservoir of geothermal energy. A small part of this thermal energy, around 10%, is remnant heat which remains from the time, about 4.5 billion years ago, when the earth was formed from the dust of stars. The major part (90%) results from the energy released in the decay of radioactive elements such as uranium, thorium, and potassium which are contained in the earth. Our present understanding of the earth's structure is that the central core of the earth is solid and has a temperature larger than 5000 °C. Surrounding this core is a liquid consisting mainly of iron, with a mean temperature reaching 4000 °C. At a depth of about 10 km below the surface of the earth temperatures are still as high as 300 °C. About 99% of the mass of our planet has a temperature greater than 200 °C. The average geothermal temperature gradient near the surface of the earth is 3.3 °C/100 m.

While geothermal energy is accessible everywhere on the earth, there are favored regions where temperature gradients are larger than average. In a country like France, for instance, the geothermal gradient averages 4 °C/100 m but has a broad variation, 10 °C/100 m in the Alsace region and 2 °C/100 m at the emerging of the Pyrenees Mountains. In other parts of the world, gradients can be particularly large in localities having hot springs, for example, Iceland, and in volcanic regions at the borders of tectonic plates the thermal gradients may reach 30 °C/100 m.

The geothermal gradient is not the only important factor which determines accessibility of geothermal energy. The permeability of the rocks is also an important parameter because it determines the rate at which heat can be conducted to the surface. The average energy flux for conduction of geothermal heat to the surface is 60 mW/m². We are fortunate that it is not much larger;

Our Energy Future: Resources, Alternatives, and the Environment
By Christian Ngô and Joseph B. Natowitz
Copyright © 2009 John Wiley & Sons, Inc.

otherwise the earth's surface would be much hotter. Each year more than 100,000 TWh of heat energy is conducted from inside the earth to the surface. The total outward flow is around 30 TW. The geothermal flux (energy per unit area) is however only about one-thousandth of the incoming flux of solar energy incident on the earth.

Although geothermal energy is not a renewable source in the real sense, it can be considered as a virtually inexhaustible resource of energy. It can be tapped, from underground water reservoirs, for instance, as long as the equilibrium between the amount of energy removed from that reservoir and the replenishment of heat flowing into the reservoir is maintained. In some places, the underground drift velocity of heat might be too slow and the amount of energy extracted from the reservoir might not be fully recharged on a practical timescale. This would limit the useful life of that energy source.

Geothermal energy can be used directly or transformed into electricity when the temperature of the source is sufficient. The current global share of the installed capacity for electricity production and for direct use is shown in Figure 8.1.

The installed capacity for direct use ($\approx$77%) is much larger than that for electricity production. However, the final yields are different. This can be seen in Figure 8.2 showing the annual output of geothermal energy. Because of a smaller yield (about 30% for direct use compared to about 74% for electricity production), the direct-use thermal energy drops down to $\approx$57% of the total final energy production.

The quantity of electricity produced from geothermal energy is shown in Figure 8.3 for different regions of the world.

The quantity of energy produced by direct use is shown in Figure 8.4 for different regions of the world.

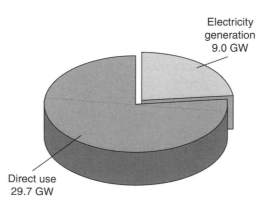

Electricity
generation
9.0 GW

Direct use
29.7 GW

Figure 8.1. Global installed capacity of geothermal energy, 2005. The share between electricity generation (in gigawatts of electricity) and the direct use (in thermal gigawatts) is shown. *Source*: Data from World Energy Council, www.worldenergy.org.

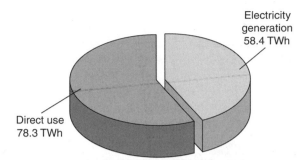

Figure 8.2. Global annual output of geothermal energy 2005. The share between electricity generation (in terawatt-hours of electricity) and the direct use (in thermal terawatt-hours) is shown. *Source*: Data from World Energy Council, www.worldenergy.org.

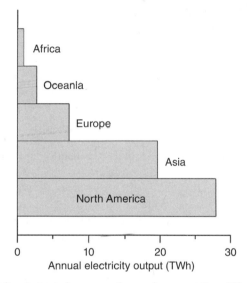

Figure 8.3. Electricity output from geothermal energy for different regions of the world, 2005. *Source*: Data from World Energy Council, www.worldenergy.org.

8.1. AVAILABLE IN MANY PLACES

The accessibility and utility of geothermal power resources depend on the intrinsic nature of the local geology; the stratification, constitution, and permeability of the subsoil; the geothermal gradient; and so on. Geothermal energy usually needs to be concentrated prior to being used in a heat extraction system. A heat carrier is needed to transport the energy from the underground to the system which will be used to provide energy to the consumer. Sometimes this carrier exists naturally, as is the case in hydrothermal sources, hot aquifers

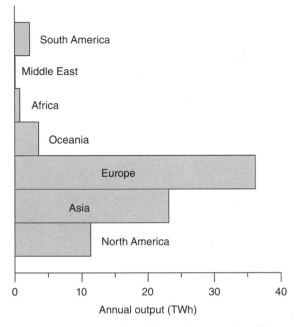

Figure 8.4. Direct-use heat output of geothermal energy for different regions of the world, 2005. *Source*: Data from World Energy Council, www.worldenergy.org.

inside permeable rock formations which are found in various areas on the earth. The carrier is the water of the aquifer.

High-enthalpy geothermal fields, that is, places delivering a good-quality heat which can be easily transformed into work, are found in regions of high geological activity. They are arranged in belts around the earth. These belts are located at the borders of tectonic plates, in regions of recent volcanic activity, or in places where the crust of the earth is thinner. For example, in the Pacific Ocean there is a belt that extends through Japan, The Philippines, Indonesia, Argentina, Central America, and Western North America.

Low-enthalpy fields, that is, those delivering a heat of lower quality as far as the production of work is concerned, exist more frequently than high-enthalpy ones. Such fields may be found in deep sedimentary basins like the Parisian and Aquitain Basins in France or along the gulf coast of the United States, for example.

Fissured or porous volcanic formations close to the borders of tectonic plates are active regions characterized by large geothermal gradients. The Pacific "ring of fire" and the large American rift provide examples. These geothermal systems transfer heat by convection phenomena. In convection, matter is transferred from one place to another. In this case, cold material close to the surface seeps downward and warms up. Simultaneously, hot matter located deep inside the earth moves toward the surface. High-temperature geothermal sources at temperatures larger than 150 °C are available at depths ranging from 1500 to 3000 m. At smaller depths, below 1000 m, medium-temperature geothermal sources at temperatures between 90 and 150 °C can be harnessed.

More or less fractured platforms in the stable continental shelf can be found at about 5000 m depth. Exploiting these, it is possible to produce water at high temperature. First cold water is injected into a well 5000 m deep. The injected fluid causes hot rocks to crack, releasing heat. The heated water diffuses and is recovered in a second well from which the heat is used to produce electricity. Heat is harnessed using either conduction or conduction and convection phenomena. A rate of flow of 100 L/s and a temperature of the outgoing fluid of 200 °C allow transportation of the heat from deep underground to the surface with a good yield. Use of this kind of "hot, dry rock resource" or "enhanced geothermal system" was first proposed in the United States. Experiments along this line have been carried out at Soultz-sous-Forêts in Alsace, France. The principle of the method is shown in Figure 8.5.

It is possible to find formations with variable porosity and heat permeability in sedimentary basins at depths ranging from 2000 to 4000 m. Because of

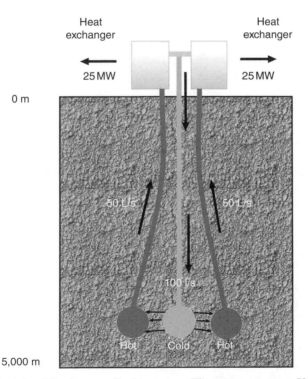

Figure 8.5. Principle of Soultz-sous-Forêts system. The temperature at 5000 m depth is about 200 °C.

existing defects and interfacial discontinuities between rocks, hot water can flow to the surface and hot springs are observed. Medium- to low-temperature geothermal resources are then available. These exist in a number of places in the world: for example, Rio Plata, in South America, the Mississippi-Missouri Basin, the Parisian and Aquitain Basins, the Boise region in Idaho, and the Beijing region in China.

Very low temperature geothermal resources may be found at relatively shallow depths (below 1000 m). The temperatures available are lower than 40 °C. They exist all around the world. The heat of these sources comes either from the sun in the first few meters of depth or from diffusion of groundwater which has been heated by hot rocks which are deeper inside the earth.

Also of note are geopressured geothermal resources which are hot aquifers containing dissolved methane. They are found under high pressure at large depths (3–6 km) and have temperatures between 90 and 200 °C. Such a formation is present in the northern part of the Gulf of Mexico. Three forms of energy could be derived from such a field: heat from the hot fluid, hydraulic work from the high pressure, and energy contained in the dissolved methane gas.

8.2. DIFFERENT USES

Geothermal sources can be classified as high (>160 °C), medium (90–160 °C), low (30–90 °C), and very low (<30 °C) temperature resources. Depending upon the temperature of the source, different types of applications can be envisioned, but the two main uses of geothermal power are in the production of heat and electricity. Heat can be produced with any of these, which means that it is available essentially everywhere on the planet. Hot aquifers are often used for direct heating of buildings or industrial and agricultural applications. Low and very low temperature sources are exploited using heat pumps for heating or air-conditioning purposes in homes or other buildings. Electricity can only be produced with high- or medium-temperature sources (high- and medium-enthalpy sources).

The first demonstration of electricity production from a geothermal source was carried out in 1904 at Larderello, Italy, by Prince Ginori-Conti. Using a small dynamo powered by geothermal steam, he was able to power five lightbulbs. In 1905, the electric lighting of both the plant and the village of Larderello was achieved. About 10 years later the project reached a commercial stage with two electricity production units of 2.6 MW each.

At the present time hydrothermal sources are the only geothermal resources employed commercially. Electricity is produced when the temperature of the source is larger than 90 °C. Depending on the depth of the hydrothermal source, it can produce hot water, steam, or a mixture of the two. The best sources are those which produce dry steam at high temperature, but they are scarce. Two big steam resources are known at present: at Larderello, Italy, and The Geysers field in Northern California in the United States. The Geysers Field produces more than 850 MW of electricity. Low-temperature sources are exploited for their heat, either directly or with heat pumps.

Research and development on the exploitation of other geothermal sources is ongoing. Hot, dry rock resources or enhanced geothermal systems are very promising with a potential of generating 200 GW in the United States and 60 GW in Europe. However, small earthquakes resulting from attempts to exploit this technology have recently been observed in France and in Germany. This is a clear indication that further studies are needed and some care must be taken in the development of these resources.

8.3. TECHNOLOGIES

The simplest way to exploit geothermal energy is to directly use the heat which is produced. The geothermal fluid can be pumped through a heat exchanger to heat air or a liquid which in turn may be circulated in a heat network to heat buildings, for example. Direct use of geothermal energy has better efficiencies, between 50 and 70%, than its use to produce electricity and should be favored whenever it is possible.

For low-temperature geothermal resources (typically below 20 °C), it is possible to use heat pumps of different sizes depending upon the applications. The great advantage of geothermal heat pumps is that they can work in either direction. In cold weather they can extract heat from the ground and supply that heat to homes or buildings. In hot weather, they can be used to provide air conditioning by removing heat from these same homes or buildings and storing it in the earth. By properly dimensioning the system (heat pump and storage), it is possible to heat in winter and cool in summer at a very low cost. Indeed, heat pumps act as energy amplifiers. Their operation requires electricity from the grid, but in the end they deliver much more energy that has been consumed, the difference coming from the energy extracted from the resource. We shall return later to a discussion of heat pump technology, which is very interesting as far as energy efficiency is concerned.

The simplest way to produce electricity from hydrothermal sources is to use dry-steam (Figure 8.6) or flash-steam systems (Figure 8.7). In dry-steam systems, the steam is extracted from the well and cleaned before it is directed through a steam turbine. Sizes of plants range from 15 to 120 MWe, the lowest power systems being part of larger modular systems.

A heat pump is a system that allows transfer of heat from a low-temperature source to a high-temperature one. Since this is not possible spontaneously, work must be performed on the system to do that. The system is then referred to as an "open" system. A heat pump requires electricity from the power grid to function. It is best understood as an "energy amplifier" since with 1 kWh electricity taken from the grid one can typically produce 3 or 4 kWh of heat (even more with high-performance devices). Conversely, heat pumps are able to work as air conditioners if the system is operated in reverse. In that case, the yield is a bit smaller than for heating. Typically 2 or 3 kWh of cooling is produced for each kilowatt-hour taken from the grid.

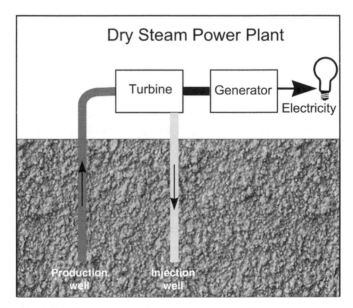

Figure 8.6. Principle of dry-steam power plant. *Source*: From www1.eere.energy.gov/geothermal/powerplants.html.

Flash-steam systems are used when the heat reservoir is mainly a liquid at sufficiently high temperature (above about 160 °C). The principle is to rapidly convert ("flash") a large part of the liquid into steam. In the simplest way this can be done by reducing the pressure of the liquid and directing the resultant steam through the turbine. In more sophisticated technologies, dual-flash cycles in which the steam is separated at two different pressures are used.

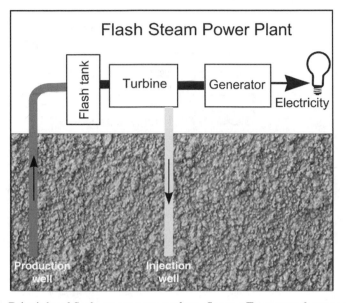

Figure 8.7. Principle of flash-steam power plant. *Source*: From www1.eere.energy.gov/
geothermal/powerplants.html.

Dual-flash systems can produce 20–30% more power than a single-flash system.

When the temperature of the hydrothermal liquids is over 350 °F (177 °C), flash-steam technology is generally employed. In these systems, most of the liquid is flashed to steam. The steam is separated from the remaining liquid and used to drive a turbine generator while the water is returned to the geothermal reservoir. The economics of most hydrothermal flash plants are improved by using a dual-flash cycle. Systems in the range 10–55 MWe have been developed. Units of 20 MW$_e$ can be used in a modular approach as is currently the case in Mexico or the Philippines.

Binary-cycle geothermal power generation plants differ from dry-steam and flash-steam systems in that the water or steam from the geothermal reservoir never comes in contact with the turbine/generator units (see Figure 8.8). Binary-cycle plants are another technology allowing production of electricity from geothermal energy. In a heat exchanger, heat from the hot geothermally heated water is transmitted to a second fluid having a much lower boiling point than the water. The secondary fluid is flashed to vapor, which then drives the turbines. With a rate of flow of the primary liquid of several tens of liters per second electricity can be produced at a competitive cost. For such systems low-temperature sources (100–160 °C) or moderate-temperature ones (160–190 °C) can be used even if they are corrosive brines or brines with insoluble components. Reinjection of the primary fluid prevents pollution and depletion of the resource.

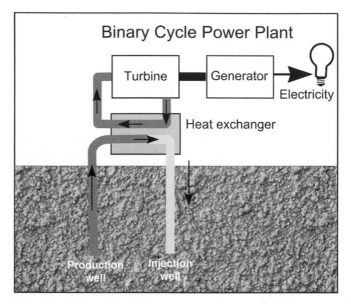

Figure 8.8. Principle of binary-cycle power plant. *Source*: From www1.eere.energy.gov/geothermal/powerplants.html.

With low-temperature resources, Rankine cycles employing organic fluids are used to produce electricity. With modular plants, operational availability up to 98% can be reached. Binary-cycle plants are also efficient for medium-temperature sources, but more sophisticated technologies are required to get the highest yield possible. Output also depends on the proportion of steam in the resource. In steam-dominated resources a geothermal cycle is first used in which the steam is directed through a steam turbine and then condensed in a heat exchanger where the energy is transmitted to an organic fluid to produce electricity as in other binary-cycle plants.

> Austria has recently used deep aquifers (2500 m depth) at temperatures of 106° and 110 °C to operate power plants of 1 MWe and 250 kWe, respectively, and produce electricity. Design and drilling costs for such plants are relatively inexpensive.

A breakthrough in the domain of geothermal electricity production may occur in the future with the use of supercritical fluids. Some reservoirs exist where the temperature is larger than 374 °C and the pressure larger than 220 bars. In this case the water is in a supercritical phase where technically there is no distinction between liquid and vapor. This phase has a much higher heat capacity than the liquid. If it could be brought to the surface under the same conditions, one could recover about 10 times more energy than with conventional reservoirs. Feasibility studies are underway in Iceland where it could be possible to extract fluids at temperatures ranging between 400 and 600 °C at 5000 m depth.

8.4. GEOTHERMAL ENERGY IN THE WORLD

Today, the use of geothermal power ranks third after biomass and hydroelectricity among renewable energy sources. In 2005, the world's geothermal energy production of electricity was 57 TWh. In the European Union, this production was 5.5 TWh. Since the 1980s, the time of the "second oil shock," there has been a strong increase in the development of geothermal energy sources for production of electricity. The installed electricity capacity is indicated in Figure 8.1. More than 20 countries are now using geothermal energy to produce electricity and 7 countries account for 90% of the world capacity (The United States is first followed by the Philippines and then Mexico, Italy, Indonesia, Japan, and New Zealand.) At the beginning of 2005, there were 43 plants producing electricity from geothermal power in the United States. There has been a strong decline in the United States in the recent years because of the decline of the giant Geysers field. At the world level this has been partly compensated by new capacities opening in several countries around the world (e.g., Indonesia, Mexico). For some of these countries geothermal production represents a very significant fraction of the total production in the country. For example, the Philippines produces 21.5% of its electricity with geothermal energy.

As far as heat production is concerned, in 2005, 80 TWh was produced from a combined installed power capacity in 72 different countries of 25.9 GW$_{th}$.

Low- and medium-temperature geothermal sources represent an installed power of 12.1 GW$_{th}$. Heat is directly extracted from aquifers. Very low temperature applications using geothermal heat pumps represent 13.8 GW$_{th}$. This corresponds to more than a million heat pumps installed all around the world. Heat produced by medium- and low-temperature geothermal sources is mostly used for baths and swimming pools (43%) and for heating of buildings (30%). There are also other applications: greenhouse heating, aquaculture, and so on. However, applications of geothermal power depend very much on the country. In France, for example, geothermal heat is used mostly for urban heating.

The installed power of the 12 countries of the European Union leading in direct use of geothermal energy is shown in Figure 8.9. The amount of energy harnessed, is shown in Figure 8.10. Comparison of the data in the two figures shows that the harnessing of the energy is more efficient in some countries than in others.

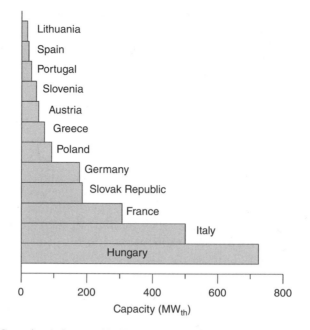

Figure 8.9. Capacity of top 12 European Union countries, 2006. Geothermal heat pumps are not included. *Source*: Data from EurObserv'ER, www.energies-renouvelables.org.

Geothermal heat pumps are becoming increasingly important in the European Union. In 2005 the installed power, corresponding to more than 450,000 units, was $5.4\,GW_{th}$. Sweden $(2\,GW_{th})$, Germany $(0.8\,GW_{th})$, and France $(0.75\,GW_{th})$ are the leading producers. Direct use of geothermal energy has seen small growth compared to electricity generation. Three countries (the United States $(5.4\,GW_{th})$, China $(2.8\,GW_{th})$, and Iceland $(1.8\,GW_{th})$] represent 59% of the world total.

In Figure 8.11 we show the number of geothermal heat pumps installed in the 12 European Union countries having the largest number and in Figure 8.12 we show the corresponding power installed.

Geothermal energy has the important advantage of being a continuous energy resource, available 24 hours a day, throughout the year. Annual percentage time availabilities between 80 and 90% can be reached. Geothermal energy is also very low in CO_2 emission. The average CO_2 emissions estimated using data obtained in a survey of 73% of the geothermal plants are $55\,g/kWh$. This corresponds to about 10% of the CO_2 emitted from a natural gas plant. This value can be significantly decreased if the geothermal fluid is reinjected into the ground and it is possible to have practically no CO_2 emission. In terms of land use, power plants using geothermal energy occupy less land than other energy resources. They require about $400\,m^2/GWh/yr$. For comparison the

Direct uses of geothermal energy

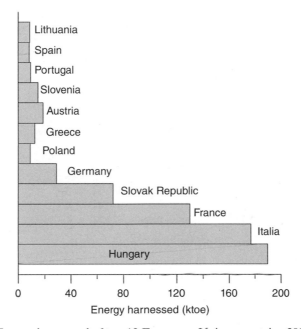

Figure 8.10. Energy harnessed of top 12 European Union countries, 2006. Geothermal heat pumps are not included. *Source*: Data from EurObserv'ER, www.energies-renouvelables.org.

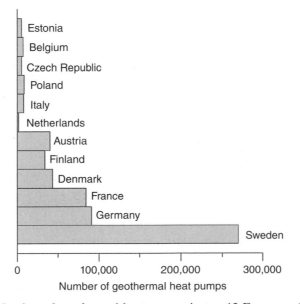

Figure 8.11. Number of geothermal heat pumps in top 12 European Union coutries, 2006. *Source*: Data from EurObserv'ER, www.energies-renouvelables.org.

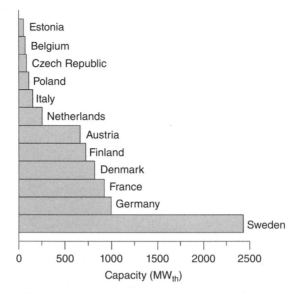

Figure 8.12. Installed power of geothermal heat pumps in top 12 European Union coutries, 2006. *Source*: Data from EurObserv'ER, www.energies-renouvelables.org.

requirements for photovoltaic energy is $3240\,m^2/GWh/yr$ and for wind power $1340\,m^2/GWh/yr$.

8.5. CONCLUSION

Geothermal energy is a sustainable energy resource which will be increasingly exploited in the future. It has the advantage of being almost pollution free, and while some parts of the world are particularly favored, it is available all around the world at any time. It is therefore a domestic energy source which can be exploited in different forms depending on the local conditions of the resource. It can be used to produce both heat and electricity, which makes it quite flexible. In order that it be more efficiently exploited, a precise mapping of sources is needed. This has still not been done in many countries. Geothermal energy usage in the world increases regularly at a level of 4% each year but is growing much faster in some countries. In France, for example, the increase was 30% in 2005. Geothermal energy, which is presently the third most used renewable energy source in the world, is expected to be an even more important contributor to our energy needs in the future.

CHAPTER 9

Wind Energy

Temperature differences between different locations on the surface of the earth and between different altitudes create pressure differentials which cause the airflows we know as wind. These temperature differences result from differential solar heating of the earth, which varies from region to region and with the time of day. Heated air has a tendency to rise and be replaced by colder air. This leads to wind, which carries energy. Part of this energy can be tapped to do useful work.

Wind has been used as an energy source since antiquity, but the new technologies developed to harness it far surpass those which existed in the past. Huge amounts of electricity can now be produced by wind turbines. With an average growth of about 30% per year over the last few years, wind energy is one of the fastest growing energy technologies. Europe is the leader in the exploitation of wind energy with 75% of the world's installed wind-powered energy-generating capacity. The first wind farms were constructed on land, but many new wind turbine installations are now being placed offshore.

9.1. ALREADY A LONG HISTORY

The first documented use of wind energy was by the Egyptians in approximately 3500 B.C. They built boats propelled by square sails to sail on the Nile. These sails functioned only with the wind at the back. Modifications of the shape and structure of the sails soon allowed sailing into the wind. With such boats it became possible to explore remote regions and to develop extensive trade with other regions.

The first machines used to extract energy from the wind on the earth were vertical-axis machines. Their first extensive use appears to have been near the end of the sixth century in the region which is now Iran and Afghanistan. In the Asian civilizations of China, Tibet, India, Persia, and Afghanistan, wind power was used mainly to pump water and irrigate fields.

Our Energy Future: Resources, Alternatives, and the Environment
By Christian Ngô and Joseph B. Natowitz
Copyright © 2009 John Wiley & Sons, Inc.

A change in the technology was made by the Arabs, who introduced horizontal-axis wind machines. During the Crusades, Europeans observed these machines and brought the idea back to Europe. About the eleventh century, horizontal-axis machines consisting of a tower and four blades were introduced in Europe. They were mainly used to grind grain for millers. For that reason they were called windmills, a term still used today. By the eighteenth century this European technology, also known under the name of "Dutch windmill," had been extensively developed in Europe and many windmills were in operation (around 8000 in Holland, 10,000 in Germany). A number of Dutch windmills were also built along the eastern coast in North America. Windmills and watermills were the first renewable energy devices used to replace animals or men and produce useful work.

The Dutch windmill consists of a tower, typically 10–15 m in height supporting four blades facing into the wind. It is estimated that about 100,000 windmills were in operation in Europe in 1820. They were essentially used to pump water and mill grain. Improvements in the structure of the blades allowed moderation of the rotational speed of the blades in high-wind conditions.

In the late eighteenth century, the use of the steam engine and other machines based on thermodynamic processes and the availability of fossil fuels led to a decreased interest in wind-powered machines. As a consequence, generation of wind energy declined gradually during the past two centuries, although some research and development in this field was continuing.

In 1920, in Denmark, windmills were coupled to batteries to produce continuous current through a grid to power lighting on some of the Danish islands.

In 1930, the U.S. government determined that midwest farms should have electricity to power radio receivers. Specific wind-driven aerogenerators were developed for that purpose. The best known was the Windcharger produced by the Sioux City Company in Iowa. At continuous current the basic model of this type of generator had a power of 200 W at 6 V.

The technology of modern windmills, also called wind turbines or aerogenerators, began at the end of the nineteenth century due to an innovation proposed by Poul La Cour, a Danish engineer. La Cour's idea was to couple an air turbine to an electrical dynamo to produce electricity. During World War I fuel shortages led to the use of many such wind turbines to produce electricity in Denmark and other countries.

Between 1925 and 1957, a large number of wind turbines manufactured by the Jacob Brothers Company, having output powers in the 2.5–3-kW range and associated with storage batteries, were used in the United States. By 1941, at Grandpa's Knob in Vermont, Americans had tested a 1250-kW wind turbine consisting of a two-blade horizontal-axis turbine coupled to an asynchronous generator. The EDF Company in France was also very much involved in the development of horizontal-axis wind turbines. Between 1950 and 1960, two big wind generators were tested in France. One, with a turbine having 30-m-long blades coupled to a synchronous generator, had a power of 800 kW. The second had a power of 1000 kW and was coupled with an asynchronous generator. At that time, production of electricity in this way was not economically competitive with other means of production.

Around 1920, Georges Darrieus, a French aeronautical engineer, developed a vertical-axis wind turbine with two blades. This design was patented in 1931. A Darrieus wind turbine having a power of 4000 kW was built in 1983 on the shore of the Saint Lawrence River in Quebec. Although it was at that time the most powerful wind turbine in the world, this vertical-axis technology turned out to be less reliable than the horizontal-axis technology and more difficult to maintain.

9.2. FROM THEORY TO PRACTICE

Wind corresponds to the motion of a mass of air. Consider a region in space in which we assume that the velocity V of the air is constant. During a time t, the air moves over a distance equal to Vt. The mass of the air per unit of time passing through a given area A perpendicular to the wind velocity is the mass of air contained in the volume, VA. If the air has a density (mass per unit volume) ρ, the total mass of the air passing through area A is just $M = \rho VA$. Since the kinetic energy is equal to $\frac{1}{2} MV^2$, the power P is given as

$$P = \tfrac{1}{2}\rho A V^3$$

We see that the power of the wind is proportional to the cube of the velocity of the wind. If the velocity doubles, the power is multiplied by a factor of 8. This sharp dependence of P upon V means that small wind velocities do not contribute much to the energy production. Note also that this rapid increase of power with wind velocity means that preventive measures must be taken to assure that high wind speeds do not damage the wind turbines.

The wind turbine industry usually takes $\rho = 1.225\,\text{kg/m}^3$ for the density of air. This corresponds to the density of dry air at sea level, at $15\,°C$, and at the standard pressure of 1 atm. In that case, if A is expressed in square meters and the speed of the wind is $V = 10\,\text{m/s}$, the power is $612.5\,\text{W/m}^2$.

However, this power cannot be completely extracted because the speed of the wind is not reduced to zero behind the wind turbine; otherwise the air would accumulate. According to Betz's law formulated in 1919 by the German physicist Albert Betz, the maximum theoretical energy that can be extracted from the wind independent of turbine design is $\frac{16}{27}$ of the incoming kinetic energy:

$$P_{\text{max}} = \frac{16}{27}\left(\frac{1}{2}\rho A V^3\right)$$

Although still a simplified version of reality, Betz's law gives a first-order estimate of realizable power. Real systems give a power even lower than the Betz limit because there are several dissipative effects which decrease the yield, friction of the air on the blades, reduction of the torque imparted to the blades because of turbulent phenomena, and so on. In the best cases powers around 85% of the Betz limit can be realized. This varies with the speed of the wind. There is no great hope to significantly improve this yield.

Usually small machines have a lower efficiencies than large machines. The efficiency compared to the incoming energy of the wind (59.3% for the Betz limit) is of the order of 40–50% for large wind turbines and 20–40% for smaller ones (with a power smaller than 100 kW).

The wind turbine is usually coupled to a generator and mechanical energy is transformed into electricity. The coupling can be direct or through a gearbox which steps up the rotational speed. The generator is mounted at the top of the tower together with the gearbox, if it exists, and the turbine. This is summarized in Figure 9.1.

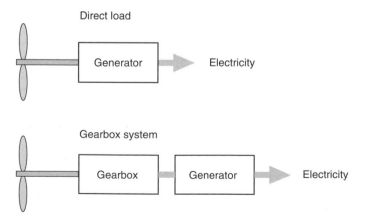

Figure 9.1. Principle of electricity generation from turbine. It can be done directly or using a gearbox system.

There are additional efficiency-decreasing steps which occur before electricity is produced. For example, the gearbox of the system has an efficiency which depends on the rotational speed. It ranges from 80–95% for large wind turbines to 70–80% for small ones. Furthermore, the subsequent transformation of the rotational mechanical energy to electricity has also a yield which is about 80–95% for large machines and 60–80% for small ones.

Small machines always have a smaller overall yield with respect to theoretical energy than do large systems. Small installations of a few kilowatts for home electricity supply have an overall efficiency of the order of 20–25% of the energy incident on the wind turbine.

Most large-size wind turbines have three blades. With blades increasing in length, carbon fiber is becoming the material of choice, replacing fiberglass. There is also an evolution toward variable-speed machines. They have the advantage that the rotor rotates more slowly in low winds, reducing noise. Also, as the wind velocity changes, they deliver a more stable output of electricity compared to older technologies. Indeed, one of the problems of wind energy is the coupling to the grid because fluctuations in the wind velocity and power output can trigger some instabilities. Progress in the wind turbine domain has reduced these drawbacks.

A wind turbine can produce energy at a minimum wind speed near 3.6 m/s but requires a speed above 5 m/s to be economically viable. A mean speed of the wind of 10 m/s is usually acceptable but the optimum speed is around 15 m/s. Because the output power varies according to the cube of the speed of the wind, it is necessary to limit the output power in the case of very high winds to prevent mechanical damage. This can be done by pitch control on the blades or by stall control. Pitch control is favored for large wind turbines because it is more efficient.

The French manufacturer Vergnet S.A. specializes in the manufacture of small two-bladed wind turbines which have the ability to resist powerful winds, gales, or storms. These wind turbines can be installed on tiltable towers which can be lowered for maintenance and during storms. This makes them well suited for use in tropical storm areas.

9.3. DEVELOPMENT OF WIND POWER

As indicated above, with the current technology, a wind speed larger than 5 m/s is necessary to economically harness wind power. At any given time, such winds are blowing in many parts of the world. Unfortunately it often happens that places where there are particularly good wind resources are also those where the population density is low and consequently where the electricity demand is small. This is the case in Patagonia, for instance. In other cases, the demand may be there but the infrastructure to carry electricity to the consumer does not exist. This is the case in some parts of Africa. In theory, even taking into account the fact that it is not possible to put wind turbines in every promising location, the available wind resource is several times that required to meet today's world electricity needs.

The main technical problems with exploiting wind as an energy resource are that the air which carries wind energy has a low density (1000 times less than water) and the winds at any given site may be irregular. Since power depends upon the cube of the velocity of the air impinging on the blades of a turbine, wind irregularity can lead to a large variation in output power. Further, if the wind is too strong, the wind turbine must even be stopped; otherwise it could be destroyed.

Interest in wind energy was dramatically renewed in the 1970s at the time of the oil crisis. Fortunately, research in this field had been carried out continuously throughout the last century. Because of that, it was possible to take advantage of the experience acquired and quickly reach an industrial scale of production for wind machines. During the last decade a particularly strong increase in the development of wind energy capabilities has taken place.

In 1990, the total world installed wind power was about 2 GW. The global cumulative installed capacity has increased rapidly since then, as illustrated in Figure 9.2, which shows the evolution from 1996 on. Between 1996 and 2001 a growth rate close to 40% per year was observed. The installed capability reached 48 GW by the end of 2004, 59 GW in 2005, and 94 GW at the end of 2007.

The annual installed capacity is shown in Figure 9.3.

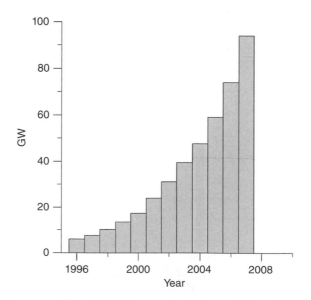

Figure 9.2. Global cumulative installed capacity between 1996 and 2007. *Source*: Data from www.gwec.net.

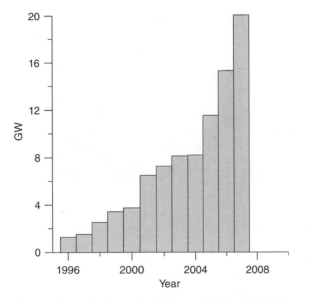

Figure 9.3. Global annual installed capacity between 1996 and 2007. *Source*: Data from www.gwec.net.

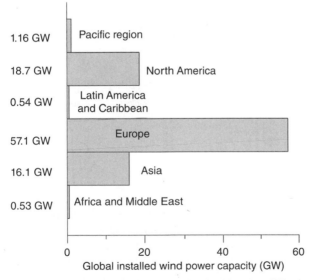

Figure 9.4. Global installed wind power capacity in 2007 for different regions of the world. *Source*: Data from www.gwec.net.

The global installed wind power capacity in 2007 is shown in Figure 9.4 for different regions of the world.

The distribution of the total installed global capacity is shown in Figure 9.5. Although Europe is still the leader in terms of cumulative installed capacity,

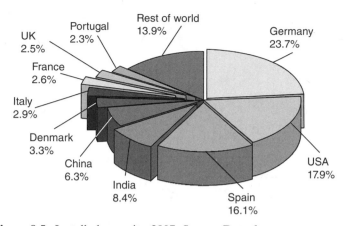

Figure 9.5. Installed capacity, 2007. *Source*: Data from www.gwec.net.

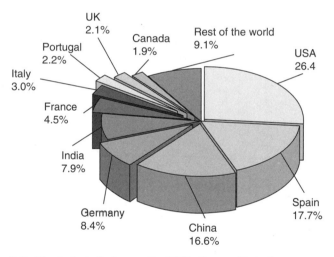

Figure 9.6. Newly installed capacity, 2007. *Source*: Data from www.gwec.net.

there are many new installations outside Europe. The distribution of installed new capacity during the year 2007 is shown in Figure 9.6. We see that the order has changed between the two plots. This is due to the fact that some countries are now making major efforts to install wind turbines. The United States leads in this category in 2007, with 26.4% of the new capacity.

The 12 European countries which lead in the installed capacity of wind turbines at the end of 2007 are listed in Figure 9.7. Presently the total installed capacity of the 27 European Union countries is 56,535 MW, of which 1080 MW is derived from offshore wind turbines.

As we see from the preceding figures, almost 60% of the present world capacity is installed in three countries. Germany leads the list, followed by the

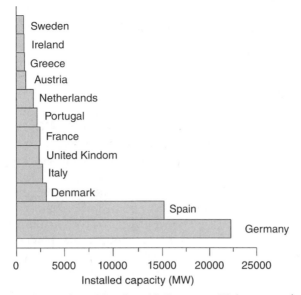

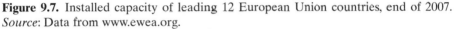

Figure 9.7. Installed capacity of leading 12 European Union countries, end of 2007. *Source*: Data from www.ewea.org.

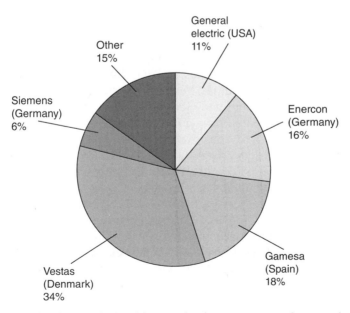

Figure 9.8. Distribution of wind turbine production among manufacturers in 2005. The market is dominated by five manufacturers. *Source*: From J. Bonal and P. Rossetti, *Energies Alternatives*, Omniscience, 2007.

SAME TECHNOLOGY BUT DIFFERENT IMPACTS

The use of wind turbines may have a varying impact on greenhouse gas emissions. Take the case of Denmark, where coal plants account for a large part of the energy production. Wind turbines have the advantage that, when the wind is blowing (typically this occurs about 30% of the time in good areas), electricity is produced without CO_2 emission. This leads to an important reduction in emissions compared to the emissions if only coal plants are used. Therefore wind turbines make good sense in Denmark.

In France, which generates 80% of its power from nuclear plants, closing a nuclear plant and replacing it by wind turbines would require building thermal plants to produce electricity when there is no wind (about 70% of the time). Indeed, one needs power plants able to start rapidly, which is not the case with nuclear plants. The best thermal plants as far as CO_2 emissions are concerned are gas plants. However, they emit CO_2 when they produce electricity. In such a case one goes from a situation where there was no CO_2 emission (with nuclear plants) to a situation in which CO_2 is emitted during 70% of the time. The situation is therefore worse as far as the greenhouse effect is concerned.

United Sates and Spain. The construction capacity of wind turbines is also concentrated, reflecting a need for high initial investments of capital. As displayed in Figure 9.8, five companies account for most of the world's wind turbine manufacturing capabilities: Vesta in Denmark, Enercon and Siemens in Germany, Gamesa in Spain, and General Electric in the United States. The strong development in some European countries is due to the fact that there are substantial subsidies from the governments and national buy-back policies at rather high prices. This is driven by both the desire for alternative energy supplies and environmental considerations. Building and operating wind turbines are therefore quite interesting financially.

The power as well as the size (diameter of the blades) of wind turbines continues to increase (see Figure 9.9). In 1980, a typical wind turbine had a size of 15 m and an output power of 50 kW. Several thousand such devices were sold. In 1990, 40-m devices with powers of 500 kW were available. In 2003, a diameter of 124 m and a power of 5000 kW were achieved. Projects of 10 MW are planned around 2010 with a size of 160 m in diameter and 20 MW is expected to be reached in 2020. This evolution will go on, but there seems to be a limit around 20 MW.

One reason to increase the size of wind turbines is the fact that in some countries there is shortage of suitable sites. At a given site, larger wind turbines have an advantage because their hubs are farther from the ground and the

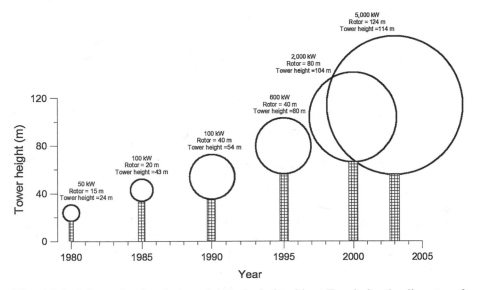

Figure 9.9. Schematic of evolution of size of wind turbines. For clarity, the diameter of the rotor is not to scale. *Source*: Data from www.ewea.org.

speed of the wind increases with altitude. Since the power depends upon the cube of the wind velocity, the power recovered is therefore larger. The wind also becomes more stable and turbulence is reduced.

Nevertheless, large turbines have a number of drawbacks. First, the forces applied on the mechanical parts increase with size and this leads to more than a linear increase of weight with size. Huge quantities of concrete are needed to anchor large wind machines to the ground. Surprisingly, the amount of concrete per unit of produced kW_e turns out to be larger that required for a nuclear plant. Transporting the different parts of large systems can be costly. Such a constraint is not so stringent for offshore wind turbines.

Between 1990 and 2004, the average size of wind turbines installed in Denmark, Germany, and the United Kingdom has increased from about 200 kW to around 2000 kW. At this date, the average turbine size in Spain was just about half of that (1100 kW). At the world level, in 2004, 23% of the operating wind turbines had a capacity between 750 and 1000 kW, almost 30% had a capacity between 1 and 1.5 MW, and more than 40% had a capacity larger than 1.5 MW. The overall electricity production efficiency of wind turbines has increased regularly over the years (about 2–3%/year over the last 15 years).

Since 2002 the investment cost of wind electricity has been about 1000 €/kW and the price is about 50 €/MWh. This cost is still larger than that associated with generation using fossil fuels or nuclear fission. During the past 15 years the cost of electricity in Europe has decreased at a rate of about 3%/ year. As a result, subsidies and special buying conditions are currently necessary for wind energy to develop. However, taking into account the fact that the cost of electricity produced by a gas plant is likely to increase in the future and a carbon emissions tax may be introduced, wind energy might soon become economically competitive.

The cost of wind-generated power depends very much on the wind availability at the site of the wind turbine. For example, for the wind farms Tararua I and II in New Zealand, the average operating time is 4530 h. In Germany, it is only 1850 h/year. The German company Eon has reported that in 2003 the electricity production of the 14-GW installed wind turbines was 18.6 TWh, that is, a yield of 15%, which is quite low.

Given the rapid increase in energy consumption in modern societies, it is very difficult to meet demands using only renewable energies. For example, Denmark has a population of 5.3 million inhabitants and more than 6000 wind turbines are installed. In 2005 wind produced about 18% of the total electricity of the country. Denmark's annual energy production from wind turbines increased by about 3.7 TWh/year between 1990 and 2002. However, during the same period of time, the total electricity demand of the country increased by about 12.7 TWh, which is larger than the total for wind electricity production.

The significant development of wind energy resources has raised new problems. Wind turbines are usually installed in windy areas where, in most cases, the population density is low. Therefore it is necessary to extend the existing electrical power transmission grid to use this resource. This can be a major expense. For example, it is anticipated that, by 2015, the German wind energy program will require about 1.1 billion euros to extend the grid to new energy production sites.

As the installed wind energy capacity becomes a significant fraction of the total grid capacity, wind intermittency may lead to instabilities in the grid. Techniques to manage this must be in place.

Wind intermittency is really an issue. Because wind electricity is not always produced at the time the consumer needs it, this can be a source of instabilities on the grid. In Denmark, the consequence of intermittency is that most of the wind-generated electricity is exported because there is no domestic need at the time it is produced. In 2004, for example, 84% of Denmark's wind-generated electricity was exported at a low price because the demand was low in the domestic market. Since wind-generated electricity is still subsidized by the state, this corresponds to a net revenue loss for the country.

Wind turbine farms also have some environmental impact. Wind farms can be noisy and they constitute visual intrusions on the landscape. This latter, the visual impact of wind turbine farms, is probably the most impressive and controversial aesthetic aspect of these facilities. While some may view them as installation art of a particular beauty and symmetry, others decry the spoilage of otherwise pristine vistas.

There are also concerns for the safety of migrating birds which may be killed by collision with the spinning blades and bats which can die from internal injuries resulting from large pressure changes existing close to the blades.

9.4. OFF-SHORE WIND TURBINES

In both Europe and the United States, offshore turbines are important emerging alternatives to the land-based ones. There are several reasons for that. In some countries the number of land sites to install wind turbines is limited. In addition, a strong opposition to installation of wind farms is sometimes encountered from local populations ("not in my backyard"). On the technical side, wind is usually more regular offshore than onshore and the average speed is higher with less turbulence. Operating times larger than 3000 h/year have been realized by Danish offshore farms. This large time of operation compensates to some extent for the higher costs of offshore wind plants compared to those onshore.

The first farm consisting of offshore wind turbines was commissioned in 1991 in Denmark in a region of the Baltic Sea where the sea bed was only a few meters deep. The wind turbines used were similar to those used inland but with reinforced foundations.

At present the world's largest offshore wind farm is located at Horns Reef on the west coast of Denmark. The installed capacity of 160 MW is composed of 80 wind turbines with a power of 2 MW. It became operational in 2002. Another Danish facility, located close to Loland Island, consists of 72 wind turbines of 2.2 MW capacity. Other projects are underway.

There are also some disadvantages of developing offshore wind turbine installations. Stronger foundations are needed because the structure must resist not only wind but also waves (the power of waves can be significant with pressures on surfaces reaching as high as 1300 t/m^2). Further, storms at sea or near coastlines tend to be significantly more violent than those over large land areas. For deep sea beds, methods currently used in oil industry can be used to anchor the turbine to the ground of the sea bed.

The connection to the electrical grid is also more expensive. If the distance to the shore is greater than about 10 km, high-voltage transportation systems are needed. This increases the complexity and cost of the technology.

As it is more expensive to install wind turbines offshore than onshore, increasingly powerful individual turbines have been developed. In 2007, the rated power of a wind turbine with a rotor 120 m in diameter and a cradle located 100 m above sea level was 6 MW. Offshore wind turbines are normally designed to operate at least 20 years and they are constructed to resist "one-in-a-hundred-year" storms.

Since Europe has many densely populated areas close to the sea in regions where wind can be harnessed, the European Union has evidenced a very strong interest in development of offshore wind turbines. Projects are underway in the Netherlands, Germany, the United Kingdom, Belgium, Ireland, and Sweden. Offshore wind farm projects have a total installed power often larger than 100 MW and some projects in Germany exceed 1200 MW. In Europe at the beginning of 2007, there were 386 offshore wind turbines operating. This corresponds to a total installed power of 785 MW. Investment costs have been of the order of 2000–3000 €/kW. This is higher than for inland wind turbines (between 1000 and 1300 €/kW). Electricity produced is therefore more expensive (between 100 and 130 €/kWh).

In the future, one can expect the development of new-generation wind turbines to be erected in places far from the shore and in deep sea beds. Since they will be located far from the coast, other ways to store or transport the energy are possible: hydrogen production, for example.

9.5. CONCLUSION

Wind energy is a major renewable resource. Well-developed technologies to harness wind already exist and future improvements can be expected. Over a little more than the past decade, spurred by the expected future decline of the availability of cheap oil as well as the fact that there is no associated emission of greenhouse gases, there has been a strong development of wind power technology. Although it is presently economically competitive only in certain specific situations, the price of wind power has decreased notably during this decade of development. It should soon reach a competitive status.

CHAPTER 10

Nuclear Energy

Most of the energy we use on the earth comes initially from nuclear fusion reactions occurring inside the sun. Apart from geothermal energy all of the other energy sources discussed in this book may be traced to this energy source. Solar energy is due to fusion reactions taking place inside our star, which is a gravitationally confined thermonuclear fusion reactor. We exist because those nuclear reactions exist.

Although nuclear reactions are the energy source which powers our sun, mankind's use of nuclear power sources on the earth is relatively new. These sources have been exploited on earth for only a little more than half a century. In contrast, fossil fuels have been extensively used for more than two centuries.

10.1. BASICS OF NUCLEAR ENERGY

The study of nuclear phenomena began in 1896 when Henri Becquerel discovered natural radioactivity. In 1934 Frederic and Irene Joliot-Curie carried out the first artificial transmutations and in 1938 nuclear fission was discovered by Otto Hahn and Fritz Strassmann. Immediately after, in 1939, Frederic Joliot, Hans Alban, Lev Kowarski, and Francis Perrin showed that it was possible, using fission, to initiate a chain reaction to produce energy: The principle of the nuclear reactor was born.

In 1896 Henri Becquerel noticed that uranium ore was spontaneously emitting an invisible radiation which was able to expose a photographic plate. In 1898, Pierre and Marie Curie isolated two other elements from uranium ore. These elements, polonium and radium, were even more radioactive than uranium.

Our Energy Future: Resources, Alternatives, and the Environment
By Christian Ngô and Joseph B. Natowitz
Copyright © 2009 John Wiley & Sons, Inc.

An atom consists of protons, neutrons, and electrons. The protons and neutrons, collectively known as nucleons, are confined in a small central core of the atom, called the nucleus. Most of the mass of the atom is accounted for by the nucleus. Chemical bonds formed by the much less massive electrons, which are outside of the nucleus, may lead to the formation of bound groups of atoms known as molecules. Chemical reactions involve reorganization of the electrons and the formation of or breaking of chemical bonds. The energies involved in elementary chemical processes is of the order of electron-volts, eV ($1\,eV = 1.6 \times 10^{-19}\,J$). Combustion (the reaction with oxygen) of elements or molecules can lead to a significant release of thermal energy. In the reaction of a carbon atom with an oxygen molecule, $C + O_2 \rightarrow CO_2$, for example, $4\,eV$ of energy is liberated. Thus atoms have a noteworthy energy density.

In contrast, nuclear reactions correspond to processes in which nucleons are reorganized. The energies involved are of the order of mega-electron-volts, a million times larger than in chemical reactions. In the case of the fission process, for example, where close to 240 nucleons are involved, the energy released is about 200 MeV. Thus, nuclear fuel has a high energy density compared to that of fossil fuel. For comparison, 1 g of oil has an energy content of 0.012 kWh while 1 g of fissile material contains 23,000 kWh. Renewable energies are about one million times less concentrated than the fossil fuels—oil, coal, or gas. As a consequence of the large energy concentration in nuclei, wastes which are produced in nuclear processes are also a million times more compact than those coming out from fossil fuels. Even though some of these wastes can be particularly dangerous, their small volume can mitigate the problems associated with storing them safely.

Nuclear reactions allow production of considerable quantities of energy with a small amount of fuel and at a competitive price. However, some people are afraid of radioactivity and nuclear energy. This probably goes back to the first use of nuclear weapons in Japan in 1945. Furthermore, the fact that radioactivity is invisible is an additional concern.

10.1.1. Atoms and Nuclei

All objects of nature, whether they are alive or inert, are made out of atoms. An atom is formed from three types of particles: protons and neutrons, which constitute the nucleus, the atom's core, and electrons, which are external to the nucleus. The masses of neutrons and protons are around 1840 times larger than that of electrons. Each atom is characterized by its number of protons, Z (the atomic number), and its mass number A, which is the total number of nucleons (protons plus neutrons) contained in the nucleus. The number of neutrons is thus equal to $A - Z$. The neutron is an electrically neutral particle discovered in 1932 by James Chadwick. The proton carries an elementary positive charge, $+e$, equal in magnitude but opposite in sign to that of the electron. Protons and neutrons are bound together in the nucleus because of the nuclear interaction, which is strong but of short range [of the order of femtometers ($10^{-15}\,m$)].

The dimensions of atoms are a few angstroms ($1 \text{ Å} = 10^{-10} \text{ m}$), that is, 10^5 times larger than that of the nucleus. Most of the volume of an atom is therefore empty space. Almost the entire mass of an atom is concentrated in the nucleus, whose dimensions are a few femtometers ($1 \text{ fm} = 10^{-15} \text{ m}$). As a consequence, the density of the nucleus is enormous: approximately $2 \times 10^{17} \text{ kg/m}^3$ or $200 \times 10^9 \text{ kg/cm}^3$.

The electric charge of the nucleus of an atom of atomic number Z and of mass number A is $+Ze$. To compensate for this positive charge and get a neutral atom from the electric point of view, there are Z electrons with negative elementary charge $-e$. The nucleus and the electrons are bound together by the electromagnetic interaction, which is more than 100 times weaker than the nuclear force but has an infinite range. A particular atom, X, consisting of Z electrons and a nucleus containing A nucleons (Z protons and $A - Z$ neutrons) is typically denoted by the symbol $_Z^A\text{X}$ (sometimes only by ^AX since the symbol X gives implicitly the atomic number Z). For example $_{82}^{208}\text{Pb}$ is the lead atom, which contains 82 electrons. There are 208 nucleons in the nucleus, 82 protons, and 126 neutrons. The chemical properties of an atom are determined by its number of electrons, which is equal to its atomic number Z.

Atoms having the same number of protons but different numbers of neutrons are called isotopes: They have the same chemical properties but different nuclear properties. For example, there are three isotopes of hydrogen:

- The most abundant one is the hydrogen isotope, $_1^1\text{H}$, with a nucleus consisting of a single proton. It constitutes 99.9% of the hydrogen found in nature.
- Deuterium is the hydrogen isotope, which has a nucleus made of one proton and one neutron ($_1^2\text{H}$). It is much rarer than the previous one. It accounts for 0.015% of natural hydrogen.
- The third isotope, $_1^3\text{H}$, known as tritium, has a nucleus containing one proton and two neutrons. It is unstable with respect to radioactive decay and does not exist in nature except in traces produced by cosmic rays.

10.1.2. Radioactivity

Many natural or artificial nuclei are radioactive. They have an extra energy and can spontaneously disintegrate to become more stable. There are three main types of radioactivity: α, β, and γ. See the Appendix at the end of the chapter. Emission or capture of an electron, e^-, or emission of a positron, e^+ (the antiparticle of the electron), is known as beta decay and leads to isotopes of adjacent elements. Emission of an alpha particle, a $_2^4$He nucleus, leads to a daughter nucleus two atomic numbers lower. Gamma decay, the emission of a photon, an energetic quantum of light, leaves the same nucleus in a less excited state. Heavy nuclei can also break in two pieces in a process known as fission. Collectively these types of disintegrations of nuclei are known as radioactive decay. They are schematically represented in Figure 10.1 (see Appendix for more details about the mechanism). Radioactive isotopes can occur naturally or can be artificially produced. More than 2000 artificial radioactive isotopes have been created. But most of the radioactivity present in the environment is of natural origin. The unit of measure of radioactivity is the becquerel (Bq). It corresponds to one disintegration per second whatever the emission is. Natural radioactivity from an average human is 8000 Bq. This unit has replaced an older one, the curie: $1\,C = 3.7 \times 10^{10}$ Bq, or 37 billion becquerels. Radioactivity decreases gradually as a function of time. Natural radioactivity is less intense now than it was at the birth of the earth, 4.5 billion years ago. The average radioactivity of the earth's crust has decreased from 5850 Bq/kg at that time to 1330 Bq/kg now. The uranium activity has decreased by a factor of 2 and that of ^{40}K has decreased from 4230 to 370 Bq/kg.

The energy released during radioactive decay is divided between the various partners found after the process.

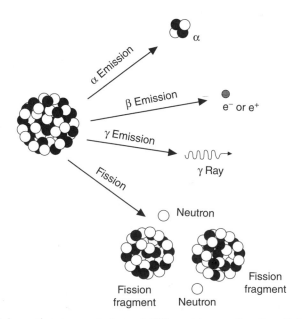

Figure 10.1. Schematic representation of different types of radioactivity. Protons are in black and neutrons in white.

In beta decay neutrinos or antineutrinos are emitted. Neutrinos and antineutrinos are elementary particles with no charge and a negligible mass. They interact with matter only by the so-called weak interaction. For that reason they are almost invisible for us. In order to capture half of an incoming neutrino flux, one would need a wall of thickness of one light year ($\approx 10^{16}$ m). Postulated by Fermi in 1930 to explain beta decay, this particle was not observed experimentally until 1956.

For the typical kinetic energies observed in radioactive phenomena, α particles, electrons and positrons (e^- and e^+), as well as γ rays, it is observed on average that

- α particles are stopped by a few centimeters of air, a sheet of paper, or the skin;
- β radiations are stopped by a few meters of air, an aluminum foil, or 1 cm of human tissue;
- γ radiation can cross several hundred meters of air or more. It can pass through the body of a human being and requires ~10 cm of lead to be stopped.

While any radioactive nucleus can eventually decay, a particular radioactive nucleus does not age: It always remains young. Its probability of disappearing is the same as long as it exists. This is not the case for humans. The probability to die increases as a human becomes older. As a consequence, radioactivity is a random process. Each nucleus has a given probability to decay. This probability is characterized by its half-life, $T_{1/2}$, which is the time at which half of them will have decayed. After one half-life, a radioactive source containing initially N radioactive nuclei has $N/2$ of these nuclei remaining. After two periods ($t = 2T_{1/2}$) there remain $N/4$ radioactive nuclei, and so on. A phenomenon that does not depend on the past is known as Markovian.

Potassium is an alkaline element present in nature and essential for the human body. In a human body there is about 165 g of potassium depending upon the person. A small quantity of the radioactive isotope ^{40}K is present in small quantities in natural potassium. It beta decays with a half-life of 1.3 billion years, which results in a radioactivity between 4400 and 5700 Bq.

10.1.3. Energy and Mass

At the birth of the universe, energy was transformed into matter. The current scientific evidence is that this occurred about 14 billion years ago. It is now known that matter can also be transformed into energy. Einstein showed, in 1905, that the energy E of a body of mass m at rest was equal to

$$E = mc^2$$

where c is the speed of light ($\cong 300,000$ km/s). A variation of energy ΔE corresponds to a mass variation Δm and vice versa. This means that there is an equivalence between mass changes and energy changes which can be expressed by the equation

$$\Delta E = \Delta m c^2$$

Expressing the mass energy equivalence in common units we can say that

$$1\,\text{kg} = 5.609 \times 10^{29}\ \text{MeV} = 2.5 \times 10^{10}\ \text{kWh} = 25\ \text{TWh}$$

Thus, the mass of an atom, made up of electrons and a nucleus, is lower than that of the free nucleus and electrons. The difference represents the binding energy, which plays the role of a cement between the electrons and the nucleus. For atoms this is a rather small difference. The binding energy of a hydrogen atom is 13.6 eV. This corresponds to a mass change of 2.4×10^{-32} g when the atom is formed from its constituents.

The fission of the ^{235}U happens in nuclear reactors and releases approximately 200 MeV. This corresponds to a reduction in mass of 3.6×10^{-25} g between the initial state and the final state. This value is small, but the mass change becomes noticeable if one considers the mass required to meet the annual energy production needs of a country. France, for example, produces around 450 TWh of electricity per year using nuclear power. This quantity of electricity corresponds to a transformation of a mass of 18 kg of ^{235}U into energy. In fact, the transformed mass must be larger because the efficiency of the power stations, due to the Carnot principle, is 33%. Therefore, two-thirds of the energy produced by the nuclear reactor is lost in heat and only one-third goes into electricity. In total, around 54 kg (119 pounds) of matter is transformed into energy each year.

As we indicated above, energy and/or mass changes in chemical and nuclear transformations may be discussed in terms of changes in binding energies which occur during the process under consideration. Figure 10.2 depicts the evolution of the binding energy per nucleon as a function of mass for the naturally occurring nuclei. The most stable nuclei, those with the maximum binding energy per nucleon, are in the vicinity of ^{56}Fe (the iron isotope with $Z = 26$ and $N = 30$) where the binding energy is close to 8.7 MeV/nucleon. The curve

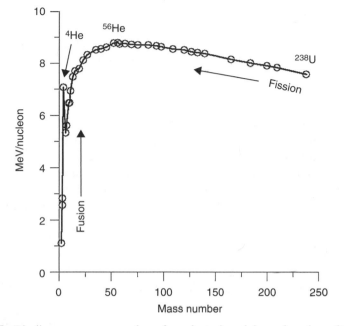

Figure 10.2. Binding energy per nucleon for selected nuclei as a function of their mass.

is relatively flat around this maximum. The helium nucleus is a singular point of the light nuclei because it is particularly stable (7.1 MeV/nucleon) compared to the neighboring nuclei. This figure shows that heavy nuclei are less bound than those of intermediate mass (7.7 MeV/nucleon for uranium). Consequently, if a heavy nucleus is broken down into two nuclei of intermediate mass, energy is liberated. This is what occurs in the fission of uranium. A little less than 1 MeV/nucleon (around 200 MeV for 236 nucleons) is released. We also see that, well below iron, the fusion of the lightest nuclei will also lead to a more stable configuration and release energy, typically several mega-electron-volts per nucleon. Thus, both fission and fusion produce energy. A practical means of harnessing fusion energy production is still at the research stage. This subject is treated in more detail at the end of the chapter.

10.1.4. Fission

Some very heavy nuclei can fission spontaneously into two fragments of medium mass. During this process neutrons are also emitted. Other nuclei fission only after they have absorbed energy. For example, the ^{235}U nucleus, which is an isotope of uranium made up of 92 protons and 143 neutrons, has a high capture probability for thermal neutrons. Thermal neutrons are neutrons whose speed corresponds to the thermal motion of the surrounding medium. Although qualified as "slow" or "thermal," they have an average velocity, at 20 °C, equal to 2.2 km/s (7920 km/h). This corresponds to an average kinetic energy of $\frac{1}{40}$ eV.

The fission of ^{235}U leads to two medium-mass excited nuclei, and two or three neutrons, depending upon the event, are emitted:

$$^{235}_{92}U + ^{1}_{0}n \rightarrow ^{236}_{92}U^* \rightarrow 2 \text{ fission fragments} + 2 \text{ or } 3 \text{ neutrons}$$

The sum of the kinetic energies of the fission fragments is about 165 MeV. On the average there are 2.5 neutrons emitted. These neutrons have an average kinetic energy of 2 MeV for each neutron. They are designated "fast" neutrons. Their velocities are about 20,000 km/s. Following the fission process the emission of electrons and antineutrinos (beta decay) and photons (gamma decay) releases approximately 32 MeV. The total released energy is approximately 200 MeV (30 pJ).

Fission leads to a mass distribution of final fragments because there are many way to break the fissioning nucleus into two pieces. In the fission decay of ^{235}U the fragmentation is usually asymmetric. For example, two typical fragmentations are

$$^{235}_{92}U + ^{1}_{0}n \rightarrow ^{236}_{92}U^* \rightarrow ^{142}_{56}Ba + ^{92}_{36}Kr + 2\,^{1}_{0}n$$

$$^{235}_{92}U + ^{1}_{0}n \rightarrow ^{236}_{92}U^* \rightarrow ^{144}_{56}Ba + ^{89}_{36}Kr + 3\,^{1}_{0}n$$

As a result, the mass distribution of the fragments has two peaks, one with its maximum corresponding to the average mass of the smaller fragment, the other with its maximum corresponding to the average mass of the heavier fragment.

All neutrons are not emitted at the same time during the fission process. The majority come from the decay of the excited neutron-rich "primary" fission fragments. Most of these are emitted immediately in the fission process and are called prompt neutrons. However, certain primary fragments can undergo radioactive decay to other isotopes, which can in turn decay by neutron emission. This "delayed neutron emission" can then take place notably later than that for the prompt neutrons. It is controlled by the half-life of radioactive decay of the parent nucleus, leading to the neutron-emitting daughter. The amount of delayed neutrons is small, only 0.65 per thousand total neutrons for ^{235}U and 0.21 per thousand for ^{239}Pu. Nevertheless they play a key role for the control of the nuclear reactors: *If they did not exist, it would be impossible to control a nuclear reactor.*

10.1.5. Fissile and Fertile

Only three naturally occurring nuclei can be used to produce energy by fission. They are ^{235}U, ^{238}U, and ^{232}Th. Only one, ^{235}U, has a fission probability larger than the probability to capture a neutron for all available neutron kinetic energies. For this reason it is called a fissile nucleus. The two others have a higher probability for neutron capture than for fission, especially for low-energy neutrons. Neutron capture by these nuclei followed by beta decay leads to fissile nuclei: ^{239}Pu for ^{238}U and ^{233}U for ^{232}Th. Therefore, ^{238}U and ^{232}Th are called fertile nuclei.

Interestingly, only odd-neutron-number nuclei are fissile with thermal neutrons. Even-neutron-number nuclei are fertile, requiring fast neutrons, with a kinetic energy larger than about 1 MeV in order to fission. This can be understood in terms of the nucleon binding energy, as shown in Figure 10.3. Fission is favored for nuclei having the smaller binding energies (light grey in the figure) while neutron capture has a larger probability than fission for nuclei having a high neutron binding energy (dark grey).

10.1.6. Chain Reaction

If one fission decay is induced by neutron capture in a fissile material, the neutrons which are produced in the reaction can themselves be captured by a fissile nucleus and the process can be repeated (Figure 10.4). Usually, not all the fission neutrons produce additional reactions because some can be

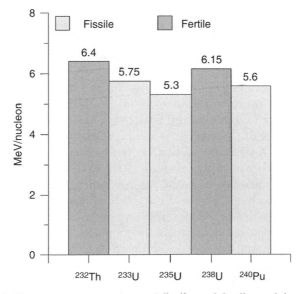

Figure 10.3. Binding energy per nucleon of fissile and fertile nuclei used to produce energy.

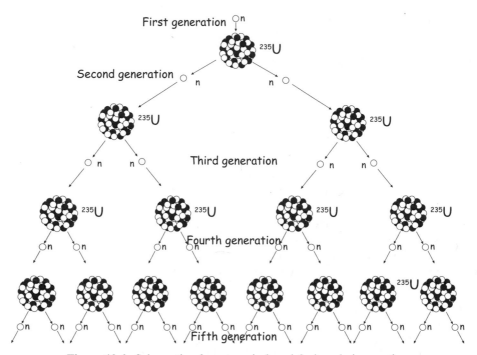

Figure 10.4. Schematic of neutron-induced fission chain reaction.

captured by fissile nuclei without leading to a fission, be captured by other nuclei present in the medium, or escape from the medium. However, if the released neutrons of one fission reaction can produce at least one additional fission reaction, a chain reaction involving multiple generations of reactions can be triggered.

The neutron multiplication factor k is defined as the ratio between the number of fissions in generations $n + 1$ to those in generation n. If $k < 1$, the medium is said to be subcritical and the fission reactions stop spontaneously. If $k > 1$, the number of neutrons grows exponentially and the medium is said to be supercritical. As the time separating two generations of neutrons is very short, there is quickly a divergent evolution. The system behaves like an energy amplifier: A small quantity of initial energy generates a lot of energy. This occurs in an atomic bomb where the fissile medium is confined as much as possible to generate the largest possible energy before it is released.

In a nuclear reactor, one controls the population of neutrons to extract energy continuously and k is equal to unity in stationary operation. It must be only very slightly higher than 1 to start the reactor or to increase its output power. All this is summarized in Figure 10.5.

The value of k depends on the geometry of the medium (if it has small dimensions, more neutrons will be able to escape to the outside), on its

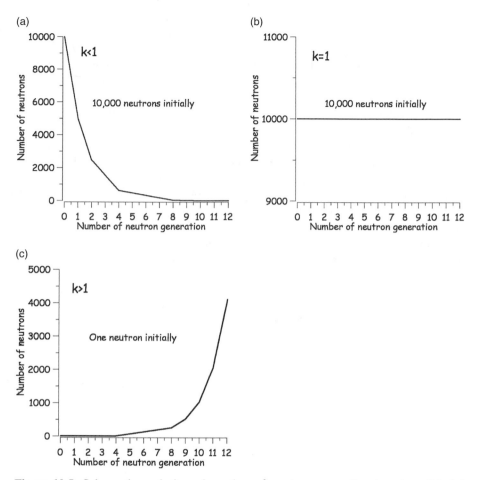

Figure 10.5. Schematic evolution of number of neutrons according to value of k. (a) For $k < 1$ there are initially 10,000 neutrons. (b) For $k = 1$ there are also 10,000 neutrons. (c) For $k > 1$ there is initially 1 neutron.

composition (some nuclei can strongly absorb neutrons and behave like poisons with respect to the propagation of the reactions), and on the homogeneity of the system (the position of various materials in the core of the nuclear reactor influences directly the properties of the neutron distribution). The capture of neutrons by a nucleus can lead, in certain cases, to a fissile isotope. As we have noted, such is the case with ^{238}U and ^{232}Th. Both nuclei are fertile nuclei from which one can manufacture nuclear fuel. This process is known as breeding. Starting from a mixture of fissile and fertile nuclei, one can produce energy and can transform part of the fertile nuclei into fissile ones. At the end of the operation there are more fissile nuclei than at the beginning.

10.1.7. Critical Mass

Although two or three neutrons are emitted for every fission reaction, they will not all induce a new fission reaction. Some of them will escape from the medium or be captured without leading to fission. If the neutrons are lost at a higher rate than they are produced, the chain reaction cannot be sustained. Given a fissile material, the point where the chain reaction becomes self-sustaining corresponds to the existence of what is called a critical mass. Achieving a critical mass depends on several factors: the nature of the fissile material, its composition, its purity, its shape, and so on. A spherical geometry is the best shape because it minimizes the surface area for a given mass. Consequently neutron leakage is the smallest. It is also possible to decrease the leakage using a neutron reflector at the surface.

Figure 10.6 shows the magnitude of the critical masses of some fissile materials having a spherical shape. The critical mass depends strongly on the purity of the material. For example, the critical mass of 20% enriched ^{235}U is 400 kg while it is 600 kg for 15% enriched ^{235}U. Because of the very high density of uranium ($19.2 \, \mathrm{g/cm^3}$), 50 kg of uranium makes a sphere whose diameter is only 17 cm.

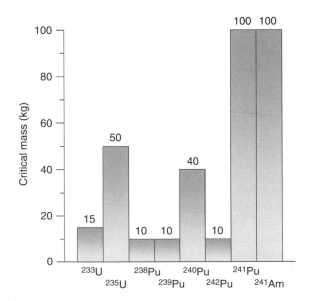

Figure 10.6. Critical masses for metal spheres at normal density. For plutonium this corresponds to a particular crystalline configuration (the α phase). *Sources*: Data from www.nti.org.

10.1.8. Nuclear Reactors

A nuclear reactor is a device in which a fission chain reaction is initiated and controlled in a steady state in order to produce energy. The fuel which is normally used is uranium, but there is considerable ongoing research on the use of thorium.

Uranium-235 nuclei fission with neutrons of any energy, but with slow neutrons of kinetic energy lower than 2 eV, the fission probability is very large. Uranium-238 significantly absorbs neutrons of intermediate energy and fissions only with fast neutrons whose energy is higher than 1 MeV. Because of these differences, there are two common types of nuclear reactors: thermal reactors, which are in the great majority used so far to produce electricity, and fast neutron reactors, which will be exploited more in the future.

Thermal nuclear reactors rely on the high thermal neutron capture probability of ^{235}U. As the major part of the neutrons emitted during fission is fast, they must be slowed down to assure efficient capture. For this purpose a neutron moderator is required. The neutron moderator is a medium in which the velocities of fast neutrons are reduced by elastic scattering collisions with the nuclei of the medium. A good moderator should consist of light nuclei so that the energy transfer in each collision between the neutron and a moderator nucleus is as large as possible and it should have a high transparency, that is, absorb only very few of the neutrons. Heavy water (D_2O) offers a good compromise between slowing down (on the average 35 collisions are necessary for thermalization) and transparency. This moderator, which is quite expensive, allows the use of natural uranium as the fuel. Light water is more effective in terms of slowing down (19 collisions on the average) but absorbs more neutrons and its use requires that the uranium fuel be slightly enriched in ^{235}U. Graphite is also a good choice. It slows down the neutrons less effectively, requiring on average 115 collisions to thermalize neutrons but it has high transparency. The light element beryllium has also been used in some experimental reactors. If natural or slightly enriched uranium fuel is used, the reactor has a heterogeneous structure in which moderator and fuel are separated to minimize neutron captures by ^{238}U. If the fuel is enriched with ^{235}U, the structure can be more homogeneous with the moderator closer to the fuel. Fast neutron reactors do not need a moderating material. They use a strong proportion of fissile nuclei because the probability of fission is much weaker.

Once the nominal output power is reached, a nuclear reactor must work in a mode in which the number of neutrons inducing fissions of the $n + 1$ generation is equal to those of generation n. This is the critical mode with multiplication factor $k \approx 1$. If there were only prompt neutrons, that is, those emitted in very short times, the system would not be controllable and would diverge very quickly. For the French nuclear reactors, this would happen in 2.5×10^{-5} s. With

$k = 1.0005$, the power would be multiplied by nearly 483 million in 1 s (40,000 neutron generations). There would be insufficient time to control the reactivity of the reactor. Fortunately, the existence of delayed neutrons, some of which are emitted several tens of seconds after the fission fragments are produced (11 s on average), makes it possible to control the chain reaction. This leads to a larger apparent neutron emission period and a corresponding divergence period of about 85 s. Consequently, a reactor must always be operated in a mode such that it is subcritical for prompt neutrons but can be slightly super-critical for delayed neutrons.

Very often the notion of reactivity is employed instead of the multiplication factor. Reactivity ρ is defined as $\rho = (k - 1)/k$. The system is stable if $k = 1$, that is, $\rho = 0$. The difference compared to stability is measured in units of 10^{-5} pcm (pcm = *pour cent mille* in French). Thus $k = 1.001$ corresponds to a reactivity of 100 pcm. For a thermal reactor burning uranium, it is impera-tive that $\rho < 650$ pcm in order to control the reactor with delayed neutrons and to remain under criticality for prompt neutrons. For a reactor burning plutonium, it is necessary that $\rho < 360$ pcm, which is more constraining. A fast neutron reactor is therefore operationally more sensitive than a thermal one.

The moderation of the reactivity of the reactor is done with control rods which contain elements absorbing neutrons strongly (e.g., cadmium, boron). Various other complementary methods are also used. For pressurized water reactors, for example, water containing boric acid, which has a huge neutron capture probability for thermal neutrons, is used.

10.1.9. Natural Nuclear Reactors: Oklo

The first man-made reactor was CP1 (Chicago pile number 1). It was con-structed by Enrico Fermi and his team and initially operated on December 2, 1942, in Chicago, Illinois. Thirty years later, in 1972, it was discovered that nuclear reactors have, in fact, previously existed on earth—around 2 billion years ago. Routine measurements of the isotopic content of uranium hexafluo-ride samples from Oklo, Gabon, carried out in Pierrelatte, France, revealed an anomalous proportion of ^{235}U. The percent abundance was found to be 0.7171 % instead of the normal 0.7202 %. Later, for some samples, the ^{235}U isotopic content was found to be as small as 0.5 %. Further investigations carried out by the CEA (the French Atomic Energy Commission) resulted in the conclu-sion that the only explanation for this anomaly was that a nuclear fission chain reaction had occurred in this Oklo uranium deposit. Other evidence supported this original hypothesis and since then it has been concluded that natural "nuclear reactors" functioned 16 different times on this site. There reactors functioned 1.9 billion years ago at very small power for 10,000–80,000 years. Evidence for an additional natural reactor was later found in Bangombé, about 30 km southeast of Oklo.

Modern-day natural uranium contains three isotopes (Figure 10.7): ^{238}U (99.2744%), ^{235}U (0.7202%), and ^{234}U (0.0054%). They are all radioactive with a half-life of 4.5 billion years for ^{238}U, 704 million years for ^{235}U, and 246,000 years for ^{234}U. Consequently the share of ^{235}U in natural uranium was larger in the past. It was close to 17% when the earth was formed, about 4.5 billion years ago, and was around 3.5% two billion years ago, near the time when the natural nuclear reactors at Oklo functioned. This latter enrichment of ^{235}U is precisely the order of magnitude of the enrichment of uranium fuel currently used in pressurized water nuclear reactors.

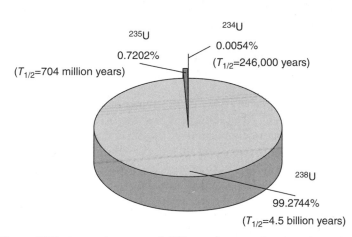

Figure 10.7. Apportionment of different isotopes in natural uranium.

These reactors worked underground at a depth of several kilometers. In these conditions, the temperature and the pressure were close to those used in pressurized water nuclear reactors (350–450 °C, 150–200 bars). South of Oklo, the chain reaction occurred at 500 m depth at a temperature of 250 °C and a pressure close to 50 bars. These working conditions are similar to boiling reactors, which we will present in Section 10.2. Other such natural reactors probably have existed in other places where rich uranium ore deposits are found (e.g., Canada, Australia), but no evidence of that has yet been found. Oklo is also of interest because it provides information on the long-term geologic transport of nuclear waste in the ground.

10.1.10. Conclusion

In this section we have given a brief introduction to the physics of the nucleus and its use to produce energy using the fission process. One advantage of nuclear energy is that it is a concentrated source of energy. In 1 g of fissile material there is as much energy as in 1.9 tons of oil. In the next section we will go into the details of harnessing this energy and the economic conditions of nuclear power generation.

10.2. USES OF NUCLEAR ENERGY

A nuclear reactor is a device in which controlled fission reactions are carried out at a steady rate in order to produce thermal energy. This thermal energy can then be used to supply heat, for industrial or home heating, or to produce electricity. While small reactors have been developed for specific purposes (e.g., ship or submarine propulsion), most nuclear reactor applications are associated with the production of electricity. Typically, heat produced from fission reactions is used to power steam turbines which are coupled to alternators to generate electrical power. In this section we focus on that particular application of nuclear power.

Nuclear reactors may be classified according to the energies of the neutrons used to trigger the fission process:

- Thermal reactors employ slow ("thermalized") neutrons to induce the fission process with ^{235}U fuel. The great majority of nuclear plants currently in operation utilize thermal reactors. Thermal neutrons are obtained by slowing down those emitted during the fission process (which are initially much faster). The thermalization occurs through collisions with a moderator material as described above. For ^{235}U the probability of thermal neutron capture and subsequent fission decay is very high. For ^{238}U, the much more abundant isotope of natural uranium, the thermal neutron capture probability is small and does not lead to fission.

- Fast neutron reactors employ the neutrons emitted by the fission process directly without use of a neutron moderator. Because the probabilities for fast-neutron-induced fission are small compared to those for thermal neutrons, these reactors require either enriched uranium or the use of ^{239}Pu as the fuel. Currently very few reactors of this type exist, but they are expected to be widely used in the future because they exploit much better the energy content of natural uranium. Indeed 99.3% of natural uranium is in the form of the ^{238}U isotope. This isotope can be used with fast neutron reactors

10.2.1. Different Technologies

A thermal nuclear reactor consists basically of a reactor vessel containing a core in which nuclear reactions take place, a moderator to thermalize the neutrons, and a heat transfer fluid which transmits the heat produced in the core to other devices which will use it. The fluid also serves as the coolant to prevent overheating of the core. Most thermal nuclear reactors in operation use water as a moderator and as a coolant (Figure 10.1).

Water-cooled nuclear reactors belong to two families: PWR (pressurized water reactors) and BWR (boiling water reactors). Pressurized water reactors are also called VVER (*Vodo-Vodyanoi Energetichesky reactor*) in the Russian version. In both cases the coolant, which is water, absorbs the thermal energy released in fission reactions and is heated.

In a PWR, water in a primary loop circulates in the core and is heated by the energy released by fission reactions. The water remains in liquid form at a temperature slightly above 300 °C because the circuit is closed and remains under pressure (around 150 bars). The thermal energy of the primary circuit is transferred to a secondary loop also containing water as saturated steam. Typical temperatures of the secondary circuit are around 275 °C and pressures are about 60 bars. The heat exchanger where heat is transferred from the primary circuit to the secondary one is called the steam generator. Finally the hot steam is used to drive a steam turbine, which in turn drives an electric generator, which produces electricity which goes to the grid. The steam leaving the turbine is condensed into liquid in a condenser.

In a BWR, There is only one water circuit. This operates at a lower pressure than in the case of a PWR, typically around 75 bars. At this pressure, the boiling point for water is near 285 °C. The design is simpler because no secondary circuit or steam generator is needed. The water circuit in the upper part of the reactor core contains both water ($\approx$85%) and steam ($\approx$15%). As in the case of a PWR, the steam is used in a turbine to drive an electric generator.

A PHWR (pressurized heavy-water reactor) uses heavy water (D_2O) as a moderator. Heavy water is expensive but, since it absorbs fewer neutrons than water (H_2O), one can use natural uranium as a fuel. This avoids costly enrichment operations. Heavy water is kept under high pressure to raise the boiling point and allow extracting a maximum of the heat released from the reactor core. Pressurized heavy-water reactors have been developed primarily in Canada where they are known under the name Candu. One of the advantages of Candu reactors is that the fuel rods can be changed without stopping the reactor. Consequently these reactors can operate with a very small service downtime compared to that required for other technologies.

A LWGR (light-water graphite reactor) is water cooled and uses graphite as a moderator. This reactor was developed in the former Soviet Union to produce both power and plutonium. Called RMBK (Reaktor Bolshoy Moshchosnti Kanalnyi), they are refuelable onload. Because of their large size, they were not constructed with containment buildings, which makes them particularly unsafe in the case of an accident. Furthermore, under certain circumstances they have inherent operational instabilities. Chernobyl has four plants of this type. The dangers associated with this type of reactor became well known to the public when one of these had a serious accident on April 26, 1986. We will return to a discussion of this accident.

A GCR (gas-cooled reactor) is a nuclear reactor cooled by CO_2 gas and moderated by graphite. It operates at a higher temperature than a PWR,

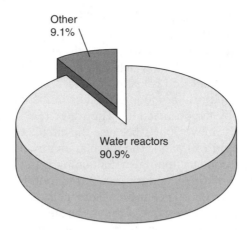

Other
9.1%

Water reactors
90.9%

Figure 10.8. Share of types of nuclear reactors between water reactors and others technologies.

leading to a better energy yield. They exist primarily in Great Britain (called Magnox in their original design) but are starting to be shut down. Due to their large size, decommissioning costs can be high.

Finally, a FBR (fast breeder reactor) is a fast neutron reactor which can belong to different technologies depending upon the cooling fluid. This fluid must not absorb too large a fraction of the neutrons, which restricts the choice of possible coolants. Sodium, lead, or lead–bismuth and helium can be used for that purpose. In this type of reactor there is no moderator since fissions are induced by fast neutrons. The term *breeder* comes from the fact that such a reactor is designed to breed fuel, producing more fissile material than it consumes. Superphenix, which was built in France and operated since 1985, was a fast neutron reactor using liquid sodium as the coolant. In December 1998, its operation was terminated for political reasons. From a purely economic point of view this was unfortunate. There were two new nuclear cores ready to be used. If they would have been burned in Superphenix, they would have provided a quantity of electricity equivalent to the demand of the City of Lyon for a period of 15 years.

Nuclear reactors using water as both moderator and coolant dominate the market of electricity production, as illustrated in Figure 10.8. In Figure 10.9 we show the share of the different families of nuclear reactors producing electricity which are presently in operation.

10.2.2. Selection Process

The first high-power nuclear reactor was developed in the United States using natural uranium as the fuel, graphite as the moderator, and air as the coolant (CP1, Oak Ridge National Lab, 1943). At that time the world was in the middle of World War II and the primary interest in developing nuclear

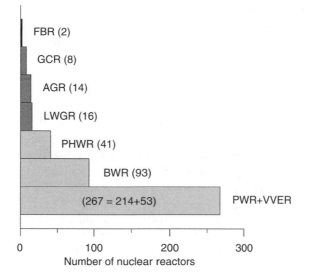

Figure 10.9. Share of different technologies of nuclear reactors: FBR = fast breeder reactor; GCR = gas-cooled graphite-moderated reactor; AGR = advanced gas-cooled graphite-moderated reactor; LWGR = light-water cooled graphite-moderated reactor; PHWR = pressurized heavy-water cooled and moderated reactor; BWR = boiling water reactor; PWR and VVER = pressurized water cooled and moderated.

reactors was not energy production but rather to synthesize plutonium for nuclear weapons. Starting in 1943, large reactors were built in Hanford, Washington. These were cooled by water from the Columbia River. Their goal was also to produce plutonium. Near the same time, in 1944, a heavy-water reactor was developed at Argonne National Laboratory in Illinois. In the late 1950s Canada, which has large uranium reserves, began development of the Candu pressurized heavy-water reactor systems.

For the first nuclear reactors, natural uranium was the only choice for use as a fuel. This dictated the use of thermal neutrons to fission ^{235}U, which is present in only a small proportion (~0.7% isotopic abundance) in natural uranium. The neutrons emitted during fission had to be thermalized quickly in order that they not be captured in the more abundant ^{238}U isotopes. The moderator could not be normal water because the ^{1}H isotope has a high neutron capture probability. The only practical choices for a moderator were heavy water, graphite, or possibly beryllium. Heavy water obtained by electrolysis was both scarce and expensive. The world's largest stockpile was in Norway. Thus graphite, produced in large quantities in the United States for electric furnace electrodes, turned out to be the best choice because it was readily available.

In parallel to natural uranium reactor development, scientists and engineers in the United States were developing the technologies for isotopic enrichment of uranium. The availability of enriched uranium (from 1 to 93%) allowed development of a new generation of reactors in which water could be employed as both moderator and coolant. Indeed, the ^{1}H isotope, which accounts for 99.985% of the mass of hydrogen in normal water, is the most efficient neutron moderator. Because water has a relatively low boiling temperature at atmospheric pressure, in such reactors a temperature above the boiling point is needed in order to get a decent yield due to the Carnot principle. Thus water is typically used at high pressure (75–150 bars). The advantages of water reactors is that they are more compact, have a better thermal yield, and use the fuel very efficiently. However, because water is quite corrosive for ordinary steels, development of PWRs employing liquid metals such as sodium, lead, and mercury has been extensively studied. The industrial development has been further spurred by the fact that this type of reactor was adopted for propulsion needs by the U.S. Navy because of its compactness.

In the period following World War II, the Soviet Union developed a reactor using slightly enriched uranium fuel moderated by graphite and cooled by boiling water circulating in pressure tubes. This finally gave birth, in 1954, to the Obninsk prototype, which can be considered the first true power reactor in the world. It was nevertheless oriented toward plutonium production and is the ancestor of the RMBK family to which the Chernobyl reactors belong.

Because they had not mastered uranium enrichment at that time and did not wish to be entirely dependent on the United States for their fuel supply, Great Britain and France developed reactors moderated by graphite and cooled by gas under pressure (first using air, later CO_2). These reactors were designated Magnox in Great Britain and UNGG (Uranium Naturel Graphite Gaz) in France.

After the war, during the period going from 1950 to 1965, a very wide variety of possible designs for nuclear reactors were considered and often tested, although usually on a small scale. Most of them functioned correctly. Various possibilities for fuel, coolant, and moderator were explored, but under safety conditions which would not be acceptable today. It is notable that no major accident occurred during those developments.

As early as 1966, prototypes of the HTR (high-temperature reactor) in Great Britain and a molten salt reactor experiment in the United States, both of which are now being promoted as the future of nuclear energy (Generation IV initiative), were tested. See Table 10.1.

The basic concepts of most reactor types in operation today or that are proposed for future exploitation were already developed before the 1960s. After that rich period, a selection occurred among the different possibilities driven mostly by economic, technological, and safety reasons. Some of the approaches were abandoned because of their size, technological problems in their construction, difficulties associated with their operation, and so on. Safety considerations have become increasingly more important in modern times.

TABLE 10.1. Order of Magnitude of Average Thermal Power per Unit of Volume and Yield for Different Reactor Types

Reactor Type	Thermal Power (MW/m^3)	Yield (%)
Magnox, UNGG	1	30
AGCR	2	41
PWR, VVER	100	33
BWR	50	33
Candu	12	29
FBR	500	40
HTR	8	>45

Source: From B. Bonin, *Energie nucléaire*, Omniscience, to be published.

10.2.3. Why Nuclear Energy?

There are several advantages to the use of nuclear power when the situation makes it possible. First, it is an energy source which delivers low-cost electricity. Furthermore, this cost is stable over several decades. Nuclear fuels do not emit CO_2 during operation. Even if one takes into consideration the entire fuel cycle, including CO_2 emitted at each stage of the nuclear process, for example, extraction and processing of the ore, enrichment of the uranium, and construction of the reactor and reactor housing, the total emissions are below that emitted when fossil fuels are used to supply the same amount of energy. The total emissions can vary, of course. Obviously, if uranium enrichment is done using electricity produced by fossil fuels, there is a larger contribution to CO_2 emission than if it is done with electricity provided by nuclear power.

10.2.4. Uranium Resources

Nuclear fuel is a highly concentrated energy source which is able to deliver large powers. However, as for any energy-producing system, the question of resource availability and power generation capabilities are important issues. The earth is finite and the amount of uranium which can be found and be economically exploited is an important issue since it determines the extent to which nuclear energy can be made available.

Uranium is a common metal which can be found in rocks and seawater. The average concentration in parts per million in the earth's crust is 2.8 ppm (2.8 g/t). Sedimentary rocks contain less (2 ppm) and granite more (4 ppm). Orebody, which is a mineralized mass from which the metal is economically recoverable, typically has a larger uranium concentration. A high-grade ore can contain 2% uranium (20,000 ppm) and a low-grade one 0.1% (1000 ppm). In Australia, for example, orebody can contain 0.5 kg/t while in Canada it can reach 200 kg/t. Seawater has a small concentration of uranium (0.003 ppm) and might be interesting as a source if the price of uranium were to greatly increase. This would depend on the difference between the energy required to extract uranium from the sea and that realized by using that uranium in a nuclear reactor.

In order to estimate uranium needs based on present-day consumption let us take the year 2005 as an example. In 2005, The 440 nuclear reactors in operation had a total thermal power of $1100\,GW_{th}$ (about one-third is transformed into electricity due to the Carnot principle). They functioned on average 7000 h/year and produced about $7700\,TWh_{th}$. This required around 64,000 tons of natural uranium. Thus 5 GWd/t (gigawatt day per ton) of useful energy was extracted from the uranium, a very small amount of that which is potentially available, ~900 GWd/t.

Naturally, estimates of the useful reserves of nuclear fuel must also depend upon the way the uranium supply would actually be exploited. As seen above, present-day thermal neutron reactor technology harnesses only a small fraction of the energy available from natural uranium. If we keep the present technology, based on thermal nuclear reactors, it is estimated that uranium fuel will be available at an acceptable economic price for the next century assuming the same consumption rate as today and the known supplies. If the price of fossil fuels strongly increases during that time, some additional uranium ore deposits could then be used economically. To obtain a sustainable nuclear power supply for a long duration requires that different technologies be employed.

Employing fast neutron reactor technology, which makes the much more abundant ^{238}U isotope useful as a fuel, would extend this period more than 10,000 years. Finally, using thorium, which is about 2.5 times more abundant than uranium, would make an additional large increase in the nuclear fuel reserves. A noteworthy point is that, with the new technologies, the energy which is not extracted in today's reactors could be extracted in the future.

The present reserves of nuclear fuels are likely to be underestimated because not much recent effort has been devoted to searching for new sources since, for a long time, the price of uranium was so low that exploration was not economically interesting. Things are now changing. The price of uranium ore has increased dramatically and there is a greatly increased renewed interest in finding new deposits.

While the cost of uranium resource is expressed in dollars per kilogram, the spot prices are reported in dollars per pound of the oxide U_3O_8. The correspondence between the different units is

$$1\,kg\,U_{nat} \Leftrightarrow 1.17\,kg\,U_3O_8 \quad \text{and} \quad 1\,kg\,U_{nat} \Leftrightarrow 2.6\,lb\,U_3O_8$$

With these equivalences, \$130/kg of natural uranium corresponds to \$50/lb of U_3O_8.

The price of U_3O_8 was around \$6/lb in 1972. It increased rapidly beginning at the time of the Arab oil embargo and reached about \$43/lb at the time of the Three Mile Island accident. Following that accident it decreased sharply. Between mid-1980 until about 2000, the price of U_3O_8 remained, on the average, below \$10/lb, which is five times smaller than the economic limit of \$50/lb (\$130/kg) for natural uranium. Starting in about 2004, a strong and continuous rise of the price of uranium occurred (\$130/kg of U_{nat} was even reached by the end of 2006). As this is being written the price has reached ~\$100/lb.

Statistics on uranium resources, production, and demand are usually found in the so-called *Red Book*, published every other year by the OECD Nuclear Energy Agency (NEA) and the International Atomic Energy Agency. In that book, uranium resources are classified in three categories:

1. The reasonably assured resources, which are classified according to their cost of exploitation (40, 80, and \$130/kg). They are known to exist because the mines are known or because rock samples have been analyzed.
2. Estimated supplementary resources inferred from a few sample analyses. These resources, which are sometimes added to the preceding ones, are not taken into account by the United States because they are not definitely proven to exist.
3. Speculative resources, which are estimated to exist based upon geologic investigations or inferred from the favorable geologic landscapes in certain areas.

These resources are called "conventional" in the sense that uranium is the main product of extraction operations. There are also unconventional uranium resources where uranium is obtained as a by-product during the extraction of other materials. The conventional uranium resources in which we are confident correspond to about 4.7 million tons at a price lower than \$130/kg. The speculative (third-category) conventional uranium resources are estimated at about 10 million tons. In total, reserves are estimated on the order of 14.7 million tons of uranium at a cost less than \$130/kg. Since one-third of the countries of the world, including Australia, which is the leading country for uranium resources, do not report their data on speculative resources, real resources may be higher.

In Figure 10.10 the shares of the known recoverable resources of uranium are displayed. Australia, Kazakhstan, Canada, the United States, and South Africa have more than half of these resources.

The annual demand on natural uranium is presently about 60,000 tons per year. Currently, only 36,000 tons comes from orebodies. The remaining part is supplied from nuclear weapons dismantling and from reprocessing of spent fuel (on average there is 0.7% of ^{235}U and 0.6% of ^{239}Pu in used uranium fuel coming out of PWRs) and the use of MOX [mixed-oxide (uranium and plutonium) fuel]. The IEA expects that there will be a tight supply situation for a long time and possibly a uranium shortage around 2012.

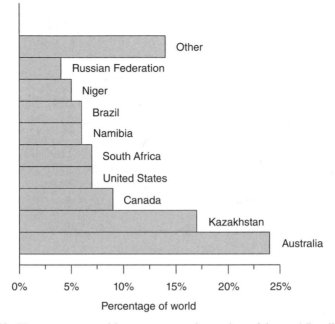

Figure 10.10. Known recoverable resources of uranium (about 4.7 million tons). *Source*: Data from www.world-nuclear.org.

In conclusion, there is enough uranium to meet anticipated nuclear energy needs. At a price lower than $130/kg of natural uranium, reasonable assured uranium resources are estimated to be on the order of 3.3 millions of tons according to the American standard and to 4.7 millions of tons if estimated supplementary resources are included. It is worth noting that, around the world, about 1.2 million tons of depleted uranium comes from enrichment operations which contain about 0.3% of ^{235}U—strategic stockpile for the future. There is therefore not much of a problem for about a century. However, to have longer term sustainable development of this energy source, it is necessary to develop fast neutron reactors which will allow much more efficient harnessing of the energy contained in natural uranium. For the future, there are also millions of tons of uranium contained in phosphates and billions of tons contained in the sea.

10.2.5. Fuel Cycles

A nuclear fuel cycle refers to all activities which occur in the production of nuclear energy. It includes mining of the ore, enrichment, fuel rod fabrication, energy production, waste management, and disposal and decommissioning of facilities. The cycle is often discussed in terms of two segments:

- The front-end fuel cycle, which encompasses the preparation of the fuel and the period in which the fuel is used in the reactor
- The back-end fuel cycle, which comprises fuel management after it has been used in the reactor

Front-End Fuel Cycle

Uranium ore is extracted using methods similar to those used for other metal ores in either open or underground pits. Uranium deposits have been identified throughout the world, but today four regions contain most of the orebodies:

- The Saskatchewan province in Canada has the richest uranium ore with a uranium concentration greater than 100 kg/t. An average enrichment of 20.7% U_3O_8 is observed in the cigar lake project.
- The Republic of Niger has mines with average U_3O_8 concentrations ranging from 2 to 5 kg/t.
- In Central Asia (Kazakhstan, Uzbekistan), the concentration drops down just below 1 kg/t.
- Australia has the largest reserves in the world (around 37% of the known reserves) but at a small concentration (about 0.5 kg/t).

Uranium deposits in the United States are low grade (typically between 0.05 and 0.3% of U_3O_8). The ore is usually extracted by mechanical methods but sometimes in situ leach mining methods are used to mine uranium.

After mining, the uranium ore is further concentrated to levels ranging from 75 to 98%. The resulting product is dissolved and precipitated to give the so-called yellowcake, which is the basic product for the nuclear industry. Its calcination leads to products containing about 75% of uranium in the form of U_3O_8 or UO_2.

Most of the reactors in use need enriched uranium at a grade ranging from about 3.5 to 4.5% of ^{235}U depending on the economic optimizations. For the enrichment process it is necessary to convert U_3O_8 to uranium hexafluoride (UF_6). Enrichment operations are normally performed by either gaseous diffusion or gas centrifugation, although other methods are possible. Gaseous diffusion is very energy intensive while gas centrifugation is more economic. The enriched UF_6 is then usually converted to UO_2 powder, which is then processed into pellets to be used as fuel. For those reactors which do not need enriched uranium (e.g., Candu), U_3O_8 is converted directly to UO_2. The fuel of PWRs consists generally in ceramic pellets of enriched uranium which are obtained from pellets of UO_2 fired at a high temperature in a sintering furnace. Cylindrical pellets are then stacked into tubes of corrosion-resistant alloy (e.g., based on zirconium). These tubes are sealed and should remain tight so that no radioactive product escapes when the fuel inside the reactor is being used. They are called fuel rods and are assembled into bundles. In a nuclear reactor core there are many bundles arranged in a regular pattern of cells. Each cell is formed by either a fuel rod or a control rod, which allows control of the reactor power. The cells are generally surrounded by a moderator and a coolant, which in most cases is water.

During the operation of the nuclear reactor, fuel is consumed by the fission process and the fuel rods must be changed periodically. Replacing old fuel generally requires shutting down the reactor. Only part (typically one-third) of the old fuel is replaced by fresh fuel at each cycle. A fuel rod typically remains four to five years inside the nuclear core, and during that time its envelope must prevent radioactive material from being dispersed into the moderator or coolant. For that reason preparation of good fuel rods is technologically demanding. Since the burn-up rates of old bundles of nuclear fuel and fresh bundles is different, bundles are normally rearranged inside the reactor core to optimize the reactor operation.

However, some reactor designs (e.g., Candu and RMBK) can be refueled without being shut down. This is because they have small individual pressure tubes containing the fuel and the coolant. Each tube can be isolated and refueled separately. For Candu, for example, about 2% of the tubes are refueled each day. Fuel reloading is better optimized than for PWRs or BWRs, but the process itself is more complicated. There are hundreds of pressure tubes to be refueled and specialized fueling machines are needed to do this job.

Back-End Fuel Cycle

Back-end fuel cycles are classified in two main categories:

1. The *open cycle* is a once-through fuel cycle where the spent fuel that issues from the reactor is treated as waste. Open cycles also have an advantage in terms of cost because there is no need to perform expensive and complex reprocessing operations; on the other hand, they entail more complex waste management procedures and lose the possibility of using fissile materials contained in the waste to produce energy. This procedure has been standard in the United States for nonproliferation reasons but this will probably change in the future. Studies are underway to develop spent-fuel reprocessing methods that make the recovered plutonium usable for nuclear fuel but unusable for nuclear weapons.

2. In the *closed cycle* the spent fuel discharged from the reactor is reprocessed. The idea is to decrease the amount of waste and to reuse the part of the spent fuel which has not fissioned. For that purpose the spent fuel is partitioned into uranium, plutonium, minor actinide nuclei, and fission products. The uranium and plutonium can be used for the fabrication of new fuels and recycled back into the reactor. At present the rest of the spent fuel is treated as high-level waste. In the future, minor actinides mixed with uranium fuel could be burned either in a thermal reactor (which is not the most efficient way) or in fast neutron reactors.

There could also be dedicated reactors to transmute selected isotopes previously separated from the spent fuel. Closed cycles have the advantage over open cycles in terms of resource utilization and for long-term disposal since the amount of final waste to take care of is much smaller. In the long term, if reprocessed fissile and fertile materials are used to produce energy, the final cost of electricity produced in both the open and closed cycles is about the same.

Plutonium is needed to start fast neutron reactors. About 10 t of ^{239}Pu, which is present in about 14 t of plutonium, is needed to start a 1-GW$_e$ fast reactor. This represents the quantity of plutonium produced by a PWR during approximately 50–60 years of operation.

The fuel discharged from the nuclear reactor core (spent fuel) is usually stored at the reactor site, usually in a water pool. This interim storage allows the fuel to cool down as it is highly radioactive when it is removed from the core reactor because it contains many short-lived nuclei formed during the fission process or during the neutron irradiation which occurs inside the core. After several years of cooling, spent fuel can be moved to dry storage. In France, for example, this interim storage close to the reactor facility takes place for about half a year. The spent fuel is then moved to a reprocessing site where it is stored from three and eight years.

Reprocessing is essentially based on the PUREX (plutonium and uranium recovery by extraction) method. This is basically a liquid–liquid extraction method which extracts uranium and plutonium from the spent fuel. Other complementary methods may be used for more specific separations. The main goal is to obtain end wastes with volumes and radioactivity levels which are more easily manageable.

During the reprocessing fuel rods are sheared and dissolved in nitric acid. The shell of the spent fuel constitutes a structural waste which must be disposed of. The liquid solution contains typically 200 g/L of uranium, 2.5 g/L of plutonium, 6–7 g/L of fission products, and traces of minor actinide nuclei. With the PUREX method it is possible to separate uranium, plutonium, and fission fragments mixed with minor actinides. Reprocessed uranium still contains ^{235}U (on the order of 0.9%). Plutonium can be used to prepare MOX fuel. At present, fission products and minor actinides are considered as end wastes and vitrified.

There are only a small number of reprocessing units in the world. Their capacities are shown in Figure 10.11.

In the future, pyrometallurgical processes could also be used with the new generation of reactors. They are more suitable for metal fuels than for oxide

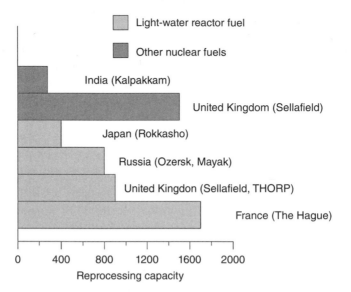

Figure 10.11. World reprocessing capacities, 2006. *Source*: Data from OECD/NEA. *Nuclear energy data, Nuclear Engineering International Handbook*, 2007.

fuels. They have several advantages: They do not use water; they can be almost immediately applied to high burn-up fuel; and they recover all the actinides, which can then be employed in reactors.

A major consideration with nuclear energy is waste management. The amount of waste depends very much on the back-end fuel cycle used. In the open cycle, the amount of waste to be taken care of is much larger than in the closed fuel cycle. Reprocessing allows separation of fissile and fertile products which can be used in the future, leaving only the final waste products which have to be taken care of. Most of the radioactivity (about 99.5%) of the spent fuel belongs to fission products. The timescale associated with management of fissile nuclear materials extends over several decades. There is no need to rush the care of used fuel since the longer we wait, the smaller will be the radioactivity.

10.2.6. Safety

Zero risk does not exist. Safety can and should always be improved. In nuclear energy generation, the first goal must be to design a reactor which is as safe as possible. This means decreasing the probability of a serious accident as much as possible. In the case of an accident in which the nuclear core is involved, the reactor should be designed in such a way that all the radioactivity is kept inside and there is no consequence for the nearby population or the environment. For that the containment vessel must be designed to withstand high pressures and temperatures.

The EPR (EPR originally denoted European pressurized reactor and now means evolutionary pressurized reactor) belongs to the so-called 3+ generation of reactors built by the French company Areva taking into account all of the experience in matters of risk obtained over several decades. It is a 60-year-service-life 1600-MW$_e$ PWR in which safety is considered a top priority. To compensate for the extra costs associated with security improvements its power is larger than usual. In the end, the cost of electricity is similar to or a bit lower than that of previous nuclear reactors. In a severe accident where the core melts, the corium (material of the nuclear core) is passively collected, retained, and cooled in a special area located below but inside the containment building.

Between the time of the first nuclear reactor (CP1 started by Enrico Fermi in 1942) and present-day reactors tremendous progress has been made in terms of safety.

By the end of 1942, Enrico Fermi and his team had started the CP1 in downtown Chicago. It had only two safety systems. The first one was a series of shutdown rods which were gravity operated and inserted by cutting a rope. The second shutdown system consisted of buckets containing a cadmium sulfate solution (cadmium is a strong neutron absorber). The buckets were located on top of the reactor and could be emptied into it in case of emergency.

The next step in Western nuclear reactor safety was to locate the nuclear reactor inside a leak-proof and pressure-resistant containment vessel. For Western reactors it was also required that the primary cooling circuit be inside the containment vessel. The first reactor built with these new safety specifications was the SR1 reactor at West Milton, New York State, in 1950. Safety requirements in Eastern countries were not as stringent and a complete leak-proof and pressure-resistant containment vessel was not required. For example, the VVER 230 series of Russian reactors had a leakage of about 25% each day, a value much larger than the 0.2% typically obtained in Western nuclear reactors.

Since the average European population density is larger than the United States by a factor of 7, further constraints have been adopted in Europe for reactor building. Japan and South Korea are in a similar situation. This has led to different levels of security that the public has sometimes misunderstood.

Before 1979, the date of the Three Mile Island accident, three major developments occurred in the approach to nuclear energy safety. The first one was to evaluate the need for defense against nonnatural external events such as plane crashes on nuclear plant sites or possible sabotage with explosives. The second was the Rasmussen report, published in 1975, which recommended evaluating the risks of all conceivable accidents that could be imagined in terms of probability. The great advantage of this report was to foster a methodical scientific approach to nuclear safety. The third improvement was the application of a quality control approach in nuclear engineering.

By evaluating any unexpected event, even a small one, it is possible to gain experience that may be used to anticipate and prevent future incidents or accidents. Presently we have experience equivalent to more than 10,000 reactor-years. After the Chernobyl accident, an international scale already existing in some countries was employed to report the different events occurring in the nuclear industry: the INES (International Nuclear Event Scale), which categorizes eight levels of unexpected events occurring in a nuclear plant. The lowest one (0) corresponds to a deviation without any safety significance. The three lower levels (1, 2, and 3) are termed incidents. The four highest levels (4, 5, 6, and 7) are termed accidents and are more serious in terms of safety. The INES is displayed in Figure 10.12.

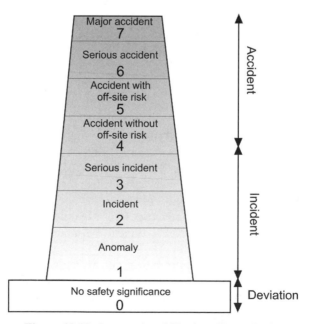

Figure 10.12. International Nuclear Event Scale.

Three Mile Island is a flat island several square kilometers located on the Susquehanna River in the United States. Two nuclear reactors (TM-1 and TM-2) were installed there. They supplied a power of 1700 MW, sufficient to meet the demands of about 300,000 U.S. families. The accident occurred at about 4 A.M. on March 28, 1979, in the TM-2 power plant unit. This accident was due to a human failure. What was learned from this accident has been of great importance to the improvement of subsequent nuclear safety.

In March 1979 there was an accident at the Three Mile Island (TMI) nuclear plant in the United States. It started when a valve on top of the pressurizer did not close as expected. As a consequence there was loss of coolant. This incident was amplified because the technical staff made an incorrect analysis of the event and made wrong decisions. After the TMI accident, new standards of nuclear safety requirements were introduced. The first was the requirement for *"defense in depth," the installation of* several independent protections designed to stop the progress of an accident. If one fails, the next one is activated to stop the evolution of the accident. The second is the development of a *"safety culture"* among the technical staff.

The Chernobyl accident (1986 in what is now the Ukraine) was rated 7 on the INES. There was a large external release of radioactive materials (equivalent to more than 10,000 TBq of ^{131}I). This was distributed over a very large area with long-term environmental consequences. In 1957 (in Russia) there was a serious accident at the Kyshtym reprocessing plant accident with a release of radioactive material of the order of thousands to tens of thousands terabecquerels of ^{131}I equivalent. It was rated 7 on the INES. In 1957 in the Windscale pile accident. In the United Kingdom, there was also a release of radioactive material equivalent to hundreds to thousands terabecquerels of ^{131}I.

The Chernobyl accident released 12 kg of ^{139}Pu into the atmosphere. This value can be compared to the 4.2 tons of ^{239}Pu been released in the atmosphere during all aerial atomic explosions to test the atomic bomb.

Seriously taking into account the possibilities of accident and developing methods to prevent them have made the nuclear energy industry a much safer industry than many others. There are, for example, many more casualties in coal mines than in the nuclear industry. However, the nuclear industry remains such a concern for the average population that any casualty is still regarded as having a much greater importance than casualties occurring in other industries.

10.2.7. Nuclear Waste

Because nuclear energy is a concentrated form of energy, the weight or volume of nuclear waste produced by nuclear energy sources is also relatively small. However, these wastes are dangerous because of their radioactivity. Each radioactive element is characterized by its half-life, which is the time after which half of an initial population of radioactive nuclei has disappeared. Some radioactive nuclei have short half-lives and consequently a high radioactivity because many of them disintegrate per unit of time. Others have a long lifetime and a small radioactivity level. All intermediate situations are also possible. The problem is that very often short- and long-lifetime nuclei are mixed together and this mixture can have both a large activity and a long lifetime. The best way to deal with this would be to perfectly separate nuclei according to half-life and manage each of these fractions. The shorter the period of a radioactive element, the more radioactive it will be, but it will also cease to be a problem in a shorter time. In contrast, long-lived nuclei are not very radioactive. This is the case for plutonium, for example, which has a half-life of 24,000 years. Other isotopes can live even longer. Usually one considers short-lived nuclei to be those with a period smaller than 30 years and long-lived nuclei to be those with a period larger than about 30 years.

It is interesting to review the amounts of radioactive waste produced in a country such as France, which draws a large fraction of its power from nuclear sources. Each year in France 50,000 tons of radioactive waste is produced, corresponding to a volume of 20,000–25,000 m^3. About 75% comes from the nuclear power industry. The remaining part, 15%, comes from non–nuclear industry activities like hospitals, universities, research, or maintenance of nuclear weapons. To compare this amount of waste with other kinds of waste, note that French hospitals produce about 700,000 tons of waste per year, commonplace industrial wastes are 130 million tons, dangerous industrial wastes are 18 million tons, household wastes are 30 million tons, and agricultural wastes are 600 million tons.

Wastes coming from spent fuel are highly radioactive. Technological wastes coming from handling of these are less active but may occupy a large volume.

There are several different classifications of nuclear waste. These can be different in different countries. We will not consider these variations in detail. Basically wastes are classified as follows:

- High-level waste (HLW) is that produced by nuclear reactors. It contains fission products and transuranic elements. It is highly radioactive and thermally hot. Typically, in a developed country, HLW accounts for more than 95% of the total nuclear waste radioactivity.
- Intermediate-level waste (ILW) has a lower radioactivity than HLW. It comes from metal reactor fuel cladding, chemical sludge, contaminated material from reactor decommissioning, and so on.
- Low-level waste (LLW) comes from hospitals, industry, research, and the nuclear fuel cycle. It includes any matter coming from an "active area" and may sometimes have the same radioactivity as it had before entering the active zone.
- Very low level waste (VLLW) includes rubble or scrap metal which was not directly in contact with highly radioactive materials but was in restricted areas.

Table 10.2 indicates the order of magnitude of the initial activities of selected radioactive wastes.

Assuming an open fuel cycle ("one-through" mode) and one thousand 1-GW$_e$ nuclear reactors operating in the world, one would need to open one disposal facility like Yucca Mountain every three or four years. Therefore, it is extremely desirable to evolve toward the closed-cycle operation mode in order to decrease the volume of nuclear waste to be managed.

10.2.8. Conclusion

Nuclear energy is a rather young energy source, having been used a little less than half a century. Its main advantage is that it produces low-cost electricity.

TABLE 10.2. Order of Magnitude of Radioactive Activities of Different Types of Waste

Type of Waste	α Activity (GBq/m^3)	β and γ Activities (GBq/m^3)
Very low level waste	10^{-2}	10^{-2} to a few 10^{-1}
Low-level waste	<1	1–10^2
Intermediate-level waste	10^3	10^6
High-level waste	10^6	10^8

Note: For comparison, the natural radioactivity of the human body is on the order of 10^5 Bq/m^3 = 10^{-4} GBq/m^3, the activity of uranium ore is typically around 500 GBq, and that of granite rock is around 2.7×10^{-3} GBq.

The small price of natural uranium relative to the total cost of production means that the final cost remains relatively stable as a function of time. This provides a competitive advantage for industries and businesses of a country which employs this energy. Since it is a concentrated source of energy, high power, which is needed for some industries, is available. However, nuclear power plants require higher initial investments than are needed for power plants employing other fuels, gas, for example.

Because of increasing energy demand, the need for a more secure energy supply, and the oil shock, in the 1970s France decided to strongly support development of nuclear energy. Today about 80% of electricity in France is produced by nuclear power. If fuel-fired plants had been developed in the 1970s instead of nuclear reactors, today it would be necessary to buy oil to run them. Assuming the cost of a barrel of oil to be \$80, each inhabitant of France would have to pay an additional 1000€ per year to buy the oil necessary to produce the electricity used today in France. The total amount is comparable to the education budget of that country.

For nuclear energy to be a sustainable energy resource, it is necessary that fast neutron reactors, which have the advantage of better using the potential energy contained in natural uranium, be developed. The usable fuel reserves will then be multiplied by two orders of magnitude. Developing fast nuclear reactors which work at high temperature will allow an increase in the efficiency of electricity production from about 30% now to roughly 50%.

The use of nuclear energy can also help decrease CO_2 emissions because no CO_2 emissions occur during reactor operation. On the other hand, emissions can occur in associated operations such as transportation of the radioactive fuel or waste, uranium enrichment using electricity from fossil fuels, and making concrete for the containment buildings.

The large timescales associated with the development of nuclear power facilities imply that nuclear energy usage cannot be quickly introduced and requires long-term advance planning. Other concerns with the use of nuclear energy include the possible impact of radioactivity on human heath, nuclear incidents or accidents, proliferation risks, and the problem of nuclear waste storage.

10.3. THERMONUCLEAR FUSION

Humanity's dreams of an essentially inexhaustible source of energy might well be realized if functional thermonuclear fusion reactors can be built. But this technology is still in the research stage and will probably not produce electricity on a practical scale before the next century. In this chapter we discuss the fundamental nature of nuclear fusion reactions and the practical considerations that shape the current research efforts.

10.3.1. Nuclei: Concentrated Sources of Energy

The combustion of oil, gas, or coal is a chemical reaction. A chemical reaction corresponds to a reorganization of the atoms involved in that reaction. During this reorganization chemical bonds are broken and formed. The energy released in combustion of fossil fuels reflects the difference in bonding energies between the reactants and products of the reaction.

A nuclear reaction corresponds to a reorganization of the neutrons and protons from which the reactant and product nuclei are constituted. The energy changes involved in nuclear reactions reflect the changes in the binding between neutrons and protons in the nuclei involved. Since the nuclear force which holds nuclei together is about 1 million times stronger than the Coulomb force which binds atoms, the energy changes in nuclear reactions are about a million times higher than those observed in chemical reactions. While both chemical and nuclear reactions may release or absorb energy, only those reactions that release energy are useful to produce energy.

There are two types of nuclear reactions that can provide exploitable energy on a large scale. The first is neutron-induced nuclear fission in which the capture of a neutron by a heavy nucleus is followed by a subsequent breaking into two medium-mass nuclei and a large energy release. We have presented this mechanism in the first two sections of this chapter. In the second type, known as nuclear fusion, two light nuclei merge into a heavier nucleus. For light nuclei this also leads to significant energy release. In both cases the energy released in the process comes from the binding energy differences between the initial and final nuclei. One understands why by looking at Figure 10.2, which depicts the binding energy per nucleon of stable nuclei as a function of their mass numbers (the numbers of neutrons plus protons in these nuclei). It grows from 1 to 7 MeV per nucleon between $A = 2$ (deuterium) and $A = 4$ (helium). It is about 8.5 MeV per nucleon for nuclei with mass number A between 30 and 120. It decreases to 7.5 MeV per nucleon for $A > 220$. This indicates that fusion of the lightest nuclei or fission of the heaviest will both lead to more stable nuclei and energy will be released in the process. The amount of energy liberated per nucleon is larger for fusion than for fission. One kilogram of natural uranium can produce $\approx 10^5$ kWh in thermal reactors or $\approx 10^7$ kWh in fast reactors. In the sun the thermonuclear reactions provide $\approx 1.8 \times 10^8$ kWh/kg of matter.

But energy release alone is not sufficient. It is necessary that, once started, the reactions are self-sustaining. Thus, for example, in the burning of fossil fuel, ignition provides the initial energy to start a fire that then can continue energy release as long as the fuel and oxygen required for the combustion are accessible.

In order to be useful for energy production a nuclear reaction must also induce additional, self-sustaining, reactions. In addition, the rate of reactions and energy release must be controllable. As we have seen in Section 10.1, this becomes possible for fission reactions because about 2.5 neutrons are emitted in each reaction and some of them contribute to the nuclear chain reaction that powers a reactor. This chain reaction can be moderated using neutron absorbers. The technical goal for nuclear fusion is to recreate self-sustained fusion reactions such as those that power our sun (or a hydrogen bomb) but in a controllable manner. So far this goal has not been achieved. When it is, the goal of a virtually inexhaustible energy supply will have been realized. The question of economic viability will then be primary.

10.3.2. The Sun

It is only relatively recently that the mechanism generating energy in the sun has been understood. In 1938, the physicist Hans Bethe explained this energy as resulting from nuclear fusion reactions. For this work he received the Nobel Prize in 1967. The net process generating this energy may be written as

$$4 \, ^1\text{H} \rightarrow 4 \, \text{He} + 2 \, e^+ + 2 \, ^0\nu + 26 \, \text{MeV}$$

This equation reaction is the sum of simpler nuclear reactions, the first one being the fusion of two ^1H nuclei (protons). Individually, a reaction of this type is rare because it is controlled by the weak interaction. On the average, approximately 5 billion years is required before any two particular protons fuse. However, the number of protons is very large and therefore the rate of reactions is sufficient to provide a large amount of energy at the earth's surface. The sun is thus a nuclear fusion reactor confined by its own gravitation. Each second, a little less than 5 million tons of matter is transformed into energy. On the earth it is not feasible to reproduce the process exactly as it occurs in the sun. The light-nuclei fusion reactions that will be used to produce energy on the earth are therefore very different from the one taking place in the sun. The means of confinement of the reacting medium, the plasma, where the fusion takes place will also be very different.

10.3.3. Fusion of Light Nuclei

The most interesting reaction for an earth-based fusion reactor is the (D,T) fusion reaction between deuterium (D = ^2H) and tritium (T = ^3H), two heavy isotopes of hydrogen. It is written

$$\text{D} + \text{T} \rightarrow \, ^4\text{He} + \text{n} + 17.6 \, \text{MeV}$$

The released energy appears as kinetic energy of the helium nucleus, 3.5 MeV, and the neutron, 14.1 MeV. Using this reaction one can produce as much energy from a 1-g mixture of deuterium and tritium as can be realized from 8 tons of oil, that is, approximately 83,000 kWh. However, there remains the necessity to recover the energy released to produce electricity. The large kinetic energy of the neutron makes the absorption of this energy more difficult than in the case of nuclear fission.

In a fission reactor the neutrons which induce the reactions are electrically neutral and go easily into the uranium nucleus. In the (D,T) fusion reaction, the two particles which interact are positively charged. Because the nuclear interaction is of short range, it is necessary to overcome the Coulomb barrier between the nuclei to bring the interacting nuclei close enough to react. Distances smaller than 10^{-15} m must be reached. A way to overcome Coulomb forces and bring nuclei close enough so that they fuse is to bring the reaction mixture to high temperature. The nuclei will then move with sufficient kinetic energies to overcome the Coulomb barrier. For the (D,T) reaction, temperatures higher than 100 million degrees are required. For comparison the temperature in the center of the sun is 15 million degrees. It is also necessary to confine this very hot plasma long enough so that these nuclei can interact with each other. Finally, these conditions must be maintained for a period long enough so that a part of the released energy induces further reactions in the plasma. They are difficult to realize.

10.3.4. Difficulties

To support fusion it is necessary that the product of the density, n, and the containment time, τ, is larger than 10^{20} m$^{-3} \cdot$ s. This gives, for the product $nT\tau$, the "ignition condition" (near the so-called Lawson criterion):

$$nT\tau \approx 10^{21} \, \text{m}^{-3} \cdot \text{keV} \cdot \text{s} = 1 \, \text{bar} \cdot \text{s}$$

This equation suggests two limiting possibilities for achieving a fusion reactor. The first consists of confining high-density plasmas for short times. The second is to confine low-density plasmas for long times.

The first method is pursued in the "inertial confinement" approach in which a D–T mixture is strongly compressed using either laser beams or beams of energetic heavy ions (≈ 1000 g/cm^3; 50–100 million degrees). The laser technique is being pursued at the National Ignition Facility at the Lawrence Livermore National Laboratory in the United States, scheduled to operate in 2009. The French Laser Megajoule project under construction near Bordeaux is scheduled for completion in 2011. The goal of these facilities is to carry out experiments dedicated to simulate microscale nuclear explosions. In these facilities, a D–T mixture contained in a small capsule will be compressed and heated by bombardment with high-power synchronized laser pulses. Under irradiation, both the density and the temperature of the mixture increase strongly and the fusion process is initiated.

A second way to produce thermonuclear fusion is to use confined low-density energetic plasmas (with the number of particles about 10^{20} particles/m^3). The basic principle is to confine a D–T plasma in a magnetic structure and to heat the mixture until it reaches a sufficiently high temperature to initiate fusion reactions.

10.3.5. A Bit of History

The first experiments on confinement of hot plasmas with magnetic fields date back to 1938 in the United States. Since 1946 considerable effort has been devoted to this subject. This research remained confidential until 1958. At that time earlier results were declassified and international collaborations were established. The difficulties of the problem were largely underestimated and it was not until 1968 that an important step was taken by Russian scientists who introduced a particular magnetic structure for plasma confinement: the tokamak. This structure proved to be much more efficient in confining the plasma than any other solution and constituted a real breakthrough. The stellarator, another magnetic structure, offers a possible but less powerful alternative. Due to the currently existing research on tokamaks, considerable progress has been made in confinement studies. In 30 years, the achievable product $nT\tau$ has increased by a factor of 1000. The progress achieved so far corresponds to a growth rate which is a little higher than that observed in semiconductor development (Moore's law). Nevertheless much remains to be done before a practical operating device can be employed as an energy source.

To achieve a controlled thermonuclear fusion plasma, the essential next step toward a practical fusion reactor, seven parties—Europe, the United States, the Russian Federation, Japan, the People's Republic of China, the Republic of Korea, and India—have joined together to pursue the ITER project. ITER means "the way" in Latin. Initially proposed by Mikhail Gorbachev at the Geneva Summit of 1985, ITER is being constructed in Cadarache, France. The total cost will be approximately $10 billion over 30 years. The International Fusion Materials Irradiation Facility (IFMIF), a related facility housing a high-intensity generator of 14 MeV neutrons, will carry out materials studies in support of the ITER project. Located in Japan this facility will account for approximately 15% of the total financing for ITER.

10.3.6. Thermonuclear Fusion in Tokamaks

In order to produce energy continuously the plasma must remain hot and confined. The energy released in a D–T reaction is shared between the α particle (helium nucleus) and the neutron. The α particle carries 20% of the energy, the neutron 80%. The α particles produced remain confined in the plasma by the magnetic field of the tokamak. They transfer their energies to the plasma as they suffer collisions with the other particles. Neutrons, which are uncharged, escape. They are stopped in breeder materials surrounding the vacuum chamber of the tokamak. If the fraction of the energy released by fusion reactions and trapped in the plasma is not sufficient to sustain the process, it is necessary to provide additional energy for this purpose.

Energy may be lost from the plasma by several mechanisms. A large part is removed by transport phenomena of the particles and heat radiated to the outside. The plasma also radiates energy as Bremsstrahlung and synchrotron radiation photons originating from energy losses by electrons in the plasma. Impurity atoms removed from the walls of the confinement vessel can absorb energy when the atomic levels of these atoms are excited. This last phenomenon can lead to a rapid loss of the plasma confinement (disruption phenomenon).

The amplification factor Q is defined as the ratio between the power provided by fusion reactions and the external injected power. If $Q > 1$ the fusion reaction liberates more energy than is injected. The break-even point for the operation of a fusion reactor corresponds to the situation where $Q = 1$. The aim is to reach ignition, the condition in which the power provided by the fusion reactions compensates for any losses. In this condition, there is no need to provide extra energy. Once lit, the plasma burns and continues to do so as long as there is fuel. Ignition corresponds to $Q = \infty$ since the external power is zero.

10.3.7. ITER: New Step Towards Mastering Fusion

As indicated above, ITER is a very large research instrument being built and exploited within the framework of an international collaboration. Its goal is not to produce electricity but to control the conditions under which it is possible to make a thermonuclear fusion plasma. ITER will be 8 times larger than the JET (Joint European Torus) international fusion reactor located in England. The power of ITER is $500\,MW_{th}$ and the plasma will be confined for a maximum time of $400\,s$. Whereas the gain of the JET reactor was 1, ITER will have a gain $Q = 10$. This is still insufficient to produce electricity at an industrial scale. To achieve that, it will be necessary to reach $Q = 40$ to maintain the temperature of the plasma at 150 million degrees. The characteristics of the ITER are indicated in Table 10.3.

Phase 2 of the project envisages the construction of the reactor DEMO, with a power of $2000\,MW_{th}$, in which it will be possible to obtain a gain of 40 during a time compatible with electricity generation and to study in more detail the production of tritium fuel in a lithium blanket, the extraction of energy, and so on. Finally, if the results correspond to those expected, PROTO, a first industrial prototype, with an electric output of $1000\,MW_e$ could be built. Taking into account the duration of each project (several decades) and their costs, it is difficult in the current context to expect fusion-powered production

TABLE 10.3. Main Characteristics of ITER

Parameter	Value
Main radius	6.2 m
Small radius	2 m
Plasma volume	837 m^3
Magnetic field	5.3 T
Plasma current	15×10^6 A
Fusion power	500 MW
Neutron flux	0.5 MW/m^2
Power amplification (Q)	10
Maximum confinement time	400 s

of electricity on a large scale before the next century. The development time might be shortened if it were concluded, at the international level, that the urgency for this new source of energy warranted the commitment of the necessary effort and resources to an accelerated program.

10.3.8. About Fuel Reserves

The operation of a fusion reactor of 1000 MW$_e$ would require 100 kg of deuterium and 150 kg of tritium per year. The fuel resources needed are abundant and well distributed around the planet. The deuterium concentration in seawater is 33 g/m^3, which corresponds to reserves of about 4.6×10^{13} tons. These reserves would be sufficient to provide about 10 billion years of energy at the rate of the earth's current electricity consumption. Since in another 5 billion years our sun will have exhausted its own fuel and our planet will be destroyed, we can consider the reserves of deuterium to be inexhaustible. On the other hand, tritium, which is a radioactive nucleus with a period of 12.3 years, does not exist in sizeable amounts on the earth. It is synthesized starting from lithium (^{6}Li). Producing the necessary amount of the short-lived tritium isotope in the reactions of neutrons with lithium nuclei would require 300 kg of ^{6}Li. fortunately, lithium is very abundant. The average lithium content in the earth's crust is about 50 ppm of ^{6}Li. The reserves are about 12 million tons available at a price lower than \$5/kg. As for nuclear fission, the price of producing the fuel for thermonuclear fusion will represent only a small part of the price of a kilowatt-hour of generated electricity. This amount of lithium corresponds to an approximately 5000-year supply of energy. These reserves are of the same order of magnitude as those of uranium, which will last 10,000 years if used in fast reactors. There is also lithium in seawater (0.17 g/m^3, or approximately 230 billion tons). In a future fusion-driven economy the major question may well become whether it is energetically and economically interesting to exploit this resource.

10.3.9. Longer Term Possibilities

There are other reactions that might be considered for future fusion reactors. For example, the D–D reaction

$$D + D \rightarrow T + p + 4 \, \text{MeV} \quad \text{or} \quad {}^{3}\text{He} + n + 3.37 \, \text{MeV}$$

is interesting because it uses only deuterium, an essentially inexhaustible resource, as discussed above. However, the achievement of controlled D–D fusion requires higher temperatures than does the D–T reaction.

Another possible reaction is

$$D + {}^{3}\text{He} \rightarrow {}^{4}\text{He} \, (3.6 \, \text{MeV}) = p \, (14.7 \, \text{MeV}) + 18.3 \, \text{MeV}$$

It produces only charged particles which are easily captured to extract energy. Unfortunately ${}^{3}\text{He}$ is very rare and exists in large quantities only on the moon where it is deposited by the solar wind.

As a final example we note the reaction

$$P + {}^{11}\text{B} \rightarrow 3 \, {}^{3}\text{He} + 8.7 \, \text{MeV}$$

While elemental hydrogen and boron are abundant, the temperatures necessary to realize this reaction are much higher than for the others we have considered.

10.3.10. Safety and Waste Issues

The plasma of a fusion reactor contains very little matter (a few tens of milligrams of D–T mixture per cubic meter). A divergence of the reaction is hardly possible because any uncontrolled disturbance of the medium leads to a fast cooling of the plasma and termination of the reaction.

The tritium used as fuel is radioactive. It will be produced in a lithium-containing blanket surrounding the plasma chamber. This tritium must be carefully confined and managed because it is capable of diffusing quickly into most materials.

The products resulting from the D–T reaction are α particles and neutrons. The α particle, produced in its ground state, is not radioactive. The free neutron decays with a half-life of ~11 min. This is long compared to the time necessary for it to be captured by another nucleus. At the end of the operating lifetime of the fusion reactor, the structural materials will be activated but the half-lives of the radioactive nuclei produced are relatively short. After about 100 years the majority of the activated materials will only be very slightly radioactive. This is clearly an advantage for fusion reactors in comparison to fission reactors, which generate wastes having much longer decay times.

10.3.11. Conclusion

The energy releases in nuclear fusion reactions are about a million times higher than those observed in the burning of comparable masses of fossil fuels. However, even if fantastic progress were to be made in the domestication of thermonuclear fusion, application on an industrial scale is still far away. Much research and development remain to be done. The international ITER project, which will extend over a period of 30 years, will make it possible to test all components essential to the development of a controlled fusion device. It represents an indispensable step in the development of practical thermonuclear fusion reactors capable of producing electricity.

APPENDIX

In radioactive α emission, also called α decay, the initial nucleus of atomic number Z and mass number $A(\,_{Z}^{A}X)$ emits a helium nucleus $_{2}^{4}He$ made of two protons and two neutrons:

$$_{Z}^{A}X \rightarrow \,_{Z-2}^{A-4}Y + \,_{2}^{4}He$$

Alpha decay is mainly observed for heavy nuclei. The major part of the released energy is found in the form of kinetic energy shared between the α particle $_{2}^{4}He$ and the daughter nucleus $_{Z-2}^{A-4}X$. Since the α particle is much lighter than the daughter nucleus, it takes the bulk of the kinetic energy available.

There are two types of β radioactivity: β^{-} and β^{+}. In β^{-} radioactivity, an electron e^{-} is emitted together with an antineutrino $\bar{\nu}$. This antineutrino is an elementary particle of very low mass and charge and carries part of the energy released during the reaction in the form of kinetic energy:

$$_{Z}^{A}X \rightarrow \,_{Z+1}^{A}Y + e^{-} + \,_{0}^{0}\bar{\nu}$$

The final nucleus has an atomic number one unit larger than the initial nucleus.

For β^{+} radioactivity a positron e^{+} (antiparticle of the electron) and a neutrino ν are produced during the decay according to the reaction

$$_{Z}^{A}X \rightarrow \,_{Z-1}^{A}Y + e^{+} + \,_{0}^{0}\nu$$

The final nucleus has an atomic number one unit smaller than the initial nucleus.

Similar in effect to β^{+} emission and competing with it is a phenomenon known as electron capture in which an electron of the atom is captured by the nucleus leading to a transformation of a proton into a neutron n and a neutrino:

$$_Z^A X + e^- \rightarrow {}_{Z-1}^A Y + n + {}_0^0 v$$

Here, again, the final nucleus has an atomic number one unit smaller than the initial nucleus.

Radioactive decay by γ emission corresponds to the emission of electromagnetic radiation (γ photons are very energetic light) of very short wavelength by a nucleus which is in an excited nuclear state. It is the nuclear analog to the emission of radiation by an atom when an electron moves from a high atomic energy level to a low one:

$$_Z^A X^* \rightarrow {}_Z^A X + \gamma$$

The mass and the atomic number of the nucleus remain unchanged.

CHAPTER 11

Electricity: Smart Use of Energy

Electricity is not an energy source but rather a convenient and flexible way to transport energy. Energy is needed to produce it. It is difficult to imagine a modern home without electricity. There are now so many electric motors or electric appliances in each modern home that electricity has become necessary to modern life.

Unfortunately electricity is not available to everybody in the world. There are still about 1.6 billion people in the world who have no access to electricity. These are usually the same people who have insufficient access to food and other necessities. The comparison between the populations with and without electricity in different regions is shown in Figure 11.1. One unfortunate consequence of living without electricity is that school children cannot study when the sun is down. This is more common in southern countries where day and night have almost the same duration all year long and is a strong hindrance to development.

> About 1.6 billion people in the world are without electricity. Most live in rural areas ($\approx$1.3 billion, or 80%).
>
> The affected areas with the largest number of people are South Asia, which has 580 million people without electricity in rural areas and 126 million in urban areas, and sub-Saharan Africa, where the population without electricity in rural areas corresponds to 438 million people and in the urban areas to 109 million.
>
> In South Asia, the percentage of electrification is 52% while it is only 26% in Sub-Saharan Africa. In China and East Asia, the electrification percentage reaches 89%, but there are still 182 million people in the rural population and 41 million people in the urban population without electricity.

Our Energy Future: Resources, Alternatives, and the Environment
By Christian Ngô and Joseph B. Natowitz
Copyright © 2009 John Wiley & Sons, Inc.

In North Africa and in the Middle East the percentage of electrification is 86%, yet there remain 30 million people without electricity in the rural areas and 17 million in the urban areas.

Latin America has a high percentage of electrification (90%), but there are still 38 million people living without electricity in rural areas and 7 million in urban areas.

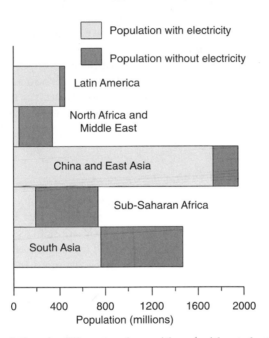

Figure 11.1. Populations in different regions with and without electricity, 2005. *Source*: Data from *World Energy Outlook*, IEA, Paris, 2006.

11.1. RAPID DEVELOPMENT

Electricity is a rather new energy vector. One of the first applications of electricity was for lighting. The first electric light was installed in 1850 on the Pont-Neuf in Paris. Observers were astonished at the high intensity of the light provided. Even so, at the beginning, electricity had difficulties in establishing itself as the preferred source of energy for consumers. During the second part of the nineteenth century there was strong competition between gas and electricity for primacy in lighting applications. Lighting using alcohol lamps, a technology strongly pushed by Germany, was also in competition with gas and electricity.

Initially, light obtained from electricity was more expensive than that produced by the other methods. Its advantage was mainly for use in market niches, theaters, for example, because electric lights did not produce smoke and were less of a fire hazard. It was only at the end of the nineteenth century that incandescent lights replaced arc lamps that electricity started to be commonly used by private individuals. Electricity developed most rapidly in the United States because the country was new and older energy sources were less developed than in Europe.

At the beginning of the twentieth century, the energy cost to produce light varied strongly according to the technology used. This is seen in Figure 11.2, which shows costs of different lighting systems in France in the year 1900. We see a difference of a factor of 7 between the lowest and the highest price. Since the cost of lighting was a nonnegligible part of the energy budget of private individuals, finding a means to lower the cost became important.

In the figure, "gas mantle" refers to a technology introduced by the Austrian scientist Carl Auer von Welsbach. It is sometimes referred as the "incandescent gas mantle" or "Welsbach mantle." The mantle is made of oxides which emit bright white light when heated by a flame. Modern camping lamps work on this principle.

Electric appliances (e.g., refrigerators, washing machines, dishwashers) exist in much greater numbers today than before, but they need less energy. In France the number of such electric appliances (called "white products") increased from 60 million in 1978 to 100 million in 1992 while the corresponding total energy consumption dropped from 8.33 toe in 1978 to 8 toe in 1992.

In the period between 1973 and 2004, the world's electricity production increased by a factor of 2.86, from 6130 to 17,531 TWh. The average annual growth rate of electricity production over this period was larger than the total for all energy production: 3.4%. Electricity consumption has grown much faster in emergent countries than in developed countries. Between 1973 and 2004, the rate of electricity production increased by 4.9%/year in non-OECD countries and by 2.7%/year in OECD countries.

In 1973, 72.9% of the electricity production was in the OECD countries while this share dropped to 58.2% in 2004. In 2005, electricity production had reached 18,292 TWh. The total final consumption in 2005 was 15,021 TWh. This corresponds to an average power of 1.7 TW. The distribution losses were 1597 TWh. The bad news with respect to environmental concerns is that a large part of electricity (65.8% in 2004) is generated from burning fossil fuels. This leads to large amounts of CO_2 emission as well as emission of other pollutants.

The rapid development of electricity usage in France is illustrated in Figure 11.3 showing the electricity consumption from 1945 to 2007. Just after World

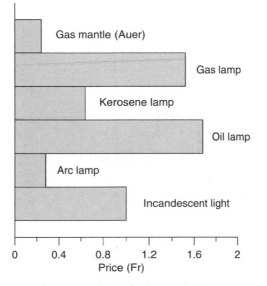

Figure 11.2. Comparison (in French francs) of cost of different energy sources producing same amount of light, 1900. *Source*: Data from *Electricité*, Eyrolles, 2005.

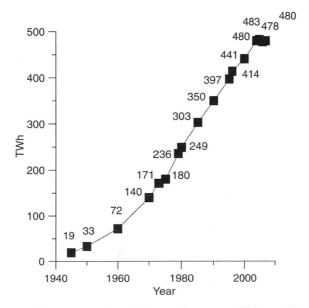

Figure 11.3. Electricity consumption in France between 1945 and 2007. *Source*: Data from *Electricité*, Eyrolles, 2000, and DGEMP, www.industrie.gouv.fr.

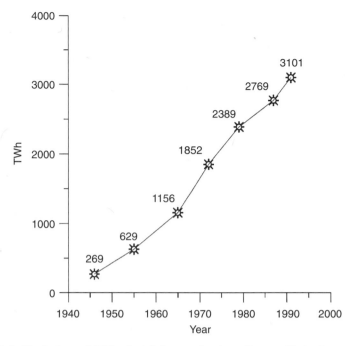

Figure 11.4. Evolution of U.S. electricity production. *Source*: Data from *Electricité*, Eyrolles, 2000.

War II the electricity consumption was very low. It has been multiplied by a factor of 25 between that time and today.

Electricity has also had a remarkable growth rate in the United States. In 1940, U.S. electricity production (269 TWh) was already a little more than half that of France's current production (about 480 TWh). Since 1940, U.S. electricity production has been multiplied by a factor larger than 10. The increase is displayed in Figure 11.4.

The great advantage of electricity is that it can be produced from any primary energy source. Figures 11.5 and 11.8 illustrate that and show the advantage that electricity has in this regard compared to heat or cold production and to transportation, which is almost entirely dependent on oil. Furthermore, electricity can be used to produce heat or cold and can also be used to meet transportation needs with electric vehicles.

11.2. ENERGY SOURCES FOR ELECTRICITY PRODUCTION

While at the world level a variety of sources are exploited for production of electricity, two-thirds of the global electricity production is from fossil fuels

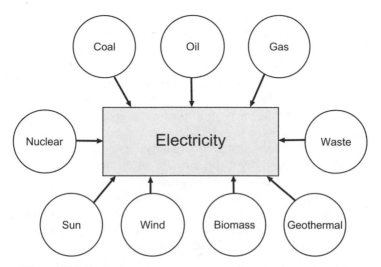

Figure 11.5. Primary energy sources used to produce electricity.

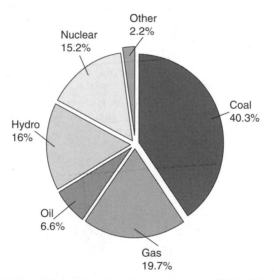

Figure 11.6. World electricity generation by fuel source, 2005. "Other" includes geothermal, solar, wind, combustible renewable, and waste. *Source*: Data from *Key World Energy Statistics*, www.iea.org, 2007.

and just over 40% is produced from coal. Coal is the largest single source. This is illustrated in Figure 11.6. Since CO_2 is emitted when fossil fuels are burned, electricity generation is a large source of greenhouse gas emissions. The other sources of energy for electricity production (nuclear, hydro, and other renewables) do not produce CO_2 during operation.

Among renewable energy sources, hydropower is the dominating one in electricity production (2809 TWh in 2004). In second place is biomass and waste, with 227 TWh. In this latter case this is probably not the best way to use this resource since it can be exploited in more valuable applications.

In Figure 11.7 the distribution of electricity generation from renewable primary sources other than hydropower is displayed. While biomass leads, wind energy has quickly become an important source of electricity due to its strong development over the last decade.

In 2004, the world's total installed capacity for electricity production was 4054 GW. Installed capacity and the amount of electricity produced are two different things because efficiencies of production are different for different sources of energy. Figure 11.8 gives the shares of installed capacity which are accounted for by different primary sources. For nuclear energy this share is quoted for electric power generation and not for thermal power generation, which would correspond to a figure three times larger. For hydropower the global capacity is equal to 851 GW, of which pumped energy storage represents 79 GW.

The share of installed capacity for power generation by other renewable energies is shown in Figure 11.9. These sources provide about 2% of the global power production. In this grouping wind energy now represents half of the production from renewable energy sources other than hydropower.

Electricity is used in many activities. The distribution of electricity consumption among different domains is displayed in Figure 11.10 for the year 2005. More than 90% of electricity is consumed in industrial use, in residential use, and for commercial and public services. Between 1973 and 2004 electricity consumption in the OECD countries increased by a factor of 2.3.

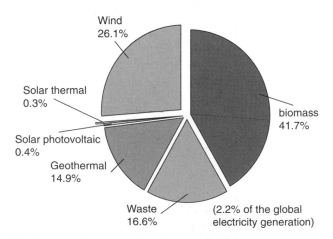

Figure 11.7. World electricity generation by renewable energy sources except hydro, 2005. *Source*: Data from *Key World Energy Statistics*, www.iea.org, 2007.

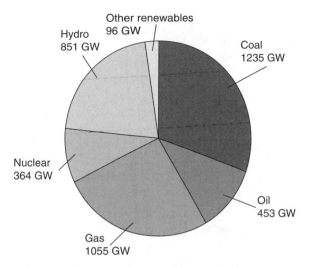

Figure 11.8. Share of installed capacity for power generation, 2004. *Source*: Data from *World Energy Outlook*, IEA, Paris, 2006.

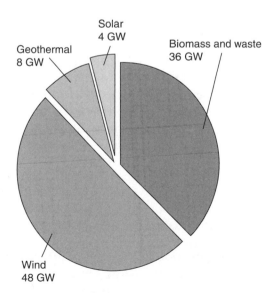

Figure 11.9. Installed capacity for electricity production by renewable energies other than hydro, 2004. This total capacity represents 96 GW. This is a small part of the global capacity of 4054 GW. *Source*: Data from *World Energy Outlook*, IEA, Paris, 2006.

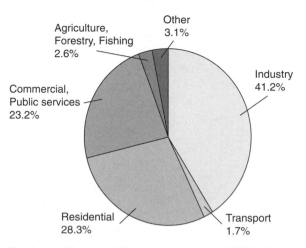

Figure 11.10. Total world electricity consumption, 2005. *Source*: Data from www.iea.org.

11.3. NO UNIQUE SOLUTION

The best way for a country to produce electricity depends upon that country and its available resources. Different countries have chosen different solutions to meet their electricity demands.

The IEA distinguishes between gross electricity production, which is measured at the alternator level of the power station, and net electricity production, which is gross electricity production minus the electricity consumed within the station. Net electricity production is the quantity of electricity leaving the power station. The difference between the two depends upon the type of power plant. It is about 7% in case of conventional thermal stations, 6% for nuclear plants, and 1% for hydro stations. In the case of hydropower stations the amount of electricity produced from pumped storage is included.

In Figure 11.11, net U.S. electricity production in 2004 is broken down by source. The production was dominated by fossil fuels. Renewable energies other than hydro remain very small. As a consequence, power generation leads to large CO_2 emissions.

A completely different situation is observed in France where almost 90% of electricity production is made by energy sources which are CO_2 emission free (Figure 11.12). Nuclear energy, which produces 78% of the electricity, was chosen in the 1970s after the first oil shock to obtain some independence from oil. In view of recent increases in the price of oil and of the perceived needs to reduce greenhouse gas emissions, this appears to have been a good choice.

In contrast to France, almost two-thirds of the electricity generated in Germany is derived from fossil fuels (Figure 11.13). Germany has large coal resources which are used to produce electricity. Coal accounts for 79% of the total fossil fuel contribution.

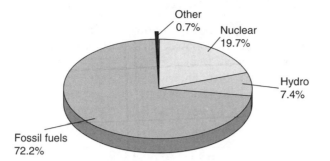

Figure 11.11. Net U.S. electricity production, 2004. "Other" includes other renewable energy except hydro. *Source*: Data from *Electricity Information*, OECD/IEA statistics, Paris, 2006.

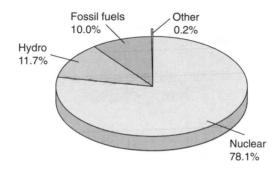

Figure 11.12. Net electricity production for France, 2004. "Other" includes other renewable energy except hydro. *Source*: Data from *Electricity Information*, OECD/ IEA statistics, Paris, 2006.

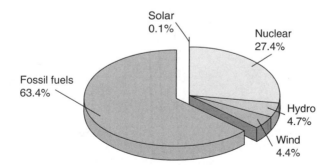

Figure 11.13. Net electricity production for Germany, 2004. *Source*: Data from *Electricity Information*, OECD/IEA statistics, Paris, 2006.

   In Denmark 60% of the electricity is derived from coal and fossil fuels generate more than 80% of the electricity (Figure 11.14). Wind provides 17.2%, but this small country is close to having saturated the land available to it for windmill installation. Installing windmills in the sea remains an option.

When it is available, hydropower is an optimum choice for production of electricity. However, hydropower cannot usually meet the demand of a whole country and needs to be complemented by other means of production. For example, in Sweden, which has large hydropower resources, 40.2% of the electricity is produced by hydropower. Hydropower and nuclear energy together produce more than 90% of Sweden's electric power. The split between different sources is displayed in Figure 11.15.

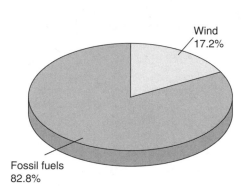

Figure 11.14. Net electricity production for Denmark, 2004. *Source*: Data from *Electricity Information*, OECD/IEA statistics, Paris, 2006.

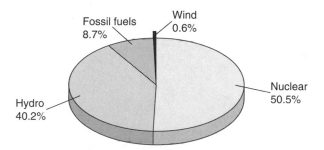

Figure 11.15. Net electricity production for Sweden, 2004. *Source*: Data from *Electricity Information*, OECD/IEA statistics, Paris, 2006.

Hydropower is also very important in Canada and produces close to 60% of the electricity of the country (Figure 11.16).

Finally, Figure 11.17 shows the sources of production of electricity in China for the year 2005. In China coal dominates power generation. The present rate of construction of new coal-fired plants in China is approximately one a week.

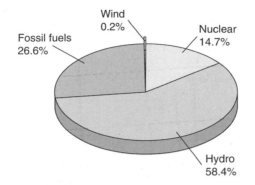

Figure 11.16. Net electricity production for Canada, 2004. *Source*: Data from *Electricity Information*, OECD/IEA statistics, Paris, 2006.

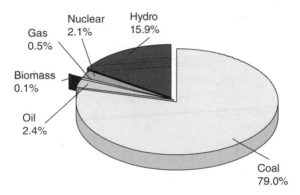

Figure 11.17. Electricity production in China, 2005. *Source*: Data from www.iea.org.

11.4. FROM MECHANICAL ENERGY TO CONSUMER

Mechanical energy generated by a primary energy source (fossil fuel, nuclear, wind, etc.) is transformed into electricity using an alternator. The mechanical energy rotates the moving part of the alternator, which is called the rotor. The rotor consists of a magnet which rotates inside the stator, which is the fixed part of the alternator and consists of a set of conductor coils on an iron core. The frequency output of an alternator depends upon the number of poles of the magnet and on the rotational speed. The efficiency of an alternator is high, of the order of 95%, and can reach almost 99% for large units of 1000 MW. Even with such a high efficiency, the amount of heat generated by the small losses corresponds to a large dissipated power, of the order of 1 MW, and cooling of the different parts of the alternator stator is required.

A great deal of recent progress has been made in the generation of electricity from coal. Before World War II, 800 g of coal was needed to produce 1 kWh of electricity in a coal-fired plant. Today, half of this quantity of coal is enough.

Electricity must be transported from the place where it is produced to places where it is consumed. This is done on a wire transmission network. In the early 1900s, Thomas Edison tried to develop a DC transmission system but did not succeed in transmitting electricity for long distances. Nikola Tesla developed the AC power transmission and distribution technology that we still use today. Nikola Tesla was also the inventor of polyphase transformers.

Power lines can transport alternating current over long distances at low cost provided the voltage is high enough. Transporting energy in the wires leads to losses due to the Joule effect, that is, dissipation as heat. Losses are smaller at higher voltages. For an alternating current change of voltage can easily be done using transformers. The transformation efficiency from one voltage to another is close to 100%. Transformers are able to raise and lower the voltage between the production site and the site of consumption. Different standards for the transmission voltage and current frequency have been chosen in different parts of the world. In the United States the standard is 120 V, 60 Hz. In Europe it is 220 V, 50 Hz.

In France there are about 100,000 km of high-voltage power lines. Of these 47% correspond to very high voltage (400 and 225 kV) and 53% to high voltage (63 and 90 kV). Several types of pylons are used to support the wires. Beaubourg pylons for the 400-kV power lines are 50 m high and weigh 45 t. For the 225-kV power lines they are 42 m high and weigh 20 t. The energy losses due to the joule effect are of the order of 2.5% in France. This is about 12 TWh per year or equivalent to the production of one and a half nuclear plant.

Current is mostly transported in the form of three-phase current. The advantage of three-phase current is that it is less costly in terms of generators and wires. The distribution is more efficient than for a single-phase alternating current. Furthermore many industries require three-phase power. For very large distances, the problem of AC lines is that they radiate energy because they act like antennas. This why, in some cases, direct current is used to transport power. There are such examples in the west coast of the United States.

The ability to transport electricity over large distances spurred the development of centralized power generation systems rather than local power

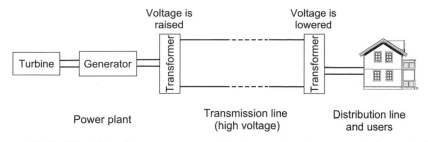

Figure 11.18. Principle of a power system showing production, transmission, and distribution of electricity to end users.

generation systems. Centralized generation has existed since the end of the nineteenth century. The sketch of a power system going from the power plant to the consumer is shown in Figure 11.18.

It is costly and difficult to store large quantities of electricity and to make it immediately available on demand. Therefore, strong attempts are made to balance the supply and the demand on a real-time basis.

11.5. IMPACT ON ENVIRONMENT

Since fossil fuels, especially coal, are employed for a large part of electricity production, one of the largest environmental impacts of electricity production results from CO_2 emission. The details of the impact of CO_2 and of other pollutants on the environment are addressed in other chapters. Here we simply note that, in our opinion, the impact is such that the development of CO_2 free-energy sources for electricity production, that is, renewable energies and nuclear energy, should be accorded high priority.

Another impact of electricity production is the thermal energy released into the environment during power production. Thermal power plants need water for cooling. If the thermal plant employs an open cooling system, the water which is used for cooling is released at a larger temperature than it had at intake. For a 600-MW fossil fuel plant the rate of water flow is about $10 \, \text{m}^3/\text{s}$, leading to an increase in temperature on the order on $12 \, °C$. In the case of a nuclear reactor of the PWR type, a 1300-MW_e power plant needs $42 \, \text{m}^3/\text{s}$ of water for cooling and produces an increase of the water temperature between 11 and $15 \, °C$. The impact on local temperature changes on aquatic life and vegetation is a complicated issue. Both positive and negative impacts have been observed.

Transporting, distributing, and using electricity has relatively little direct impact on the environment. Studies of the possible effects on human health of electromagnetic fields associated with power transmission lines have not produced conclusive evidence for such effects. This it is mostly the visual impact, which can be environmentally significant because poles or pylons and wires are needed. Placing the wires underground could minimize the visual effect. This is possible but generally expensive and is done only in some particular cases. Since this solution can involve huge costs, the real issue is to decide whether or not the money is more usefully spent elsewhere.

11.6. COST

The cost of electricity depends on the energy source and technology used to generate it. In the case of centralized production of large quantities of electricity, hydropower and nuclear power are the most competitive energy sources. In Europe prices at the power plant on the order of 3 eurocents per kilowatt-hour can be obtained. Hydropower can even be cheaper in some cases. Gas-turbine combined-cycle plants were very competitive earlier in this century, but increase in the cost of natural gas make this technology presently less appealing than some others.

Delocalized electricity production based on renewable energies may be competitive in some cases and should become more and more competitive in the future. The advantage of such delocalization is less in developed countries where a distribution grid already exists than in developing countries with no grid in place. In this latter case, depending on population density, it may be cheaper to use delocalized sources of energy (solar or wind) than to build a grid and use centralized power generation.

The price of electricity depends upon the type of consumer and on the country. In the OECD area, prices for industrial consumers vary significantly from country to country. According to the IEA, the lowest price is in Norway (4.3 U.S. cents per kilowatt-hour in 2005) while it is the highest in Italy, with 16.1 U.S. cents per kilowatt-hour in 2004. On average the price for OECD industrial consumers was 6.5 U.S. cents per kilowatt-hour in 2005. Electricity for residential consumers has an even greater range of prices. The prices per kilowatt-hour go from 7.1 U.S. cents in Norway to 29.5 U.S. cents in Denmark. The average in the OECD countries is 11.3 U.S. cents per kilowatt-hour. Prices in 2005 increased by 2.5% compared to 2004.

In Figure 11.19 the average price in 2007 is shown for selected countries. A large variation is observed.

The cost given for electricity is usually an average cost, but the price of electricity can vary quite a lot over a 24-h period, sometimes by more than a factor of 10. It can also depend upon the season. The reason is that the production of electricity is managed to meet the demand because storing large quantities of electricity is not easy and is costly. Storage of large quantities of energy using pumped energy storage would be very useful with nuclear power plants, for example, where it is difficult to rapidly change the output power over a large range. In this case electricity produced during night, when the demand is low, can be stored and recovered for use during peak hours.

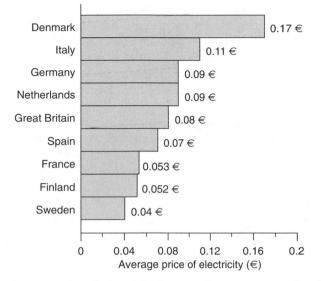

Figure 11.19. Average price of electricity in some European countries. *Source:* Data from NUS Consulting Group, 2007.

The capital coefficient of a given economic domain is equal to the ratio between the amount of money which is necessary to invest in order to produce a given value of goods and the value of the goods produced. For example, if $2 of capital is needed to produce $1 of goods, the capital coefficient is equal to 2. Electricity is a capital-intensive industry. The capital coefficient is of the order of 11 as evaluated in a British study. It is pretty much the same for the gas and water industries. For transportation and communications the capital coefficient values are a little less, around 4. Normal commercial operations are less capital intensive with capital coefficients on the order of 0.7.

11.7. CONCLUSION

Electricity is a convenient energy vector which becomes ever more important in modern societies. With the expected increases in the use of heat pumps, the introduction of plug-in hybrid vehicles or the use of electricity to produce hydrogen by electrolysis for second-generation biofuels as well as other potential new uses, we can expect that the demand for electricity will continue to increase at a speed greater than that of the total primary energy demand.

When a population density is large enough, centralized power plants are a good solution to meeting electricity demand because they allow a very economic production of electricity. Producing power with big plants necessitates having a grid to distribute electricity to the end users. Although significant improvements to aging networks are clearly needed, such a grid already exists in modern countries. In centralized power production, the total installed power capacity can be smaller than in the case of individual installations because access to many consumers allows a smoothing of the electricity demand. Typically between 7 and 10 times less installed power is required in centralized power generation systems compared to delocalized production systems. When the individual units of a delocalized system can be started and stopped on demand, this is not much of a problem except for the cost of the initial investment.

In countries with a small population density and with no grid, producing electricity in a decentralized manner is the best solution because it is quite costly to build or extend the grid. In this case storage issues are of utmost importance when intermittent energy sources are used. Where a grid also exists, one may also consider selling any extra electricity produced to the grid. However, in many situations, especially in small countries where the time is the same over the entire country (no time lag), it will be difficult to sell the electricity because most of the people need it during the same time period.

CHAPTER 12

Energy Storage: Weak Point of Energy Supply Chain

Oil, gas, coal, and wood contain energy which can be released by burning and be used to heat or do work. Thus, all of these can be considered to be vehicles for storing energy. In that same way, a liter of gasoline can be thought of as a storage system for 10 kWh of energy that can be used in a car engine to propel an automobile. The fossil fuels are storage vehicles for energy that was released from the sun millions of years ago. Renewable biomass is a similar storage vehicle for solar energy released more recently and captured in the photosynthesis process. Energy is also intrinsically stored in uranium nuclei from which one can produce heat and then electricity in a nuclear power plant.

Other energy sources, direct solar energy or wind energy, for example, may be available only intermittently. Direct solar energy is received only during the daylight hours and its intensity varies with time of the day and season. Windmills provide electricity only when the wind is blowing above a certain velocity. The ability to store the energy produced by such sources is essential for meeting our energy demands. We should store energy when it is being produced in excess and use it as required to help meet later energy needs, much in the same way that we store food to meet our future nutritional needs. To do this requires the development of efficient and flexible energy storage technologies. This is more difficult than might first be thought. In this chapter we present an overview of the current status of energy storage systems.

Modern civilization relies increasingly on electricity, a convenient energy vector to carry energy from the power plant to the consumer, and there are now so many devices requiring electricity to operate that it would be difficult to live without it. Citizens of developed countries have come to expect it to be available on demand. A major problem with electrical power is that it cannot be stored in the grid. Absent some means of storage, the production rate of electrical power at any given time should be equal to its rate of consumption. This requires a delicate balance. On a daily basis, electricity demand

Our Energy Future: Resources, Alternatives, and the Environment
By Christian Ngô and Joseph B. Natowitz
Copyright © 2009 John Wiley & Sons, Inc.

can vary dramatically with time. As an example of this, the diurnal variation of electricity demand in France is illustrated in Figure 12.1. Demand is large in the morning, when people get up and begin working, and in the evening, when they switch on their lights and their television sets. In contrast, at 3–4 A.M. the electricity demand is at a minimum because most people are sleeping and most industries or service facilities are not functioning.

In France, the installed electric power (maximum power capability) was equal to 116 GW in 2005. This was required to ensure a reserve capacity which could meet peak demands anytime during the year. The total electricity production in 2005 was 549 TWh, corresponding to an average power requirement of about 63 GW. Typical households have a distribution capability of 6 to 12 kW of electrical power. This power capability is necessary if a number of appliances are operated at the same time but, on average, the power needed is about 1 kW.

Consumers use considerably more electricity when the weather is very hot or very cold. Large heating or air-conditioning needs lead to large power demands and the power system must be dimensioned to meet these demands. This again implies the need to oversize the installed power capabilities in order that peak demands can be met. One means of mitigating this problem is the

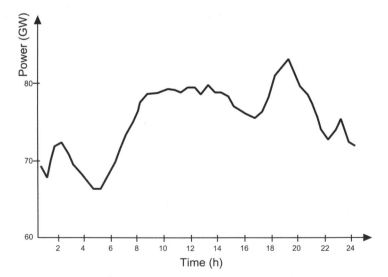

Figure 12.1. Typical evolution of required power during one day in France. The minimum power requirement occurs about 5 A.M. and the power requirement is largest near 7 P.M. *Source*: Data from www.rte.fr.

construction of peak-demand fossil-fired power plants that can be brought into operation quickly when needed. Unfortunately, in addition to requiring great investments, such plants have another serious drawback. They emit a lot of CO_2 and other pollutants.

Since it is not easy to modulate the output power of many power plants, storage of electricity provides an attractive alternative to meet peak demand. Storing large quantities of electricity would allow the leveling of power generation and decrease the installed power requirements. From the point of view of a power-generating facility this would introduce important economic efficiencies, allowing it to produce and store electricity during the off-peak hours and sell it during the peak hours.

Under certain conditions, it is also important that energy be storable as heat (or energy deficits as refrigeration) because these forms of energy cannot be easily transported over long distances. Such capabilities also have important implications in the global energy consumption picture. For example, hot-water tanks using electricity allow leveling of electricity production since the water may be warmed during the night when the electricity demand is normally lower.

12.1. ELECTRICITY STORAGE

Storage capabilities for electrical power are needed for a variety of applications at all steps of the energy supply chain, that is, production, transport, and distribution. They are particularly important in remote places powered by intermittent renewable energy sources which are not connected to the electrical grid as well as for stand-alone applications (e.g., photovoltaic systems in isolated places) or for systems connected to the electrical grid but requiring an uninterruptible power supply.

What are the main reasons to store electrical energy? The first is to minimize the total amount of installed power that is needed. Storing large quantities of electricity allows power-generating facilities to store electricity during off-peak hours, when it is cheap and the demand is low, and used to meet daily fluctuations and peak demands (Figure 12.2). Another advantage is that power plants can run at almost constant output power, reducing in this way the wear on their components. This is particularly useful in the case of nuclear plants which usually provide constant baseline power and are not able to respond quickly to rapid changes in demand.

The second reason is that, at any given time, the electricity supply must exactly match the demand. Any interruption in the supply may lead to outages. Computers are particularly sensitive to power fluctuations and many systems are now controlled by computers. Serious breakdowns of sophisticated control systems may occur with a nonnegligible impact on the economy. In the United States, which has an aging electrical power grid, it is estimated that the cost of electrical outages reaches about $80 billion each year, although two-thirds

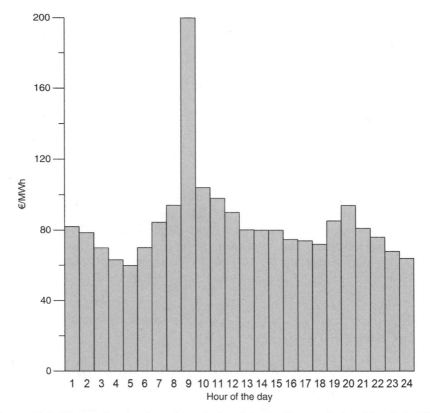

Figure 12.2. Fluctuation in the price of electricity in France during March 6, 2008. There is almost a factor of 3 difference between the lowest and the highest prices. *Source*: Data from www.powernext.fr.

of the interruptions last less than 5 min. Electricity storage capabilities supported by sophisticated control systems can mitigate short-duration power fluctuations and can provide a back-up source of energy during longer interruptions.

Some industries, for example, manufacturing plants for integrated circuits, require uninterruptible power supplies as well as stable voltages and frequencies. They would be much less vulnerable to interruptions if suitable electricity storage systems were available.

A third consideration is the rapid pace of development of wind and solar energy sources. These energies are renewable but inherently intermittent and subject to large fluctuations. Storage of electricity allows a smoothing of these variations in output. Also, electricity unused at a given time can be used at later times to meet peak demands.

Such storage is possible but often difficult and expensive when large quantities of electrical power are needed. Physical and chemical restrictions combine to make progress in the development of the storage capabilities needed to meet consumer demands slow. As a result the storage of electricity is really the weak link in the energy supply chain.

Unlike the case for coal, oil, or gas, it is not easy to store large quantities of electricity. For most storage purposes it has to be transformed into another form of energy: potential, chemical, and so on. An illustration of the problem is as follows: Using pumped hydro techniques to store the amount of energy equivalent to the average consumption of a refrigerator used in France (380 kWh/year) would require the transport of almost 1400 tons of water to a height of 100 m, even more if we include a reasonable estimate of the efficiency of transformation.

12.1.1. Characteristics of Electricity Storage

In storage of electrical power, energy density, the amount of energy stored in a given volume or a given mass, is the key parameter. The energy that can be recovered from a storage system depends upon this energy density but can also depend upon how the system has been charged and how it is discharged.

When electricity storage systems are used, the overall efficiency of the storage and recovery operations determines the economical viability of the process. This overall efficiency is the product of the efficiency corresponding to the storage step in which electricity is used to fill the storage system and the delivery efficiency of the storage system. The latter is determined by the ratio between the energy the system provides and the energy that it contains. Depending upon the system, the yield may vary between 50 and 90%. The timescale associated with the charging and discharging operations can also be of importance. An important concern is the self-discharge of some systems in which the energy content decreases as a function of time even if they are not used.

Gasoline has an energy density of about 41 MJ/kg, or 34.6 MJ/L. The energy density of TNT (the explosive) is 4.2 MJ/kg, or 6.9 MJ/L. A lithium ion battery has an energy density on the order of 0.7 MJ/kg and that of a lead battery is ~0.1 MJ/kg. In the latter case, the practical recovery efficiency is in the range of 75–85%.

The number of times one can store and retrieve electricity, the cyclability of the storage device, is also important and determines the storage system's lifetime for a given application. This quantity is not always well defined because it can depend upon the depth of the discharge. For electrochemical batteries, the typical cyclability for a full charge–discharge process is of the order of 500–1000 cycles depending upon the technology. This figure can be increased if a full discharge is not reached. In hybrid vehicles, for example, the battery is never completely discharged. At most only a few tens of percent of its capacity is used. This mode of operation greatly extends the lifetime of the battery. Since the battery is an expensive component of the hybrid car, this is a particularly cost-effective strategy.

The time needed to completely charge a battery or another storage system is also a parameter to consider, especially in transportation applications. A rapid charging may sometimes be worse than a slow one, leading to greater degradation of the battery components. In such situations a more practical solution may be to replace the spent battery with a spare one and then recharge spent battery over a longer time period.

Safety of storage systems is also a very important issue. Some Li ion batteries used in cell phones or laptop computers have been known to explode. Even if the probability of such an accident is small, this is unacceptable because injuries may result. Lithium–iron phosphate cathodes are promising in terms of safety.

12.1.2. Large-Quantity Storage Technologies

For electricity storage there are currently only a few technologies capable of storing large quantities of electricity, especially for grid applications. The two well-established technologies for this purpose are pumped hydroelectric storage and compressed air energy storage. The first one is the one that is most extensively used at present. A third technology under development is thermal energy storage. This last can be coupled with heat pumps to increase the efficiency of the storage system.

Pumped Hydroelectric Storage

Pumped hydroelectric storage is a mature technology that has been used for almost 80 years. For about half a century it was the only technology available to store large quantities of electricity. The principle is displayed in Figure 12.3. During off-peak periods excess electricity is used to pump water, from a base-level reservoir to a reservoir at a higher level. This is the storage step in which the surplus electrical energy is transformed into gravitational potential energy. A large amount of water is required. When needed to produce electricity the water from the upper reservoir is allowed to flow down through a hydraulic turbine. A volume of 3600 liters of water dropping a distance of 100 m produces 1 kWh of electricity. Powers of about 1 GW can be obtained in this way. Since the electricity stored can be recovered quickly, it is a technology that is well adapted to dealing with fluctuations in demand and to meeting peak

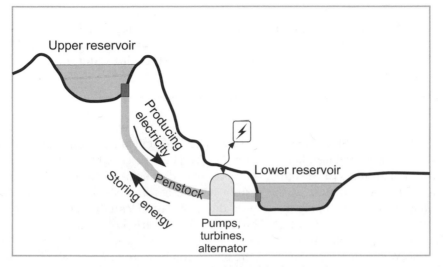

Figure 12.3. Schematic principle of pumped hydroelectric storage system.

demands at certain hours of the day. Pumped hydroelectric storage systems are costly to construct and often require long lead times. In this way they are similar to conventional hydraulic systems that are designed to produce electricity directly.

The gravitational potential energy of the stored water is $E = mgh$, where m is mass of water stored, g the gravitational acceleration constant, and h the elevation (the vertical distance between the upper and lower reservoirs). The power P is the derivative of the energy and is given by $P = (dm/dt)gh$, where dm/dt represents the rate of the water flow. The energy (power) per unit of mass (flow) is just equal to gh. This gives, with typical elevations, energy densities on the order of 0.2–2Wh/kg and powers between 10^2 and 10^3W/kg.

Pumped hydroplants can provide high power levels and long discharge times and, as already mentioned, are especially suited in situations where a rapid change in the supply of electricity is needed. During hydroplant operation the use of adjustable speed pumps and turbines allows the power output to be changed in 10–30ms. An efficiency of about 70–80% can be obtained for the cycle of pumping water into the upper reservoir and releasing it back to the lower reservoir.

The installation at Grand-Maison, France, can store $137 \times 10^6\,\mathrm{m^3}$ of water in its 2.2-km^2 upper reservoir. The head is quite high: 955m. A total potential energy of 1.3TWh is stored and can be produced by completely emptying out the upper reservoir. A flow of 27m^3/s leads to a power of 1.8GW.

The Adam Beck II hydroelectric power station in Niagara Falls, Ontario, Canada, diverts water from the Niagara River above the falls through underground pipes. A reservoir was made to store this water during the night for use during the day.

In the United States, there are 38 pumped hydroplants providing 19 GW of power. In Japan the installed power is 24 GW. The efficiency ranges from about 60% in older units to 78% in new ones. The hydraulic head varies between 30 and 650 m. A cold start of the system can be accomplished in 1–4 min. The switch between pumping (storage) and generation can be carried out in 5–30 min. Capital costs of the older facilities were below $100/kW. New facilities are expected to have a much higher capital costs (from $1000/kW to $2000/kW). In that range of costs, the use of combustion turbines (see Chapter 2) becomes an economically competitive alternative to meeting fluctuating demands.

Pumped hydro is a technology widely used in France. It is interesting to have a quick look at the effect which deregulation of the electricity market has had on the cost of this storage technology. Before deregulation, Electricité de France (EDF), owned both the power plants and the grid used to transport electricity. As an effect of deregulation, the management of the grid has been removed from EDF and taken over by a new company, Réseau de Transport d'Electricité (RTE) EDF now has to pay to transport electricity from its power plants to the pumped hydro facility in order to store it there. It must pay again to transport electricity produced by the pumped hydro facility to the consumer. In the end, this new arrangement has resulted in an increase of the cost of pumped hydro storage. Although deregulation is usually considered as a means to increase competition and reduce costs to the consumer, economies of scale may be lost, leading to a different outcome.

Capital costs of pumped hydro storage facilities are large and long times are needed for their construction. Operating costs are low. The current global capability of such facilities is 90 GW, which corresponds to about 3% of the world's electricity generation capacity. In Figure 12.4, we show the number of large pumped hydro installations with a power storage capability larger than 1 GW. They account for a total power storage capability of 57 GW.

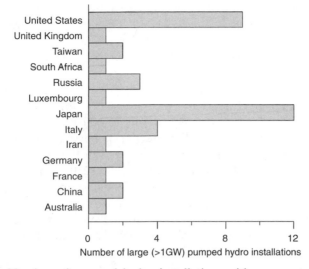

Figure 12.4. Number of pumped hydro installations with power storage capability larger than 1 GW. *Source*: Data from http://electricitystorage.org.

Compressed Air Energy Storage

Compressed air energy storage is another method to store electricity during off-peak periods. Electricity is used to compress air and store it in an airtight underground container. The air container may be an aquifer, a porous rock formation, or a cavern in underground salt domes. When needed, the compressed air is expanded through a turbine to generate electricity. Compressing air dissipates energy into heat. Existing compressed air energy storage facilities employ a diabatic storage cycle. In this case the heat produced in the compression phase is released into the environment. In the unloading phase, some heat is required because the expanding air cools down. The air is heated by an extra source of energy, for example, a natural gas-fired burner, prior to expanding in a turbine. The efficiency for these facilities is typically around 50%, of the same order of magnitude as using natural gas in a combined-cycle turbine.

There are, in principle, other ways to use compressed air energy storage. They differ in the way in which the heat (or heat loss) produced in the compression (or expansion) phase is managed. The best method would be to employ an adiabatic storage cycle in which the heat produced in the compression phase is stored and used in the expansion phase. The heat could be stored in either a solid (e.g., concrete, stone) or liquid (e.g., molten salt, oil) heat reservoir. A good efficiency (larger than 70%) is expected for such a system. However, there is at present no facility that uses the adiabatic storage technique, which is costlier to construct than the facilities presently in operation.

In a third possible method, the compression and expansion phases could be carried out isothermally, that is, at constant temperature. The required thermodynamic transformations are slow, and thus this technique may be limited to small power storage systems, but the efficiencies for such systems could be very high.

There are presently two compressed air energy storage facilities in the world. One, located in Huntorf in Germany, was commissioned in 1978. It employs two cylindrical caverns at a depth of 600–800 m. The caverns are around 200 m high and 30 m in diameter, each having a volume of about 150,000 m^3. The pressure of the compressed air is in the range of 50–70 bars. To produce 1 kWh of electricity in the output phase, 0.8 kWh of electricity is required in the storage phase and 1.6 kWh of heat from natural gas must be supplied to the air before it goes into the turbine. A power of 290 MW can be produced over a 2-h period. It is a hybrid storage system in the sense that the extra source of energy (natural gas) is needed. The basic principle of such a compressed air storage facility is displayed in Figure 12.5.

The second facility is located in McIntosh, Alabama, in the United States. It was commissioned in 1991. Its storage chamber is a 538,000-m^3 salt cavern located at a depth between 450 and 750 m. The pressure range of 45–76 bars is similar to that of the Huntorf facility. It can produce 110 MW for 26 h. To produce 1 kWh of electricity it requires 0.7 kWh of electricity in the storage phase and 1.2 kWh of energy from natural gas in the output phase. This higher efficiency of the McIntosh plant compared to the Huntorf results from the fact that it uses the exhaust gases with a heat exchanger to preheat the compressed air.

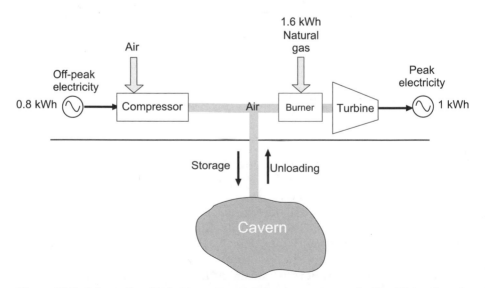

Figure 12.5. Schematic of hybrid compressed air energy storage facility. Natural gas is used to heat the air in the output phase.

The capital costs of compressed air energy storage facilities vary between about \$400/kW and \$750/kW depending upon the size of the installation and the geology of the storage chamber. For small storage units, manufactured chambers such as buried pipes could be used as storage chambers. These pipes could be buried in the subsurface and installed almost anywhere.

Thermal Energy Storage of Electricity

Huge storage capacities can be realized using thermal storage. The basic idea is to maintain two different reservoirs at different temperatures. In the storage phase, electricity is used to drive a heat pump that takes energy from the low-temperature reservoir and stores it in the high-temperature reservoir. In the unloading phase, the heat of the high-temperature reservoir is transformed into mechanical energy in a thermal engine. Since the overall efficiency for conventional thermal energy storage systems is not expected to be high, new heat pump technologies for which the global yield can be over 70% are being developed by the SAIPEM Company of Milan, Italy. This effort is currently at the project stage. Figures 12.6 and 12.7 summarize the operation of a typical installation. The medium- and high-pressure vessels are filled with a porous solid that allows circulation of gas and exchange of heat. Argon is employed as the cycling gas. The energy flows are indicated assuming a 90% yield for the compressors and expanders and an amplification factor of 3 for the heat pump.

As seen in Figure 12.6, the bottom of the medium-pressure vessel is cold while the upper part is at a medium temperature. The top of the upper part of the high-pressure vessel is at high temperature while the bottom part is at ambient temperature. During the storage phase, the solid of the medium-pressure vessel progressively cools down while the high-pressure vessel heats up. The thermal front therefore moves upward in the medium-pressure vessel

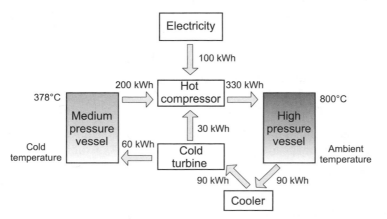

Figure 12.6. Principle and energy flow for storage phase in thermal energy storage technology. *Source*: From J. Ruer, *Saipem, Lettre energie des techniques de l'ingénieur*, No. 12, 2008.

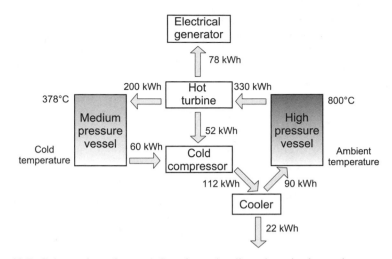

Figure 12.7. Schematic and energy flow for unloading phase in thermal energy storage technology. *Source*: From J. Ruer, *Saipem, Lettre energie des techniques de l'ingénieur*, No. 12, 2008.

while it moves down in the high-pressure vessel. In the unloading phase the thermal front moves in the opposite direction.

In Figure 12.7, the yield from the hot turbine to the electrical generator corresponds to only 24% (78/330). However, compared to the input in the storage phase (100 kWh), the yield is 78%. This is so only because a heat pump system is used to pump heat from the cold source to the hot one. Using 100 kWh from the grid, 200 kWh can be pumped from the medium-pressure vessel and 30 kWh from the cold turbine. This trick allows an increase of the yield of the thermal energy storage system leading to efficiencies similar to pumped hydro storage. With the present technology and facilities working at around 800 °C, efficiencies above 70% can be reached with a storage capacity between 35 and 50 kWh/m³. Increasing the temperature to 1000–1200 °C would increase the efficiency beyond 80% and allow storage of about 60–100 kWh/m³.

Because it is not limited to particular geographic locations with appropriate geologic features, thermal storage combined with heat pump technology is a very appealing solution to the energy storage problem.

12.1.3. Electrochemical Batteries

Batteries constitute a huge market, about $40 billion/year. Of this about $34.5 million is spent on primary batteries and about $5.5 billion on secondary batteries. In batteries, chemical energy is directly converted into electricity. A battery basically consists of an anode, a cathode, and an electrolyte. A separator containing the electrolyte separates the anode region from the cathode region to prevent the physical transport of the active materials (the oxidant and the reductant) from one electrode to the other. Batteries have been used since the middle of the nineteenth century and the lead battery, still an important component of the modern automobile, was commercially available by the end of the nineteenth century. The great advantage of batteries is that they can provide electricity at any time and any place. They have become ever more useful in the modern age, which has seen an astounding increase in the availability of portable electronic devices.

Primary batteries are those that can only be used once. The chemical energy contained in the active materials is transformed into electricity on demand. The active materials are chosen to obtain high thermodynamic potentials because the output voltage is closely connected to this quantity. Once the chemical energy is exhausted, the battery cannot be used again. Primary batteries are not primary energy sources in the sense that energy is needed to manufacture them. The amount of energy needed for battery fabrication is much larger than the amount of energy that can be recovered from them. For some batteries the ratio between the energy necessary to manufacture them and the energy that can be recovered can reach a factor of 100. In secondary batteries or accumulators, the depleted oxidant and reductant can be regenerated when the battery is charged with electricity. Obviously, this secondary battery technology is much more interesting in electricity storage applications.

Modern society is increasingly turning to portable devices (laptops, cellular phones, walkmen, etc.) for communication and entertainment needs. The energy consumption of these electronic devices increases as their power to treat information increases, requiring more energy to function. More powerful rechargeable batteries are needed to meet this demand. Electricity provided by batteries is quite expensive, but the utility and convenience of batteries is often more important than the price per kilowatt-hour.

12.1.3.1. Primary Batteries

The Leclanché cell, based on zinc–manganese dioxide, was invented in 1866. It is made of a positive electrode of carbon, a depolarizer of manganese dioxide (MnO_2), an electrolyte of ammonium chloride (NH_4Cl), and a negative electrode of zinc. The voltage is 1.5V and the volume energy density is around 180Wh/L. A metal button at the top and the metal bottom are the positive and negative contacts that are used to tap the device (see Figure 12.8).

The French engineer Georges Leclanché (1839–1882) is the father of modern batteries. He developed a battery which was more powerful and user friendly than previous ones. In 1866 he patented the Leclanché cell, made up of a carbon cathode, a zinc anode, and an ammonium chloride electrolyte. He later took out a patent for a $CuCO_3$ battery and won distinction at the universal exhibition of 1867 with a manganese battery. This latter invention was adopted by the Belgian telegraph administration and the railway system in the Netherlands.

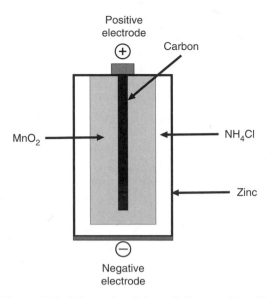

Figure 12.8. Schematic of dry cell (Leclanché cell).

Zinc–carbon batteries are less expensive batteries but also have a smaller energy density. They are derived from the wet Leclanché cell. In a cylindrical zinc–carbon cell (Figure 12.8), the outer part of the container is made of zinc. Along the symmetry axis, in the middle of the cylinder, there is a carbon rod surrounded by a mixture of manganese dioxide and carbon powder. At the cathode, the zinc is oxidized into Zn^{2+}. At the anode, MnO_2 is reduced into Mn_2O_3. Except for low-cost applications, for example, provision of batteries included in brand new electrical devices, these batteries have largely been supplanted by alkaline batteries.

Alkaline batteries have a normal voltage of 1.5 V and are widely used to power electrical devices. Alkaline batteries use potassium hydroxide (KOH) instead of ammonium chloride or zinc chloride. They are able to deliver more current and have a longer shelf life before use. The half-cell reactions are

$$Zn + 2OH^- \rightarrow ZnO + H_2O + 2e^- \quad \text{at negative electrode}$$

$$2MnO_2 + H_2O + 2e^- \rightarrow Mn_2O_3 + 2OH^- \quad \text{at positive electrode}$$

leading to the global reaction

$$Zn + 2MnO_2 \rightarrow ZnO + Mn_2O_3$$

The capacity of an alkaline battery depends upon the way the electricity is drawn from it (Table 12.1). At low power an AA battery can have a capacity of 3000 mAh. With a load this capacity can decrease to 1000 mAh. It is not easy

TABLE 12.1. Capacities, Energies, and Costs of Selected Batteries

	Alkaline				Camera Battery, 6V Lithium
	AAA Cell	AA Cell	D Cell	9V	
Capacity	1.1 Ah	2.8 Ah	17 Ah	0.6 Ah	1.4 Ah
Energy	1.6 Wh	4.2 Wh	25 Wh	4.2 Wh	8.4 Wh
Cost per kWh	$400	$120	$80	$240	$1200

Note: Assumes the nominal voltage. The voltage falls during the discharge and the real energy is lower than the nominal value. Higher costs per kilowatt-hour are common with smaller or specialized batteries, e.g., those used in hearing aids.

Source: Data from www.BatteryUniversity.com.

to recharge an alkaline battery and in practice this is not done except in special cases.

Alkaline batteries currently dominate the primary battery market. They can be stored up to 10 years and their energy densities are high.

Metal–Air Batteries Metal–air batteries are interesting devices sometimes known as "half fuel cells." The anodes are made of metal, such as zinc or aluminum. The cathodes are air electrodes made of porous carbon or metal covered with a catalyst. The electrolyte is a good conductor for OH^- ions, for example, KOH or a polymer membrane saturated in KOH. These batteries have high energy densities. Recharging them is very difficult. However, they can be refueled mechanically by replacement of the consumed metal. Rechargeable metal–air batteries are under development, but the number of possible charge–discharge cycles is small and the efficiency presently reaches only about 50%.

The Zn–air battery is a rather inexpensive metal–air battery. It is used in Europe, for example, to provide electricity to metal fences in order to keep cattle inside a field. It has a positive air electrode, a negative zinc electrode, and an alkaline electrolyte (KOH). The elementary reactions at the electrodes are

$$\tfrac{1}{2}O_2 + H_2O + 2e^- \rightarrow 2OH^- \qquad \text{at cathode}$$

$$Zn + 4OH^- \rightarrow Zn(OH)^{2-} + 2e^- \quad \text{at anode}$$

In the electrolyte

$$Zn(OH)_4 \rightarrow ZnO + H_2O + 2OH^-$$

The overall reaction reads

$$Zn + \tfrac{1}{2}O_2 \rightarrow ZnO$$

It can be used as either a primary battery or a mechanically rechargeable battery. The normal potential is 1.65 V. The specific energy is around 370 Wh/kg and the energy density is 1300 Wh/L. The voltage can be slightly reduced by adjusting the flow of air and this is sometimes done for specific applications. With appropriate design, the zinc can be replenished and the zinc oxide removed continuously from the battery. In this case it is known as a zinc–air fuel cell. Such a fuel cell has sometimes been used in experimental electric vehicles.

The Al–air battery has an even higher energy density but is less advanced in terms of development. The nominal cell voltage is 1.2 V and the energy density obtained is 1300 Wh/kg. In the future 2000 Wh/kg will probably be reached. Some possible applications are for electric vehicles, cellular phones, or laptop computers.

12.1.3.2. Rechargeable Batteries
Rechargeable batteries are particularly useful in a wide range of applications because their ability to perform over many cycles of charge and discharge provides a means of supplying large net amounts of energy in an economical and convenient way (Table 12.2).

Lead Batteries   The most widely used battery is the lead battery, invented in 1859 by the French physicist Gaston Planté. The system developed by Planté

TABLE 12.2. Capacities, Energies, and Costs for Selected Rechargeable Batteries

	NiCd (6 cells)	NiMH (6 cells)	Lead–Acid (6 cells)	Li Ion (2 packs)	Lead–Acid (for Scooters and Wheelchairs)
Capacity	600 mAh	1000 mAh	2000 mAh	2000 mAh	33 Ah
Voltage, V	7.2	7.2	12	7.2	12
Energy (per discharge), Wh		7.2	24	14.4	396
Maximum life cycle	1500	500	250	500	250
Cost per kWh	$8	$20	$8.5	$20	$1

Note: Assumes the nominal voltage. The voltage falls during the discharge and the real energy is lower than the nominal value. The cost does not include the price of electricity necessary to charge the battery and the price of the charging equipment. The lowest prices per kilowatt-hour are obtained with the older technologies. Larger batteries have lower prices per kilowatt-hour than do smaller batteries. The costs per kilowatt-hour are substantially lower than those of primary batteries because several cycles of charge–discharge are possible. Proper maintenance and use of the battery are assumed. Otherwise the costs can increase notably.
Source: Data from www.BatteryUniversity.com.

was not convenient to use and the lead battery was actually commercialized in 1881 after important advances in Luxembourg by Henri Owen Tudor (who was born the year Planté invented the lead battery and died of lead poisoning in 1928) and in France by Emile Alphonse Fauré, who improved the technology, making it easier to construct and to use. Essentially the same technology is still used today.

Even before the beginning of the twentieth century, lead batteries were used to power electric automobiles. However, internal combustion engines and diesel engines soon became more competitive and electric vehicles were replaced by vehicles using these engines. In the late 1980s interest in electric vehicles powered by lead batteries was renewed. Unfortunately the range of such vehicles is low (about 100 km for a medium-sized car) and recharge times are long, between 5 and 7 h. Nickel–cadmium and Ni–MH batteries are much better in terms of energy density and offer the possibility of increased ranges for electric vehicles. However, these batteries are more expensive.

The half-cell reactions of a lead battery are

$$PbO_2 + H_2SO_4 + 2H^+ + 2e^- \Leftrightarrow PbSO_4 + 2H_2O \qquad \text{at cathode}$$

$$Pb + H_2SO_4 \quad \Leftrightarrow PbSO_4 + 2H^+ + 2e^- \quad \text{at anode}$$

The overall reaction is then

$$Pb + PbO_2 + 2H_2SO_4 \Leftrightarrow 2PbSO_4 + 2H_2O$$

The most common application of lead batteries is as starter batteries for vehicles. A great advantage of these batteries is their low cost compared to other technologies. This is about \$100/kWh for starter batteries in cars, but it can be much more expensive for smaller lead battery units. This is much cheaper than Li ion batteries, which cost about \$800/kWh.

NiCd Batteries   The nickel–cadmium (NiCd) battery uses nickel oxyhydroxide and metallic cadmium as electrodes. The NiCd batteries have a larger energy density (around 100 Wh/L) than lead batteries. Their standard thermodynamic reversible potential of 1.35 V allows a nominal operating potential of around 1.2 V, which make them suitable to replace nonrechargeable batteries in many applications. Their great advantage is that they can be recharged. Several elementary cells can be grouped to provide a larger output voltage. For example, using six cells in series gives a battery of 7.2 V. A normal charging time is 16 h, but a rapid charge can be done in 3 h or less. At the end of the charge, a trickle current may be used to maintain the full charge until the battery is needed.

The first NiCd battery was built in 1899 by the Swedish inventor Waldemar Jungner. It was not until 1947 that a completely sealed NiCd battery of the type presently used was obtained. Nickel–cadmium battery fabrication is now a mature and robust technology. More than 1.5 billion NiCd batteries are produced each year. They can be used in applications where high discharge rates and long lifetimes are required. They have low internal resistances, allowing them to generate high currents. For that reason they are very useful for applications needing instant high powers. They can also be used over a wide range of temperatures (–40 to 60 °C). However, the energy density of a NiCd battery is not as large as that of some others and Cd is a toxic metal.

Nickel–cadmium batteries have energies per unit mass ranging from 40 to 60 Wh/kg and energy densities between 50 and 150 Wh/L depending upon the fabrication technology. The cycle durability of NiCd batteries is between 1000 and 2000 cycles and the power which can be delivered reaches 150 W/kg. The number of cycles can be greatly increased if the battery is operated at lower rates of charge and discharge. For example, in applications in space, tens of thousands of cycles have been reached. They can also operate at low temperatures, which is an advantage in some situations.

Early NiCd batteries exhibited memory effects, meaning that the battery remembered how much energy was extracted in the preceding discharges. This could lead to a drop of the battery efficiency. Such memory effects have practically disappeared today. However, other effects, also called memory effects, are observed in modern NiCd batteries. These result when the very small crystals of the active cadmium material inside the cell aggregate, leading to a decrease in the useful surface area. This may result in self-discharge or electrical short circuits in the battery.

The NiMH Battery   The nickel–metal hydride (NiMH) battery is similar to the NiCd battery but uses an alloy which is able to absorb hydrogen for the negative electrode (instead of cadmium). The positive electrode is nickel oxyhydroxide (NiOOH). The main advantage of a NiMH battery is that the energy capacity is more than 30% greater than that of the NiCd battery. The energy density per unit of mass is 30–80 Wh/kg and the energy density per unit of volume is between 140 and 300 Wh/L. The power per unit mass is between 250 and 1000 W/kg. The nominal voltage of a cell is 1.2 V.

The reactions at the electrodes are the following:

$$H_2O + M + e^- \Leftrightarrow MH + OH^- \text{ at cathode}$$

Where M is metal (an intermetallic compound) and

$$Ni(OH)_2 OH \Leftrightarrow NiO(OH) + H_2O + e^- \text{ at anode}$$

The electrolyte is KOH.

Research on NiMH batteries started in the 1970s, but the original metal hydride alloys were unstable. New alloys with improved stability were developed in the 1980s. The NiMH batteries became commercially available in the 1990s.

The NiMH batteries allow 500 and 1000 cycles at full discharge and many more if only partial discharge is allowed, as is the case in hybrid vehicles. These batteries are quite expensive. Restricting operation to partial discharges leads to a large increase in the useful lifetime. In fact, a NiMH battery should not be completely discharged because damage to the battery may occur. Overcharging should also be avoided otherwise the battery can be damaged and be potentially dangerous. Battery chargers specifically designed for NiMH batteries should be used. The fabrication technology of NiMH batteries has improved rapidly. Since 2005, NiMH batteries are being manufactured with a small self-discharge rate. About 70–85% of the energy remains after a year of storage at room temperature.

Small NiMH batteries are now widely used because in most of situations they can replace alkaline batteries and have the advantage of being rechargeable many times, which significantly decreases the net energy cost. An advantage compared to NiCd batteries is that they do not contain toxic Cd. Also memory effects are small in NiMH batteries.

Large NiMH batteries are currently used in hybrid vehicles. For example, the Prius, from Toyota, has a 1.5-kWh NiMH battery weighting 39 kg. Because these batteries are never fully discharged, they are expected to have lifetimes of more than a decade.

Li Ion Batteries Lithium ion batteries are rechargeable batteries in which lithium ions move between the electrodes. Lithium ion batteries are different from lithium batteries, which are not rechargeable and have a metallic lithium anode. In Li ion batteries the anode is a material into which lithium is inserted and from which lithium is extracted. A large choice of electrode materials exists. The choice of this material leads to different characteristics for the battery. The cathode is often made of a layered oxide, a polyelectrolyte, or a spinel.

When the battery is discharging, lithium is extracted from the anode, moves from the anode to the cathode, and is inserted into the cathode. The inverse mechanism occurs during the charging mode. Lithium ion batteries are being increasingly employed to power portable electronic devices because of their high energy densities compared to other types of batteries, such as CdNi or NiMH. They have no memory effects and they discharge only slowly when they are not used. Lithium ion battery technology is rather young, dating back to the 1970s in the research laboratories, but really available on the market only since 1991, when Sony provided the first commercial Li ion batteries.

The energy density per unit mass of a Li ion battery is typically 160 Wh/kg and per unit volume it is 270 Wh/L. The amount of power that can be delivered is on the order of 1800 W/kg. The self-discharge remains reasonable ($\approx$5–10% per month) and around 1200 charging cycles are possible. The voltage, around 3.6–3.7 V, depends upon the manufacturing technology. In Figure 12.9 the energy capacities (expressed in watt-hours per kilogram) and average voltages for selected technologies are shown. Higher voltages than the nominal ones ("charge V limit" in the figure) can be obtained. This provides more energy, but the cycle life is reduced. The lifetime is two or three years but it should be noted that Li ion batteries age even if they are not used. The best way to store an unused Li ion battery is with a charge of about 40% in a cool place.

Lithium ion batteries should be charged early and often. At 25 °C, the loss of charge after one year is 20% for a battery initially charged to 100% of its capacity and 4% if the initial charge was 40%. If they are not used for a long time, the best way to preserve the battery is to keep a charge level between 40 and 60%. Storing Li ion batteries in a cool place like the refrigerator is also a good idea. For example, at 40 °C there is 35% loss after one year of a battery initially charged at 100% while this loss is only 6% at 0 °C. Aging occurs faster at high temperature and the main problem of Li ion batteries is that they age even if they are not used.

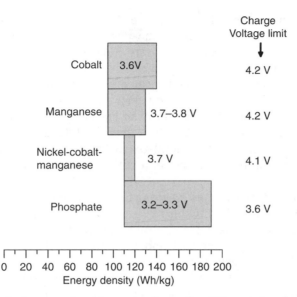

Figure 12.9. Typical energy densities and voltages for different chemical technologies in Li ion battery fabrication. *Source*: Data from www.BatteryUniversity.com.

Electrolytes used in Li ion batteries are solid lithium–salt compounds like $LiPF_6$, $LiBF_4$, and $LiClO_4$ and organic solvents such as ether. The lithium ions are not oxidized in a Li ion battery; they are just transported from one electrode to the other while a transition metal, for example, is actually oxidized (during charging) or reduced (during discharging). In the case of cobalt oxides the overall reaction is

$$LiCoO_2 + C \Leftrightarrow Li_{1-x}CoO_2 + Li_xC$$

The first Sony Li ion batteries used a lithium cobalt oxide cathode and a graphite anode. Many Asian companies now produce Li ion batteries using this technology. These batteries are used to power cell phones, cameras, laptops, and other portable devices.

In the nickel–cobalt–manganese technology, the cathode incorporates the metals in the crystal structure. The voltage of this technology is indicated in Figure 12.9. This type of cell can be recharged at 4.2 V, giving a higher energy capacity. However, the number of cycles which can be achieved could be reduced by a factor larger than 2.

Lithium ion batteries can be damaged if they are discharged below a certain value (so-called deep discharge). On the other hand, overcharging or overheating may lead to an explosion. Short circuits may also have hazardous consequences like ignition or explosion. Short circuits can arise from an internal contamination by metal particles. This has been the case several times in the past. For example, in 2007, more than 46 million Li ion cell phone batteries were recalled because of the possibility of overheating and possibly exploding. As a result of safety concerns protection devices are included in Li ion batteries to prevent overloading and deep discharges. These devices provide important protection but can add some unreliability to the Li ion battery operation since they may disable the cell if they dysfunction even though the battery itself has no problem. In addition, they require some space in the system. In terms of safety, the lithium phosphate technology is better than the metal–oxide technology. The cathode is less sensitive to high temperature and, in addition, these batteries have a longer shelf life.

Lithium ion polymer batteries belong to a technology issued from conventional Li ion batteries. The lithium salt is contained not in an organic solvent but rather in a solid polymer composite. This technology appeared around 1996. Its advantage is lower cost and the ability to shape the batteries as needed. This is particularly convenient for some applications, cellular phones, for example. The energy density in a Li ion polymer battery is about 20% higher than that of a classical Li ion battery. The gravimetric energy density is between 130 and 200 Wh/kg and the volumetric energy density around 300 Wh/L. The nominal cell voltage is 3.7 V, but it can vary between 2.7 V when the cell is discharged and 4.2 V when it is fully charged. However, care should be taken not to discharge the battery too much, otherwise it will not be possible to fully charge it again and problems may arise when it is used. Overcharging may also be dangerous, as in the case of the cobalt Li ion batteries. In regular use, more than 1000 cycles can be achieved. The lifetime is expected to be on the order of two or three years.

12.1.3.3. Flow Batteries

One of the problems of batteries is the limited amount of energy that they contain. A flow battery is one of the solutions developed to overcome this drawback. The principle is to store the reactive materials (reactants) outside the battery in separate containers which can be refilled on demand. The idea is not new and goes back to the French scientist Charles Renard, who experimented with it in 1884 to power his airship, *La France*. The airship used a propeller powered by an electric motor running on the electricity produced by a flow battery. With this airship he flew over a closed loop of 8 km above Villacoublay, France. This flight, which lasted 23 min, was the world's first round-trip by air.

It was only in the mid-1970s that the technology of flow batteries was revisited. Modern flow batteries use two electrolytes which are pumped to the cell where the electrochemical reaction takes place (see Figure 12.10). The running time is just limited by the size of the electrolyte tanks and depends also on whether they are refilled during use or not. Since thermal energy is also released in the electrochemical process, part of this energy can in principle be recovered.

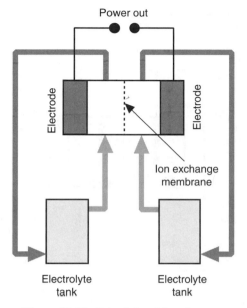

Figure 12.10. Principle of flow battery.

A zinc–bromine cell is an example of a flow battery. The reactants are zinc and bromine. Zinc is at the negative electrode and bromine at the positive one. A solution of zinc–bromine is stored in two tanks. One tank is for the positive electrode reaction, the other for the negative electrode reaction. A microporous membrane separates the two parts of the reaction cell and prevents bromine from going to the zinc electrode, which would produce self-discharge. The overall reaction is

$$Zn + Br_2 \Leftrightarrow 2Br^- + Zn^{2+}$$

The voltage is around 1.67 V per cell. Energy densities are on the order of 75–85 Wh/kg.

The two electrodes do not take part in the reaction but act as a substrate for them.

A promising flow battery is a redox flow cell based on vanadium ions, the vanadium flow battery. Vanadium is interesting because it can exist in four oxidation states. The principle of the battery is shown in Figure 12.11. The reactions taking place at the electrodes are

$$VO_2^+ + 2H^+ + e^- \Leftrightarrow VO^{2+} + H_2O$$

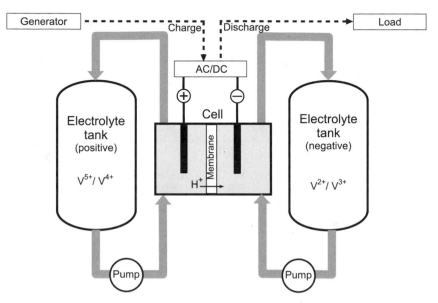

Figure 12.11. Schematic of principle of vanadium redox flow battery.

at the positive electrode and

$$V^{3+} + e^- \Leftrightarrow V^{2+}$$

at the negative electrode

The voltage in an open circuit is between 1.4 and 1.6V depending on the state of charge. The energy density is between 25 and 35Wh/kg. Installations of 250–1500kW exist in various locations in the world.

Sodium–Sulfur Battery The sodium–sulfur battery is a high-energy-density battery built with inexpensive materials. It operates at high temperature (around 300 °C) using molten reactants. It consists of liquid sulfur at the positive electrode and liquid sodium at the negative electrode. It has a high efficiency ($\approx$90%) and a long cycle life. The overall reaction reads

$$2Na + 4S \Leftrightarrow Na_2S_4$$

Both reactants are separated by a solid beta ceramic exchange membrane which conducts Na^+ ions only. The output voltage is around 2V. The principle of the battery is displayed in Figure 12.12.

The largest sodium–sulfur battery is installed in Japan. It has a power of 6MW and can provide this power for 8h. More than 30 sites in Japan are equipped with such batteries. They have a total power larger than 20MW.

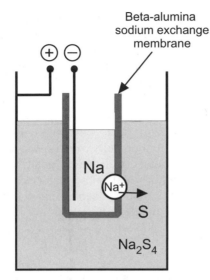

Figure 12.12. Schematic of principle of sodium–sulfur battery.

Conclusion Batteries are convenient energy storage devices. They are capable of providing electricity on demand but are expensive. Primary batteries are the most expensive. Secondary batteries have much lower costs for the electricity supplied but require charging devices and an electrical power grid as a charging source. Their use has to be planned in order that they may be recharged on an appropriate schedule.

It is interesting to quote cost estimates done in the United States, comparing different sources of energy: primary batteries (AA alkaline cells), NiCd batteries for portable use, combustion engines for midsize cars, fuel cells, and the electrical power grid. The highest cost per kilowatt-hour is that of the primary battery ($\approx$\$166) followed by the NiCd rechargeable battery ($\approx$\$7.8). The combustion engine is a cheaper source of energy ($\approx$\$0.3). The price per kilowatt-hour from fuel cells depends on the application and has some uncertainty. It ranges between \$1.9 and \$4.1 for portable use, between \$0.3 and \$0.6 for mobile applications, and between \$0.1 and \$0.2 for stationary applications. The lowest energy price per kilowatt-hour, about \$0.1, is obtained from the electrical grid (assuming a base cost of generating 1 kWh of electricity of \$0.03). The grid is clearly the most economically competitive. Detailed data can be found at www.BatteryUniversity.com. Relative costs can change from country to country depending on the tax structure. In the future batteries are expected to become increasingly important for automotive applications.

12.1.4. Supercapacitors

Supercapacitors are energy storage devices having power densities and energy densities between those of batteries and electrochemical capacitors. They are also known as ultracapacitors or electrochemical double-layer capacitors. The energy densities of commercial supercapacitors are in the range of 0.5–10 Wh/kg, but in the laboratory values on the order of 30 Wh/kg have been obtained. Even higher values are possible with different electrode materials. For comparison a standard electrochemical capacitor typically has an energy density lower than about 0.1 Wh/kg. A lead battery has an energy density of ≈30–40 Wh/kg.

In a conventional electrochemical capacitor, made of two plates separated by a dielectric material, the energy results from an excess of electrons on one plate (negative charge) and a deficit of electrons (positive charge) on the other. The capacity C is directly proportional to the surface area S of the plates and the dielectric constant ε of the material separating the plates. It is inversely proportional to the distance e between the plates:

$$C = \frac{\varepsilon S}{e}$$

In supercapacitors a different technology comes into play. As is displayed schematically in Figure 12.13, a supercapacitor is made of two porous electrodes, often activated charcoal, separated by an isolated membrane porous to ions and that have double layers of charge. The second layers are formed by ions of opposite charge to the neighboring initial layers. The thickness of each layer is very small (on the order of nanometers). Much larger capacities can be obtained in such supercapacitors, typically thousands of times greater than in normal high-capacity electrochemical capacitors. Porous electrodes have large specific surface areas, typically ≈2000–3000 m²/g, which also

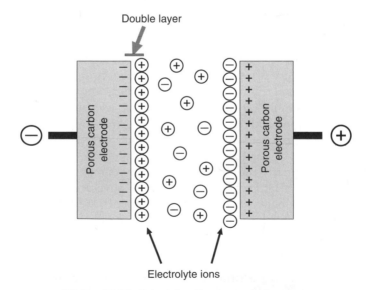

Figure 12.13. Principle of a supercondensator.

increases the capacity. It is possible to find commercial supercondensators of several thousands of farads.

Because of the small thickness of each electrochemical layer, the capacity per unit of surface are is ≈0.1 to 0.3 F/m². A consequence of the nature of the electrodes is that only low-voltage values (≈2.5 V) can be used.

The supercapacitor effect was discovered by accident in 1957 by General Electric. Standard Oil of Ohio rediscovered this effect and developed the first prototype devices in 1966, but this was not a success and a licence of the technology was transferred to NEC, which started producing supercapacitors in 1978. The development of the technology was slow until the 1990s when much progress was made in terms of materials and cost. It is now a rapidly growing commercial success mainly due to applications in the automotive area.

One of the main advantages of supercapacitors is their high power density. These are between 1000 and 5000 W/kg. This is a little smaller than for large electrochemical capacitors but much larger than batteries, which are more around 100 W/kg. Indeed, batteries have low charge and discharge times associated with the slow motion of charge carriers in a liquid electrolyte. Supercapacitors are only limited by the current going through the electrodes.

A great advantage of supercapacitors compared to batteries is the much larger number of charge–discharge cycles which can be realized, millions compared to about 1000–2000 for some good battery technologies. They are interesting in association with batteries because they can provide load balancing and part of their energy can be used to charge batteries. They are especially interesting in the automotive sector because they can be charged much more rapidly than batteries and have very good efficiencies (greater than 95%). This is particularly interesting when energy is recovered from the braking process.

In summary, supercapacitors offer many advantages as energy storage devices. They have large power densities and high efficiencies. They have 10–12 year lifetimes and can be cycled millions of times. They have a low impedance and can be charged in seconds with no danger of overcharge. The rate of charging and discharging can be very high. However, they also have some disadvantages. They have low energy densities and higher self-discharge rates than electrochemical batteries. Their linear discharge prevents the use of the full energy content. They also need sophisticated electronics to function. Finally their cost per watt is still high. The best way to use a supercapacitor is in association with a battery where it can play the role of an energy buffer between the battery and the device being powered.

12.1.5. Flywheels

Another way to store energy is in the form of rotational energy using flywheels. The basic principle is sketched in Figure 12.14. The storage device consists of a cylinder rotating in a vacuum to prevent rotational energy losses due to friction against the air. Electricity is first converted by a motor into rotational energy of the cylinder, the flywheel, which is accelerated to a high speed. When the flywheel is allowed to slow down, the rotational energy is then converted back into electricity using a generator.

The rotational energy E is given by

$$E = \tfrac{1}{2} I \omega^2$$

where I is the moment of inertia of the cylinder with respect to the rotational axis and ω is the rotational velocity. In the case of a cylinder of mass m and radius r, the moment of inertia $I = \tfrac{1}{2} mr^2$ and $E = \tfrac{1}{4} mv^2$, where $v = r\omega$ is the tangential velocity of the cylinder.

The first flywheels were made of steel and had mechanical bearings. Present-day flywheel systems have rotors made of carbon–fiber composite and use magnetic bearings. In the vacuum enclosure the rotor can spin from 20,000 rpm to over 50,000 rpm.

Flywheels have long lifetimes (decades) and require little maintenance. Between 100,000 and 10 million cycles can be realized. Charging can be accomplished in less than a quarter of an hour. The energy density is rather high compared to those of other energy storage systems, ≈130 Wh/kg. An efficiency of ≈90% is typical. The storage capacity of flywheel systems ranges from 3 to 130 kWh.

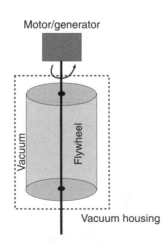

Figure 12.14. Principle of flywheel energy storage system.

Flywheels are very useful for systems where uninterruptible power is required: first, because they provide electricity and, second, because they supply power of good quality. This is well adapted to use with intermittent renewable energy systems. They are also very useful in transportation where they complement other power sources by storing energy released but not needed and recovering it later. In the 1950s buses already employed flywheel systems.

12.2. THERMAL ENERGY STORAGE

Having the ability to store thermal energy is also a major factor in energy supply considerations. Such storage can be in the form of an excess or deficit of thermal energy, that is, in the form of heat or cold. Concepts of heat and cold are usually determined relative to the ambient temperature of the environment and may differ from one country to the other. For an energy consumer, heat corresponds to thermal energy at a temperature larger than the ambient temperature and cold to thermal energy at a temperature lower than the ambient temperature.

Most produced energy is consumed in the form of heat. For example, in France, 53% of the 161 Mtoe of total energy produced is as heat. Buildings use 46% of the total heat generated and contribute to almost a quarter of the total greenhouse emissions. Heat production and utilization is a domain in which large energy savings and reductions of greenhouse gas emissions can be relatively easily achieved.

There are several reasons to store thermal energy. One of them is to store it at a time it can be easily produced and collected in order to use it at a later time when it is needed, for example, storing solar energy during the day to use at night. Indeed, as emphasized by the information in Figure 12.15, it is even more interesting to store thermal energy during the summer to be used for heating during the winter. The intensity of solar energy is a maximum in summer while heating requirements are maximal in winter. The needs associated with heating domestic water are about the same throughout the year.

Sometimes it is interesting to produce heat at a time when the energy required to produce it is cheap. This is often the case with electric water heaters in which water is warmed during the night when the price of electricity is low. This hot water can be used during the day when the price of electricity is higher.

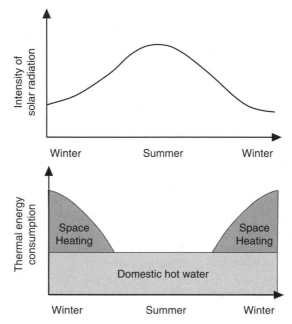

Figure 12.15. There is more solar energy in summer than in winter (top figure). However, space heating needs are maximal in winter (bottom figure). Domestic hot-water consumption is almost constant throughout the year.

Storing thermal energy allows it to be used at a different time and/or different place.

It is sometimes also interesting to store thermal energy on a large scale because it lowers effects on the environment and optimizes the use of primary energy. It is indeed sometimes inefficient to produce thermal energy just when it is needed because this usually increases the energy equipment necessary. Producing heat or cold at the right time and the right place is more interesting economically and can help reduce environmental impacts of energy generation.

Thermal energy storage also provides a way to better use intermittent renewable thermal energy sources. There are many waste heat sources at industrial sites or in urban areas which could be exploited. Using heat pumps, this waste heat could be harnessed in a very efficient way because the temperature can be raised to values useful for domestic purposes.

Storing heat also allows peak shaving in the electricity grids (e.g., with electric water heaters) or in district heating systems. In the case of cogeneration systems, it allows better use of the primary energy to produce both electricity and heat. Often electricity and heat are not needed at the same time and thermal storage is interesting in that respect.

12.2.1. Basic Heat Storage

There are basically three ways to store heat. It can be done using:

- Sensible (or specific) heat storage
- Latent heat storage which is based on phase changes in the materials employed
- Thermochemical or thermophysical heat storage

The three methods are schematically illustrated in Figure 12.16.

12.2.2. Sensible Heat Storage

Sensible heat storage takes advantage of the heat capacity of the material under consideration. The amount of thermal energy that can be stored in a given volume of material depends upon the nature of that material. Heat capacity values are given in Table 12.3 for some materials that are commonly used for thermal energy storage. Water is one of the best substances for this purpose. The heat capacity is $4.2\,kJ/kg/°C$, or $1.2\,kWh/kg/°C$. The volumetric heat capacity is about $70\,kWh/m^3$ for a temperature difference of $60\,°C$ as it is common in thermal heaters supplying household hot water.

The yield of energy storage using sensible heat is around 50%. Losses come from bad layering of the thermal energy in the storage volume and from losses

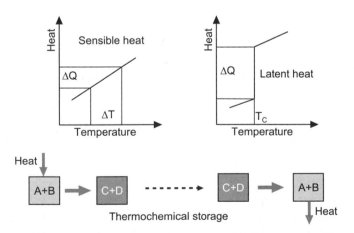

Figure 12.16. Illustration of three methods for thermal energy storage. Sensible heat produces a change ΔT in the temperature of the storage material and leads to a change of thermal energy equal to ΔQ. Some materials can undergo a change of state which requires a thermal energy: the latent heat. This takes place at constant temperature. The thermochemical reaction of A + B to produce C + D absorbs heat. These products can be stored for a long time and the rection can be reversed to produce A + B and liberate heat.

TABLE 12.3. Heat Capacities of Selected Substances

Material	Heat Capacity (kJ/kg/°C)
Aluminium	0.90
Brick	0.84
Concrete	0.88
Copper	0.39
Glass (silica)	0.84
Sand	0.84
Soil	0.80
Water	4.18
Wood	≈1.2 but increases with moisture content

of thermal energy due to thermal conductivity. It is quite an expensive method: between 200 and 450 €/MWh stored. This is much more expensive than using electricity directly to produce heat. Indeed electricity can be produced at 3 €/MWh.

12.2.3. Phase Change Materials

The other method to store heat is using the latent heat of a material. Such a material is called a phase change material. The thermodynamic phase transformation takes place at constant temperature. In a phase change going from solid to liquid (e.g., ice melting into water) energy is taken up to melt the phase change material. This energy is released when the temperature is then decreased to the point where the phase change material solidifies.

The amount of thermal energy stored per unit of mass in the case of phase change materials can be larger, but there is a constraint on the operating temperature. As far as water is concerned there are two changes of phase: from ice to liquid water and from liquid water to vapor. The thermal energy required in the different processes is displayed in Table 12.4.

Glauber salt is the sodium sulfate decahydrate (Na_2SO_4, 10 H_2O). It has a high heat storage capacity and changes phase at 32 °C. It is especially useful in the storage of solar heat for space-heating applications. It can be incorporated into tiles or in the cells surrounding a solar heater. Because of the phase change, it can store 83 times more energy than the same weight of water. It is useful between temperatures of 30 and 48 °C. This salt is inexpensive but corrosive so special containers have to be used.

TABLE 12.4. Thermal Energy Released or Absorbed for Heating Water or During a Phase Transition

State of Water	Transition	Thermal Energy (kJ/kg)	Thermal Energy (kWh/kg)
Liquid	Liquid ⇔ liquid	4.2 kJ/kg/°C	1.2 Wh/kg/°C
		420 kJ/kg/100 °C	120 Wh/kg/100 °C
Liquid ⇔ solid	Liquid ⇔ solid (0 °C)	334 kJ/kg	93 Wh/kg
Liquid ⇔ gas	Liquid ⇔ gas (100 °C)	2260 kJ/kg	630 Wh/kg

12.2.4. Thermochemical and Thermophysical Energy Storage

Thermal energy can be stored and released reversibly in various materials using thermochemical or thermophysical reactions. This can be done in place, but it is also possible to transport the materials over large distances. This is an advantage compared to transporting heat directly. Indeed, it is difficult to transport heat over large distances because of the thermal energy losses.

In thermochemical storage, a reversible chemical reaction is used to store and release thermal energy. For example, using heat, it is possible to remove water from $CuSO_4 \cdot 5H_2O$ (copper sulfate pentahydrate):

$$CuSO_4 \cdot 5H_2O + heat \rightarrow CuSO_4 + 5H_2O$$

The color of the salt changes from blue to white. Adding water to dried $CuSO_4$ gives back $CuSO_4 \cdot 5H_2O$ and releases energy:

$$CuSO_4 + 5H_2O \rightarrow CuSO_4 \cdot 5H_2O + heat$$

Thermal energy can also be stored using sorption processes: chemisorption or physisorption. In chemisorption, molecules adhere to a surface through the formation of a chemical bond while in physisorption the molecule adheres through a van der Waals bond, which is weaker. The typical energy involved in chemisorption is in the range of 50–800 kJ/mol. For physisorption the energy involved is typically less than 20 kJ/mol. Chemisorption takes place in a single layer at the surface of the adsorbing material while physisortion can occur in several layers near the material's surface.

An example of the use of the sorption mechanism is the adsorption of water vapor in a zeolite (alumina silicate) material. Placing dry zeolite material in contact with water vapor releases heat. The process is reversed if the zeolite material is heated above 100 °C. This reversible process can be repeated a large number of times. Silica gels are other efficient materials for thermal energy storage by sorption. A house with 20 m² of solar panel, a hot-water storage system of 1 m³, and a thermal energy storage system using 8 m³ of zeolite can get about 65% of its thermal energy from solar energy. About 45% of the total solar energy is used directly and the rest is energy retrieved from the storage system.

With phase change materials, volumic energy densities of the order of 100kWh/m^3 and temperatures around 40–60 °C can be obtained. Sorption materials work at temperatures a little higher than 100 °C and have volumetric energy densities near 200kWh/m^3. Thermochemical reactions occur over a wide range of temperatures (usually between 100 and 1000 °C) and have volumetric energy densities around 1000kWh/m^3.

12.2.5. Applications of Thermal Energy Storage

There are several existing thermal energy storage systems used in buildings. These are tailored to specific needs and can be quite expensive, but the existing installations provide a good means to test and evaluate the different methods employed.

For solar collectors, the most common way to store thermal energy is to use above-ground water tanks. If the solar collector is used only to provide hot water for an individual home, a storage volume of about 50–100 liters of water is sufficient for 1m^2 of solar collector. For space heating, a larger area of solar collector is required, and in this case 50L/m^2 of solar collector can provide sufficient storage. The advantage of the water tank is that it also plays the role of an energy buffer. When space-heating warm air is mixed with air coming from air-heating solar panels, pebble storage systems can also be used. This works well in passive systems where air circulates by natural convection or in active systems where the air is circulated with fans or pumps.

Heat storage is used in the German project Solarthermie-2000, which involves 7000m^2 of housing in Rostock, Germany. The goal is to have a thermal storage system capable of meeting daily and seasonal needs. There are 1000m^2 of solar roof panels and an aquifer with a volume of $20,000 \text{m}^3$ to store thermal energy. The thermal storage materials used are water, concrete, rocks, and soil. The energy necessary for heating is low: $71 \text{kWh/m}^2/\text{year}$, which means an annual total demand of 497 MWh. Over a year, 307 MWh (62%) is produced using the solar panels.

Heat storage using materials that undergo phase change is being used in Turkey where a project has been developed to heat an 180-m^2 greenhouse. Heat is produced in 27m^2 of thermal solar panels and 317 kWh per year is stored in 11.6m^3 volume of paraffin.

The German Solarthermie-2000 program has produced some interesting results regarding buildings using solar energy systems coupled to thermal heat storage systems. In north and central Europe, a small flat-plate collector solar system for domestic hot-water heating produces between 350 and 380 kWh/m^2 per year. The area needed is between 1 and 1.5 m^2 per person. The storage volume necessary is between 50 and 80 L/m^2 of collector area. The whole system can provide about 50% of the domestic hot-water needs and 15% of the total heat demand.

For an apartment building a central heating plant with diurnal storage meets part of the thermal energy demand for more than 30 apartments. Between 0.8 and 1.2 m^2 of solar collector is needed per person. About 350–500 kWh is produced per square meter of solar collector if a storage volume of 50–100 L/m^2 of solar collector is installed. About 50% of the domestic hot water can be produced and between 10 and 20% of the total heat demand can be met.

For a larger number of apartments (more than 100) it is efficient to have central solar heating with seasonal storage. Between 1.4 and 2.4 m^2 of flat solar collector per megawatt-hour of annual heat demand is required. A storage volume of 1.4–2.1 m^3 of water per square meter of solar collector is needed. One square meter of solar collector associated to such storage volumes can produce between 200 and 330 kWh/year. This can meet between 40 and 60% of the total heat demand.

The use of heat pumps can make the thermal storage process more efficient by allowing increases of the temperature. An example of using a heat pump has been described previously in the discussion of thermal storage of electricity.

The storage energy density for a water tank is small, on the order of 0.01 MWh/m^3, which means that large storage volumes are required. For seasonal thermal energy storage, larger tanks are needed. The mantle should allow for a good layering of the temperature for better efficiency of operation.

12.2.6. Underground Energy Storage

Underground storage systems that can offer the advantage of very large storage volumes are particularly suited to seasonal heat or cold storage. There are several possibilities: Aquifer thermal energy storage uses an aquifer to store thermal energy. The storage can be managed in a cyclic or continuous mode. In the cyclic mode, where either cold or heat can be produced, there are two wells, one cold and one hot, as shown in Figure 12.17. In the

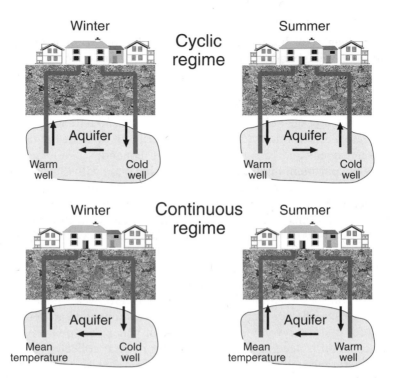

Figure 12.17. Schematic representation of aquifer thermal energy storage in the cyclic and continuous modes. *Source*: From *Thermal Energy Storage: A State of the Art*, Research program Smart Energy–Efficient Buildings, 2002–2006. http://www.ntnu.no/ em/dokumenter/smartbygg_rapp/Storage_State-of-the-art.pdf.

continuous regime one of the wells is at a temperature close to the natural ground temperature. Although the system is simpler, the limited temperature range reduces the flexibility of usage.

Another way to store thermal energy (heat or cold) is using borehole systems. Holes with a diameter of 150–200 mm are drilled to a depth of about 100–200 m. Each hole is equipped with two pipes in the open technology and with one continuous pipe in the closed technology. In the open technology, the outlet of the injection pipe is close to the bottom of the hole and the inlet of the extraction pipe is close to the top of the hole but below the groundwater table. This technology was first developed in Sweden. Such systems have been installed in Norway, mostly around Oslo. The largest borehole installation in Europe is at Lorenskog, Norway. There are 180 boreholes with a depth of 200 m. They can supply 3.2 MWh of heat during the cold season.

Thermal energy storage in caverns is also a technology allowing seasonal storage. There is a large project in Uppsala, Sweden, employing an underground cavern of 100,000 m³ of volume. It is designed to supply space heating and hot water for 550 families. The layering of the temperature in the cavern ranges from about 40 °C at the bottom to 90 °C at the top. An electric boiler backup is also installed since the solar collectors cannot provide all the energy needed by the consumers.

Ducts placed in the soil are often connected with the use of heat pumps. The ducts can be placed either horizontally, close to the ground, or in vertical boreholes. It is in this latter case that thermal energy storage is optimized. Active storage volumes can be between 10,000 and 100,000 m^3. Since the soil temperature is near 25–30 °C, heat pumps are needed to raise the temperature and produce heat for space heating and hot domestic water.

Pit storage facilities in the soil can also be used for seasonal thermal energy storage. The pit storage facility is usually located close to the surface to decrease the investment. Storage pits can be filled with water but rock is also used. Several installations exist in Denmark, Sweden, and Germany. The largest installation in Europe, a 12,000-m^3 concrete pit, is in Friedrichshaven, Germany.

12.2.7. Conclusion

Thermal energy storage is an important consideration in the management of energy production. It allows the provision of thermal energy (heat or cold) in the right place and the right time. A number of storage solutions exist but, for the moment, most of them are too expensive. Heat pumps will have to play a key role in many thermal energy storage situations because they are very efficient and can provide thermal energy at temperatures more useful for the intended applications.

CHAPTER 13

Transportation

The ability to transport people or goods from one place to another provides one of the main engines of economic development in the modern world and the energy devoted to transportation accounts for a very large part of the world's total energy consumption. For example, in the United States, almost one-third of the energy used is for transportation. At present most of this transportation is dependent on energy derived from oil. The price of oil is increasing and this resource is expected to be progressively scarcer in the future. Increases in the efficiency of use of this resource and the development of alternative energy sources to meet transportation needs have become major global priorities.

13.1. SHORT HISTORY OF TRANSPORTATION

The invention of the wheel, probably around 4000–3500 B.C. in Mesopotamia or Asia, was a major advance for humankind. It greatly facilitated the transportation of crops and other goods and materials from one place to another. The existence of the wheel allowed the development of the first carts, probably in Sumeria around 3500 B.C., in the form of two-wheeled chariots. Humans or animals provided the energy needed to propel these carts. Although the period in which humankind first went on the water is unknown; it is also very likely that it was during this period that river boats propelled with oars were first used. Traveling on waterways is complementary to traveling on land. Boats provided the means to better explore the world and to carry out extended trade. The invention of sails allowed wind to be harnessed as an energy source and provided an efficient method of propulsion.

Until the eighteenth century animals and humans were the main energy source for mass transportation. Animals were used for long time with varying degrees of efficiency. The efficiency depends on the way the animal is attached to the device. Great progress was made in these techniques in the Middle Ages

Our Energy Future: Resources, Alternatives, and the Environment
By Christian Ngô and Joseph B. Natowitz
Copyright © 2009 John Wiley & Sons, Inc.

compared to those used in antiquity. The ox was the preferred animal for plowing. Horses were domesticated around 2000 B.C. and used for transportation, but it took a long time before the energy of the animal was harnessed with good efficiency. It was not until the end of the tenth century that iron horseshoes, a major advance in protecting the animal, were introduced.

Pollution associated with transportation is not a new phenomenon. At the beginning of the twentieth century there were about 175,000 horses dedicated to transportation in New York City. Each horse produced between 10 and 15 kg of excrement each day—about 2 tons/day for the whole city. The odor and the flies which it attracted were unpleasant at best. In addition, dust contained particles of excrement, which could lead to respiratory and intestinal diseases. The excrement was often dumped in the rivers, polluting the water. In addition, about 15,000 horses died each year and some of the bodies were not immediately removed from the streets. The gradual change from transportation based on horses to use of automobiles initially resulted in a decrease of pollution.

During the nineteenth century the steam engine, which used coal as a fuel, produced a revolution in the transportation domain. The first road vehicle propelled by steam was invented in 1769 by the Frenchman Nicolas Joseph Cugnot. It could travel at a speed of about 4 km/h for a period of 15 min. In 1873, Amédée Bollée introduced the first commercial steam-powered automobile: "l'obéissante." It was able to transport 12 persons at a maximal speed of 40 km/h. Steam-powered vehicles were heavy and relatively inefficient. Therefore, this technology did not prove to be competitive with other automobile technologies emerging in that period. The first internal combustion engine used hydrogen as a fuel. It was developed in 1807 by Isaac de Rivas. The first automobile using a gasoline engine was built by Jean Lenoir in 1862. The motorcycle appeared a little later, in 1867.

The steam-powered locomotive was invented by the Englishman Richard Trevithick in 1801. Steam locomotives dominated rail transportation for about a century (from the middle of the nineteenth to the middle of the twentieth centuries). In 1783 the paddle-wheel steamboat was invented. However, it took 20 years before the first regular passenger service by steamboat was started. Credit for this development is usually accorded to the American Robert Fulton. Although these steam ships and steam locomotives had low energetic efficiencies, their availability greatly spurred economic development. Later, in 1912, the first diesel-powered boat was introduced and in 1958 the first nuclear-powered ship was launched.

The steam engine has long been replaced by more efficient technologies: the internal combustion engine and the electric motor. The first internal combustion engines used biofuels, but oil products very quickly became the dominant fuels. Electric power has been heavily used in large urban transportation systems: subways, streetcars, and trolleybuses. These technologies have improved greatly during the twentieth century due to the development of new materials and electronics.

Mankind made a number of early attempts to develop air transportation. The first aircraft (airships) were lighter than air. The zeppelin, invented by the German Ferdinand von Zeppelin, was a rigid airship. The first flight occurred in 1900 and lasted 18 min. The rigid airship technology of the zeppelins was developed during the following decade and commercial flights were started. In the 1930s there were regular flights between Germany and North America or South America. The tragic accident of the airship *Hindenburg* in 1937 helped to destroy public confidence and the commercial airship era soon came to a close.

The era of transportation using aircraft heavier than air began near the beginning of the twentieth century with several early attempts at flight: for example, Clément Adler in 1897 and the Wright brothers in 1900. The first flight of the Wright Brothers, at Kitty Hawk, North Carolina, lasted 12 s and covered a distance of 39 m. Immense progress has been made since then, especially during the last half of the century. The modern helicopter was developed in the 1940s. The first supersonic jet flight occurred in 1947. Transportation by jet plane is now a routine occurrence.

At the beginning of the industrial revolution and even rather recently, a great deal of energy was required for industrial use and transportation was less easy than it is today. Quite logically, the steel industry developed in the vicinity of coal mines and the aluminum industry developed close to sites of hydroelectric power. Today transportation is a very important part of the economy either directly or indirectly and the cost of transportation is relatively cheap. It costs about $1 to transport 1 barrel of oil from the Middle East to Europe, for example. As a consequence resources and industries are not necessarily located close to energy sources. Furthermore, different parts of a single object can be made in different places because transporting them is inexpensive. One of the major consequences of this is that work can now be relocated far away from the final consumer in places where labor is cheap.

13.2. ENERGY AND TRANSPORTATION

Transportation relies heavily on oil as an energy source. This is illustrated in Figure 13.1. The world's oil demand in 2005 was about 85 million barrels per day (about 4.9 billion cubic meters per year). About two-thirds of the oil demand is dedicated to transportation in one way or another. In 2005, 52% of the oil was used to produce fuels for transportation. Oil shortages due to

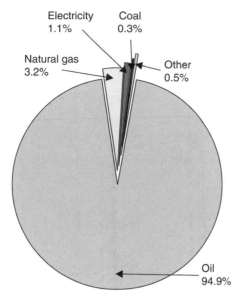

Electricity
1.1%

Coal
0.3%

Natural gas
3.2%

Other
0.5%

Oil
94.9%

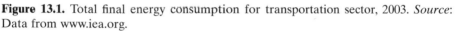

Figure 13.1. Total final energy consumption for transportation sector, 2003. *Source*: Data from www.iea.org.

either insufficient refining capacity or a real shortage of crude oil can lead to major disruptions in the transportation domain.

In 2006 oil was even more important in the United States. In that year, 95.6% of the transportation energy needs were met by oil, 2.2% by natural gas, 1.6% by renewables (biofuels), and 0.3% by electricity.

More than three-quarters of the fuel dedicated to transportation is used for road transportation, as illustrated in Figure 13.2.

Although a different classification of vehicles has been adopted by U.S. authorities, Figure 13.3 shows that the fraction of total consumption used for road transportation is even more important in the United States.

13.3. ROAD TRANSPORTATION

At the end of the nineteenth century the automobile was rapidly developing. Cars and trucks replaced horses for transportation. In 1910 the consumption of oil devoted to transportation surpassed the consumption of oil for lighting. In the early years the automobile industry was characterized by a large number of manufacturers. Initially, only rich people could afford a car. Henry Ford pioneered the manufacture of reasonably priced cars by introducing new ways of working using assembly lines. This allowed the production of cars in a more efficient and reliable way. In 1908 the Ford Model T was produced. It provided affordable personal automobile transportation. Both easy to drive and easy to maintain, it was a great success.

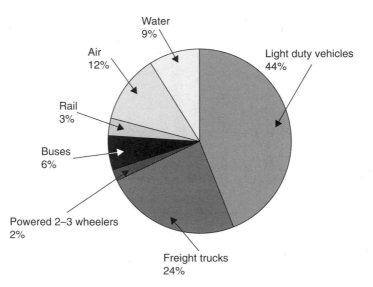

Figure 13.2. Percentage of fuel devoted to different transportation modes, 2005. *Source*: Data from www.iea.org.

At the beginning of the twentieth century France was the largest producer of automobiles. The United States was next. The number of cars produced in 1903 is displayed in Figure 13.4 for the six countries which led in this activity.

The number of manufacturers was very large at that time and increasing rapidly. In 1900 there were 30 car manufacturers in France, in 1910 there were 57, and in 1914 there were 155. The situation was similar in the United States, which had 291 car manufacturers in 1908. Currently the number of manufacturers in the world is relatively small. The recent large increase in the price of oil has destabilized car manufacturers. They had not anticipated this rise and have not invested enough in the design and production of more economic cars.

Road transportation accounts for more than 46% of the final use demand for petroleum due to the great number of vehicles on the road. The world total fleet of vehicles in 2005 amounted to 889 million.

The distribution of private cars and commercial vehicles is shown in Figure 13.5.

The share of the fleet of vehicles among the continents is displayed in Figure 13.6, showing that most of the vehicles are in Europe, America, and Asia. Rich areas have a high density of vehicles per inhabitant while underdeveloped regions have a small density. Note that Europe has a large number of private cars compared to commercial vehicles. This is quite different from America, Asia, and Africa.

The density of vehicles (including private cars and commercial vehicles) is high in developed countries and rises quickly in emergent countries.

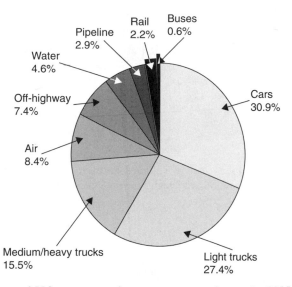

Figure 13.3. Share of U.S. transportation energy usage by mode, 2005. *Source*: Data from S. C. Davis and S. W. Diegel, *Transportation Data Book*, Edition 26, Oak Ridge National Lab, Oak Ridge, TN, 2007.

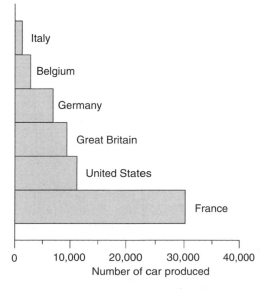

Figure 13.4. Number of car produced in 1903 for different countries. *Source*: Data from http://fr.wikipedia.org/wiki/Histoire_de_l'automobile.

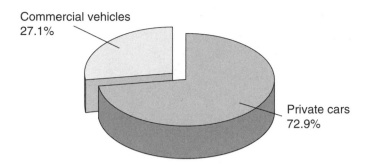

Figure 13.5. Share of global fleet of vehicles between private cars and commercial vehicles, 2005 (total 889 million vehicles). *Source*: Data from www.ccfa.fr.

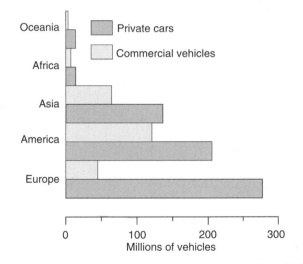

Figure 13.6. Share of global fleet of private cars and commercial vehicles among different regions of the world, 2005. *Source*: Data from www.ccfa.fr.

Figure 13.7 shows the evolution of the number of vehicles per 1000 people for selected countries between 1985 and 2006. In the United States the increase has been about 15% while in China it has been more than 650% and in South Korea close to 1200% during the same period. The number of vehicles in China in 2005 is now about the same as it was in the United States around 1915. Europe reached this level in the 1970s. Japan follows a trend very similar to the one of the European Union.

In 2006, about 69 million vehicles were manufactured in the world. Figure 13.8 shows the vehicle production associated to a country or a region

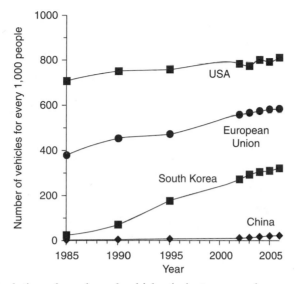

Figure 13.7. Evolution of number of vehicles (private cars and commercial vehicles) for selected countries between 1985 and 2006. *Source*: Data from www.ccfa.fr.

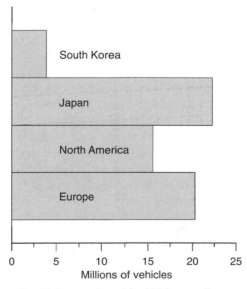

Figure 13.8 Number of vehicles produced in 2006 according to country or region to which manufacturer belongs. *Source*: Data from www.ccfa.fr.

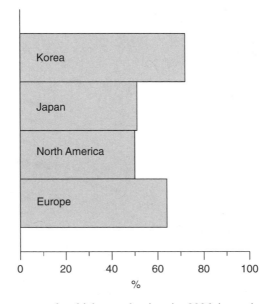

Figure 13.9. Percentage of vehicle production in 2006 in region of manufacturer. *Source*: Data from www.ccfa.fr.

to which a manufacturer belongs. China is not listed. The production in 2006 was below 0.2 million vehicles. Note that all the vehicles of a given manufacturer are not manufactured in the manufacturer's home country. The percentage of vehicles produced in the manufacturer country of origin is displayed in Figure 13.9.

In the United States, transportation accounts for about two-thirds of the petroleum use (petroleum being defined as the sum of crude oil and natural gas plant liquids). Figure 13.10 shows the percentage of petroleum used for transportation in the United States between 1973 and 2006. One observes a smooth increase averaging about 1.3% per year between 1973 and 2006. The average increase was somewhat greater in recent years, 1.6% in the last decade. Since 2003 the consumption of petroleum in the United States has been a little more than 20 million barrels per day while it was 17.3 million barrels per day in 1973. The percentage increased from 52.3% in 1973 to 68% in 2006, which means that transportation has become ever more dependent on petroleum.

13.4. SHIP TRANSPORTATION

Ship transportation was used in the past by passengers to cover large distances over the sea. Before commercial planes were available and had sufficient flight ranges, ships were the way to travel across the oceans. Presently ship transportation is primarily used for goods. It is also used to a lesser extent for recre-

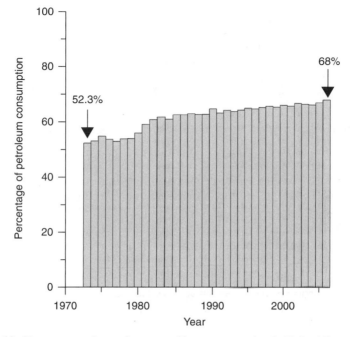

Figure 13.10. Percentage of petroleum used in transportation in United States between 1973 and 2006. *Source*: Data from S. C. Davis and S. W. Diegel, *Transportation Data Book*, Edition 26, Oak Ridge National Lab, Oak Ridge, TN, 2007.

ational trips. Many different kinds of goods can be transported: in raw form such as crude oil, coal, ores, and grains or manufactured products packaged in containers. Ship transport is also used for military applications.

Ship transportation can be used in oceans, seas, lakes, or rivers. The nature of the ship differs according to the application and the place it is used. Bulk carriers are cargo ships used to transport bulk materials such as ore, cereals, and so on. Tankers are large ships used to transport liquids such as crude oil, petroleum products, LNG (liquefied natural gas), LPG (liquefied petroleum gas), vegetable oil, wine, and chemicals. About one-third of the global ship tonnage are tankers. In the case of perishable food or materials, refrigerator ships in which the temperature is monitored and controlled are used. Containers, which can be carried on trucks, are transported over seas by container ships. For automobiles and similar items (e.g., railway cars) special ships have been designed to efficiently load and unload the cargo. Barges are flat-bottomed cargo boats which typically operate in canals, rivers, or lakes. Tugboats are used in harbors to maneuver bigger ships. Passengers are transported in ferries over short distances, sometimes even in towns, as is the case in Venice, for example. Cruise ships carry passengers for pleasure cruises. Millions of passengers per year take such trips. Sailing boats are now essentially used for recreational purposes.

Transporting goods over water is economical as far as energy consumption is concerned. This is due to the large transportation capacities of ships. For example, a 30,000-ton ship has a capacity equivalent to 750 (40-ton) trucks. About six times less energy is required to transport 1 ton of goods over 1 km with a ship than with a truck. With a power of 1 kW one can transport about 5000 kg on waterways, 700 kg by railway, and 200 kg on the road.

13.5. AIR TRANSPORT

Commercial aviation developed swiftly after Word War II. At that time, many military aircrafts were converted for use in passenger and freight transportation. Since the 1960s increasingly more efficient and comfortable aircraft have become available and, due to the development of electronics, instrumentation and control systems have simplified the flying of aircraft. The average annual growth of passenger-kilometers flown was 7% per year in the period 1975–1985. It dropped to 5.1–5.2% per year between 1985 and 2005. The average growth in ton-kilometers of freight was about 7.5–7.6% per year in the period 1975–1995. It dropped to 5.5% per year between 1995 and 2005.

In the period between 1985 and 2005 there was a larger increase in international flights than in domestic flights, as can be seen in Tables 13.1 and 13.2.

Aircraft load factors are typically between 75 and 80%. If we take into account the number of flights and the distances covered by these flights in 2006, nearly 4 trillion passenger-kilometers were flown and almost 150 billion ton-kilometers of freight were transported. International traffic represented ≈60% of the total passenger traffic and ≈83% of the air freight traffic.

In 2007, there were 831 million passengers on international flights and 1.25 billion passengers on domestic flights. For the international flights this represented an increase of 11% compared to 2006. The increase for the domestic flights was 8%. Premium flight passengers (first class and business class) accounted for 8.4% of international passengers and 5.9% of domestic passengers. For 2008, the growth rate appears to be smaller than in 2007 by about 1%. The price of oil has a very strong impact on commercial aviation. The price of oil increased about a factor of 6 between 2002 and 2008. The price of the barrel was about $20 in 2002 and it reached more than $120 in May 2008. Suddenly the cost of fuel is a large part of the cost of a flight. Therefore,

TABLE 13.1. Number of Passenger-Kilometers for International and Domestic Flights

Type of Flight	Billions of Passenger-Kilometers		Ratio 1985/2005
	1985	2005	
International	589	2197	≈3.7
Domestic	777	1523	≈2

Source: Data from ICAO environmental report, www.icao.int, 2007.

TABLE 13.2. Number of Ton-Kilometers in International and Domestic Flights

Type of Flight	Billions of Ton-Kilometers		Ratio 1985/2005
	1985	2005	
International	23.4	118.5	≈4
Domestic	10.4	24.1	≈2.3

Source: Data from ICAO environmental report, www.icao.int, 2007.

the need to build more energy efficient aircraft and to develop new, more economical fuels is now being recognized.

In fact, the energy consumption of modern aircraft has decreased significantly compared to old planes. With occupancy rates on the order of 70–80% the fuel consumption of recently built planes is on the order of 5 L/100 km per passenger. For a full Airbus A340–600 the fuel consumption rate is about 3.7 L/100 km per passenger. Roughly speaking, this means than flying over a given distance is about the same in terms of energy consumption and CO_2 emissions per passenger as driving the same distance with a car. Of course this depends upon the number of passengers in the car and the occupancy rate in the plane but the order of magnitude is the same.

A round-trip flight over the Atlantic between the United States and Europe is roughly equivalent to using a car over a period of a year. Flying between Paris and New York corresponds to an emission of about 700 kg of carbon equivalent if all greenhouse gases are included. Since new planes need less fuel than older ones to cover the same distance, this means that for a given flight less pollution is released into the atmosphere than four decades ago. During this period of time the average number of hours of utilization of the plane has also more than doubled. The average number of seats has also more than doubled.

13.6. CAR DYNAMICS

Energy is needed to move a vehicle. Typically moving a car with a single passenger requires about 20–40 times more energy than moving a bicycle. Energy is required to overcome the forces which oppose movement. These forces are aerodynamic drag, the internal friction in the drive train (except the engine and the gearbox), the inertial forces (both translational and rotational), and the tire resistive force. The relative importance of these different forces is displayed in Figure 13.11 for a medium car traveling at 100 km/h. Detailed calculations can be found in the 2004 Shanghai Challenge Bibendum report.

The largest opposing force is aerodynamic drag. At a speed of 100 km/h, it is on the order of 350 N. This force is proportional to the square of the velocity and depends very much on the shape of the vehicle. The frontal area of a passenger car is about 2 m². The frontal area can reach 9 m² for a heavy-duty vehicle. In addition the aerodynamic properties of a truck are about three

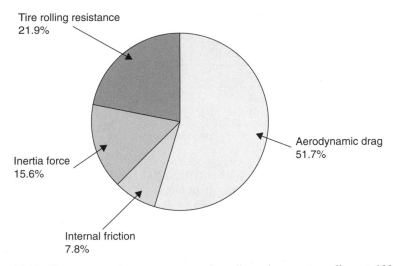

Figure 13.11. Forces opposing movement of medium-size car traveling at 100 km/h. *Source*: Data from the Challenge Bibendum, Michelin, 2004.

times worse. Driving a truck at high speed requires a lot more fuel than at a more reasonable velocity.

Inertial forces are mainly of translational origin and there is just a very small contribution from rotational inertia. The force is positive in the case of acceleration and negative in case of slowing down. It depends on several parameters, such as the type of driving, the nature of the road, and so on. It is estimated to be of the order of 100 N on average.

The internal friction forces in the drive train (except the engine and the gear box) are ~50 N almost independent of the speed.

The contact between the car and the road is through the tires, which are made of a viscoelastic material that also dissipates energy. The rolling resistance of the tires is estimated to be about 140 N. For a car the tire resistance represents about 20% of the total resistance against the motion. The share is larger for a truck reaching about 40%. It is possible to decrease the rolling resistance by a proper choice of tires. Using "green" tires instead of traditional "black" ones allows a fuel saving of between 3 and 8% depending on the nature of the vehicle. Typical fuel savings amount to about 0.25 L/100 km for a passenger car. The extra cost of such tires quickly pays for itself.

Summing up all of these contributions leads to a total force opposing the movement of about 640 N for the vehicle considered here, a relatively small one. Over a driving distance of 100 km this corresponds to an energy value of 17.8 kWh. The net calorific value of gasoline is about 43.5 MJ/kg. This corresponds to about 32.8 MJ/L or 9.1 kWh/L. With a engine of 100% efficiency, that would correspond to 2 L/100 km. However, the efficiency of the engine is roughly 30%, which means that the real gasoline consumption is about

6.6 L/100 km. This corresponds to an expected mileage of 35.6 miles per gallon for this particular car. Not surprisingly, a car consumes more fuel going uphill than downhill. A slope would add a force of about 150 N per percent of incline.

13.7. FUELS FOR ROAD TRANSPORTATION

Most of the engines used for road transportation are either spark ignition engines or compression engines. They use mostly gasoline or diesel oil as fuel.

In the spark ignition engine a spark ignites a compressed mixture of air and fuel. This takes place in the cylinders of the engine. Modern spark ignition engines have electronic fuel injection systems to ensure the best yield and the lowest pollution. Actually electronic control is now needed in modern engines to meet the restrictions that authorities have placed on the quantities of pollutants allowed to be emitted. Spark ignition engines use mostly normal gasoline or reformulated gasoline as fuel.

In compression ignition engines, air is compressed in the cylinder until it reaches the autoignition temperature of the fuel which is injected. The timing of the fuel injection and the quantity of fuel injected are critical parameters which are now managed electronically. These engines typically use diesel fuel. In the past, it was necessary to raise the temperature for cold start using an extra source of energy. Modern diesel engines can now start immediately.

The compression ratio, which is the ratio between the volume of the combustion chamber at the beginning and that at the end of the compression stage, is an important parameter related to the yield of the engine. The compression used for this is up to about 11 for gasoline-fueled spark ignition engines. It is higher, 18–22, for compression engines using diesel oil. It is about 18 for direct injection and about 22 for indirect injection when a small precombustion chamber is used.

The efficiency of a spark ignition engine increases with the compression ratio. However, if this compression ratio becomes too large, the mixture of air and fuel autoignites and the engine knocks. Knocking may damage the engine. All fuels are not equivalent in that respect. Some are better than others. The octane number was introduced to characterize the quality of a fuel with respect to knocking. Fuel with a high octane number allows high compression before knocking and increases the efficiency of the engine. In practice, there are two octane numbers used: the RON (research octane number) and the MON (motor octane number). The MON is smaller than the RON.

What is important, for a compression–ignition engine, is the readiness of the fuel to spontaneously ignite when it is sufficiently compressed. The fuel injected into the cylinder just before the maximum compression is reached must ignite within a very short time (a few milliseconds). The ability to ignite with a short delay is characterized by the cetane number. The higher the cetane number, the shorter the ignition delay and the higher the yield.

Most of the fuels used in road transportation are made from crude oil. The oil is transformed and refined into gasoline ($\approx$69% of the road transport fuels in the OECD countries) or diesel ($\approx$30%). Gasoline is derived from the light-distillate part of crude oil and diesel oil from the middle-distillate part. In the refining process various treatments such as cracking, reforming, and sulfur extraction are performed to meet the quality standards required for fuels.

Compared to diesel oil, gasoline leads to higher CO_2 emissions per unit of traveled distance but emits less NO_x and CO and is much better as far as particle emissions are concerned. This reflects the fact that diesel engines consume less for a given distance but diesel oil has a larger mass density than gasoline (see Table 13.3). Diesel oil is mainly used in heavy-duty vehicles but is also widely used for cars in Europe due to an attractive price compared to gasoline and to the lower fuel consumption for a given distance. In France, for example, 70% of the new cars sold have a diesel engine. Because of a mismatch with its refining capacity, France imports diesel fuel and exports gasoline.

The sulfur content of modern diesel oil is now small compared to diesel fuel which was delivered in the past. This reduction has been necessary to meet stronger pollution standard requirements for SO_2 emissions, which produce acid rain.

Liquefied petroleum gas is a mixture of petroleum gases (mostly propane and butane) either obtained during crude oil and natural gas extraction or as a by-product of crude oil refining. It accounts for about 0.9% of the global fuel consumption in the OECD countries. Nitrous oxide emissions from LPG are similar to those from gasoline, but other pollutant emissions are lower. In some countries LPG is more commonly used. In the Netherlands the percentage of cars using LPG is about 12%. Liquefied petroleum gas is stored as a liquid at a pressure of 6–8 bars. An LPG tank should never be completely filled. Typically 80–85% of its capacity is used in order to leave room for fuel expansion. If a

TABLE 13.3. Energy Content for Selected Fuels

Fuel	Energy Content (Low Calorific Value), MJ/L
Gasoline	31.2
Diesel	35.7
Liquefied petroleum gas (70% propane, 30% butane)	24.2
Natural gas (liquefied)	23.3
Methanol	15.6
Ethanol	21.2
Biodiesel	32.8
Dimethyl ether (DME)	18.2–19.3
H_2	8.9

For gases this is for the pressures normally used.

Source: Data from *Automotive Fuels for the Future*, IEA, Paris, 1999.

significant overpressure (larger than 20 bars) occurs, a valve allows the gas to be released. Because LPG is heavier than air and stays near ground level, LPG vehicles are not allowed to park in underground parking garages. The LPG octane number is 107.5–112 RON, better than for gasoline (90–95 RON for regular and 97–99 for super) but the energy content is about 23% less. As a consequence, a tank of LPG containing the same amount of energy as a gasoline tank has to be 2 times larger and 1.5 times heavier.

Natural gas (essentially methane) is also used as a fuel in automotive applications. It requires almost no refining before use except for the removal of water and hydrogen sulfide (H_2S). It accounts for about 0.05% of the total fuel used in road transportation in OECD countries. Some countries (e.g., Italy, countries of the former Russian Federation, Argentina, New Zealand, and the United States) have some nonnegligible fleets of natural gas–powered vehicles. Natural gas can be used either in a compressed form or as a liquid. Natural gas has a significantly higher octane number, 120 RON, than gasoline, leading to an increase of about 10% in the efficiency of the engine. Compressed natural gas is stored at a pressure ranging from 200 to 240 bars. For the same driving range, the tank is five times heavier and four times larger than for a gasoline vehicle. Liquid natural gas is stored at 4–6 bars at −161 °C. The size of the tank is about two times larger and about 40% heavier than a gasoline tank with similar energy content.

Methanol (CH_3OH) is usually produced from natural gas. Natural gas is first reformed with steam and a syngas shift is then made:

$$CH_4 + H_2O \rightarrow CO + 3H_2 \quad \text{(steam reforming)}$$

and

$$CO + H_2 \rightarrow CO_2 + H_2 \quad \text{(shift)}$$

After purification the following reactions are performed with a catalyzer:

$$CO + 2H_2 \rightarrow CH_3OH \quad \text{and} \quad CO_2 + 3H_2 \rightarrow CH_3OH + H_2O$$

Note that these reactions require hydrogen and can be used to convert CO_2 into CH_3OH. However, they require energy. Methanol can also be produced from biomass (wood alcohol). Methanol has a lower energy density than gasoline. It can be used blended with gasoline in flexible-fuel engines. It can also be used pure in heavy-duty compression–ignition vehicles. The tank needs to be about 75% larger and two times heavier than a gasoline tank. The main problems of methanol are that it mixes easily with water and is toxic. Therefore safety considerations are of utmost importance.

Ethanol is derived from biomass by fermentation processes. It is usually mixed with gasoline or converted to ETBE (ethyl tertiary butyl ether) and blended with gasoline. It can also be used pure. It has a high octane number, RON 109. Because of its smaller energy content compared to gasoline, the tank has to be about 50% larger and 65% heavier than a tank containing gasoline. Ethanol is less dangerous than methanol.

Biodiesel is esterified vegetable oil produced from different crops (sunflower, palm, soybean, rapeseed). Biodiesel can be mixed in any ratio with conventional diesel fuel or used pure. Its volumetric energy density is a little smaller than that of conventional diesel fuel ($\approx 8\%$), but its mass density is somewhat higher and the cetane number is better (51–58 instead of 48–50).

Hydrogen is also a fuel which can be used either directly or in fuel cells. This is treated in detail in Chapter 16 on hydrogen.

13.8. CO_2 EMISSIONS

Since most current means of transportation use oil-based fuels, CO_2 is emitted into the atmosphere when they are used. At the global level in 2005, emission of CO_2 in transportation applications was 24% of the total CO_2 emission (Figure 13.12). The largest contributor (45%) to global emissions remains the energy industries (electricity and heat production). Road transportation generated the largest part of the transportation-associated emissions (18% of the total emissions). A typical car still emits about 200 g of CO_2 per kilometer in normal use. This means that if 15,000 km is traveled each year, on the order of 3 tons of CO_2 is emitted. This is larger than the weight of the car itself. Although emissions given as a reference by the manufacturer are often lower than this value, these figures are measured under conditions which are more favorable than in real use. Actually they are directly connected to the official fuel consumption measured under specific conditions that an ordinary driver can usually never meet.

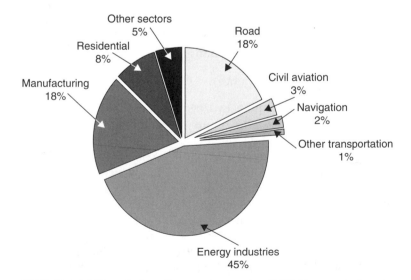

Figure 13.12. Total CO_2 emissions from energy sector, 2005. Transportation represents 24% of the global CO_2 emissions. *Source*: Data from www.iea.org, 2005.

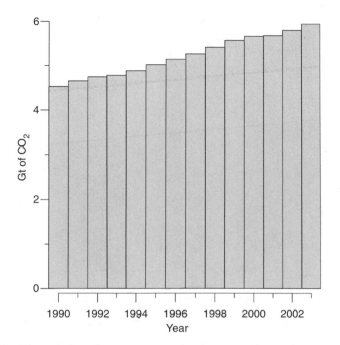

Figure 13.13. CO_2 emissions from transportation between 1990 and 2003. *Source*: Data from www.iea.org, 2005.

Even though on average new cars are more efficient and consume less fuel per unit of distance traveled than older ones, the world's total number of cars is increasing significantly. Furthermore cars tend to become heavier for safety and comfort reasons. As a result CO_2 emissions attributed to transportation needs increase as a function of time. This is depicted in Figure 13.13 for the period between 1990 and 2003.

The percentage of CO_2 emissions associated with transportation went from 22% of total CO_2 emissions of the energy-producing sector in 1990 to 24% in 2003. In 1990, the amount of CO_2 released into the atmosphere was 4.5 Gt from transportation. It was 5.9 Gt in 2003 (the global emissions were 25 Gt). This corresponds to an increase of 31%. Transportation-associated CO_2 emissions from OECD countries represent close to three-quarters of global transportation emissions. It is worth noting that the increase in the OECD countries was only 26%, indicating a strong increase of CO_2 emissions in developing countries.

Figure 13.14 gives average distributions between different emission sources in the case of energy production for the United States, France, and Germany. The data are from the IEA and may differ from other estimates. The share of transportation related CO_2 emissions is similar for the United States and France. For Germany it is much smaller. This comes from the fact that the energy sector is a very large CO_2 emitter in Germany. In France, due

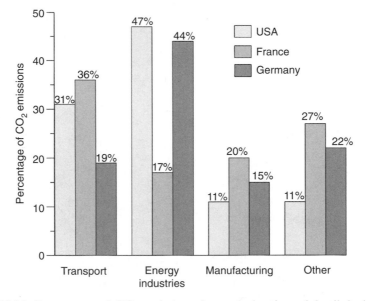

Figure 13.14. Percentage of CO_2 emissions from combustion of fossil fuels in the United States, France, and Germany as estimated by the IEA, 2003.

to nuclear energy and hydroelectricity, the energy sector is a low emitter. In the United States, the large contribution comes from the fact that people are obliged to cover large distances due to the extended area of the country.

Carbon dioxide emissions depend strongly on the means of transportation. Figure 13.15 shows an evaluation of the European Environment Agency (EEA) for CO_2 emissions in different means of passenger transportation. Sea travel and railway are the lowest emitting means of transportation while road travel is the largest shown. Air travel is not indicated in the figure but is even higher. Coaches, buses traveling outside urban areas, are more efficient than buses which travel inside urban areas and experience heavier traffic.

As far as transportation of goods is concerned, Figure 13.16 shows that, for transporting 1 ton of goods over 1 km, seaways, waterways, and railways are the least CO_2 emission intensive. Light trucks, which move goods over relatively short distances, are the greatest road transportation CO_2 emitters.

It is interesting to compare the percentage of CO_2 emissions for 2003 in the transportation area for the three countries presented above: the United States, France, and Germany. In all cases road transportation is by far the largest contributor. Figure 13.17 presents the data for the United States. Civil aviation produces a noticeable share of CO_2 emissions. This is due to the large distances covered in this country. The population of the United States was 291 million in 2003 and about 1.8 Gt of CO_2 equivalent was emitted by the transportation area. This corresponds to 6.2 t of CO_2 per inhabitant.

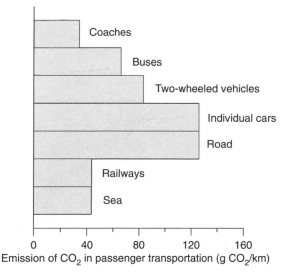

Figure 13.15. CO₂ emissions in European Union passenger transportation by different means, 2000. *Source*: Data from EEA, Copenhagen, www.eea.europa.eu, 2003.

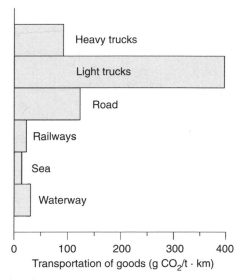

Figure 13.16. CO₂ emission in transportation of goods by different means. *Source*: Data from EEA, Copenhagen, www.eea.europa.eu, 2003.

Similar data for France are presented in Figure 13.18. Road transportation makes by far the largest contribution, but there is still a noticeable contribution from civil aviation. Railways have a very small share because a large part of the French trains use electricity, which is for the most part (90%) produced without CO_2 emissions. The consumption of electricity for railway

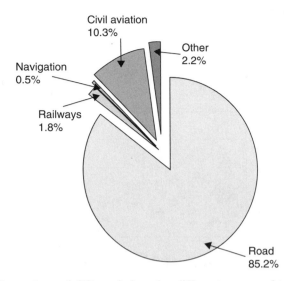

Figure 13.17. Percentage of CO_2 emissions for different means of transportation for the U.S., 2003. *Source*: Data from www.iea.org.

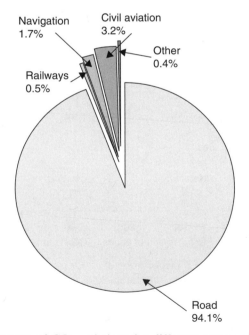

Figure 13.18. Percentage of CO_2 emissions for different means of transportation for France, 2003. *Source*: Data from www.iea.org.

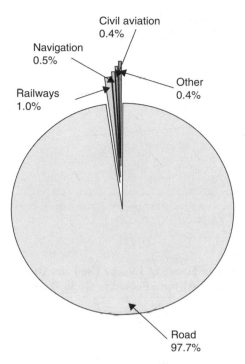

Figure 13.19. Percentage of CO$_2$ emissions for different means of transportation for Germany, 2003. *Source*: Data from www.iea.org.

transportation is about 7 TWh/year in France. The population of France was 61.8 million in 2003 and about 146 Mt of CO$_2$ equivalent was emitted in the transportation sector. This corresponds to 2.4 t of CO$_2$ per inhabitant.

The data for Germany are shown in Figure 13.19. The population of Germany was 82.5 million in 2003 and about 175 Mt of CO$_2$ equivalent, 2.1 t of CO$_2$ per inhabitant, was emitted in the transportation sector. Road transportation is again the largest contributor.

Table 13.4 shows the quantity of energy (in ktoe) and the quantity of CO$_2$ emitted (in tons) in France for a 1000-km round trip using different modes of transport. The calculations were done by the Ademe agency. Because most French electricity is produced without greenhouse gas emissions, CO$_2$ emissions associated with the high-speed train (TGV) are low. It is done for a typical average occupancy—for a car, for example, where many seats are available and, if it would be full, the values would be smaller. If it is almost empty, the values would be larger.

For France Table 13.5 shows estimates of the quantity of energy used and the quantity of CO$_2$ emitted per year for transporting a person to a workplace located 30 km from home using different means of transportation. For a car it is assumed that the person travels alone. If several persons are transported, the figures would be smaller.

TABLE 13.4. Evaluation in France of Energy and Quantity of CO_2 Emitted in a Round Trip of 1000 km

Mode of Transportation	Energy Consumed over a Year (ktoe)	Quantity of CO_2 Emitted (tons/year)
Regional airliner	66	205
Average car	47	123
Medium-distance plane	39	123
Average motorcycle	38	116
Bus	17.3	5.5
Long-distance train	8	14
TGV (high-speed train)	7	8

Note: Evaluations are done with a computer program from Ademe (www.ademe.fr; ktoe = kiloton of oil equivalent).

TABLE 13.5. Evaluation in France of Energy Used and Quantity of CO_2 Emitted per Year (215 Days of Activity) for a Person to Go to Work 30 km from Home

Mode of Transportation	Energy Consumed over a Year (ktoe)	Quantity of CO_2 Emitted (tons/year)
Car (2-liter engine, gasoline)	460	1400
Car (2-liter engine, diesel)	370	1160
Car (<1.4-liter engine, gasoline)	320	1000
Average motorcycle	185	560
Bus	100	316
Commuter train	108	280

Note: Evaluations are done with a program of Ademe (www.ademe.fr).

The carbon content of gasoline is about 640 g of carbon per liter. For diesel fuel it is about 734 g of carbon per liter. The combustion of gasoline or diesel in modern engines is almost complete ($\approx 99\%$). Therefore it is this quantity of carbon which is released when a liter of gasoline or diesel fuel is burned. To calculate the quantity of CO_2 emitted, one has simply to multiply by 3.7, the ratio of the molecular weight of CO_2 to the atomic weight of carbon. This gives

1 liter of gasoline $\Rightarrow$ 0.64 kg of carbon $\Leftrightarrow$ 2.35 kg of CO_2

1 liter of diesel $\Rightarrow$ 0.734 kg of carbon $\Leftrightarrow$ 2.69 kg of CO_2

A car consuming 7.5 liters of gasoline per 100 km emits on the average about 175 g/km of CO_2. Diesel cars need less fuel for the same distance. This usually compensates for the larger amount of carbon content per unit of volume. However, on average diesel engines emit more other pollutants and other greenhouse gases than emitted by a gasoline engine.

TABLE 13.6. Standards for Private Diesel Vehicles in Europe and Their Evolution as Function of Time for Different Pollutants

Standards	Year	CO	HC	NO_x	Particles
Euro 3	2000 (2001)	0.64	—	0.56	0.05
Euro 4	2005 (2006)	0.5	—	0.30	0.025
Euro 5	2009 (2011)	0.5	—	0.23	0.005
Euro 6	2014 (2015)	0.5	—	0.17	0.005

Note: The values are given for new models. The dates in parentheses correspond to older models which are still being produced but which would have to be improved to meet pollution standards: CO, carbon monoxide; HC, hydrocarbon; NO_x, nitrogen oxides.

TABLE 13.7. Standards for Private Gasoline Vehicles Used in Europe and Their Evolution as Function of Time for Different Pollutants

Standarsds	Year	CO	HC	NO_x	Particles
Euro 3	2000 (2001)	2.3	0.2–	0.15	
Euro 4	2005 (2006)	1	0.1	0.08	
Euro 5	2009 (2011)	1	0.1	0.06	0.005
Euro 6	2014 (2015)	1	0.1	0.06	0.005

Note: The values are given for new models. The dates in parentheses correspond to older models which are still being produced but would have to be improved to meet pollution is standards.

Automobile emission standards have been introduced in several countries to fight against pollution. Recommended limits or taxes have also been used in attempts to reduce CO_2 emissions. Table 13.6 shows the standards used in Europe for private diesel cars and Table 13.7 the standards for private gasoline cars. As the standards become more severe, improvements in engines and fuel have to be made to meet them. More sophisticated electronic devices are also being used to monitor engines and reduce pollution.

Many different pollutants are emitted during road transportation. in France it is estimated that close to one-third of CO_2 emissions and 70% of CO, which is poisonous (the CO fixes on hemoglobin and prevents oxygen from doing so), are due to transportation. Three-quarters of NO_x emissions and 60% of incompletely burned hydrocarbons are due to automotive transportation. Ozone (O_3) emitted in the low atmosphere is produced indirectly from nitrogen oxides and has a negative impact on human health. The transportation area also produces 13% of the total emissions of SO_2. These emissions lead to acid rain. Particles, emitted primarily by diesel engines, may have important consequences on human health in the long run. The smallest size particles are the most dangerous ones because they penetrate deeply into the lungs.

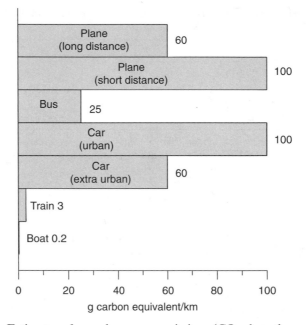

Figure 13.20. Estimates of greenhouse gas emissions (CO_2 plus other gases) for different modes of transportation. *Source*: Data from J. M. Jancovici, www.manicore.com.

A vehicle emits CO_2 but also other greenhouse gases. It is difficult to give precise estimates because it depends very much on the way the vehicle is used, how many persons are transported, and so on. For a plane the other greenhouse gases have the consequence of doubling the carbon content emitted in the atmosphere. For a train using electricity the amount depends upon the way the electricity is produced. For example, it is very different for nuclear plants and coal-fired plants. In the case of aircraft the take-offs consume a lot of fuel and short-distance flights are more energy consuming and emit more greenhouse gases per kilometer than long-distance flights. Some relevant estimates are given in Figure 13.20.

Aircraft emit gases and particles in the upper atmosphere. In 1992, aircraft emissions were estimated to contribute 3–3.5% of the global emissions. These emissions have a greater impact than comparable emissions at sea level. Several greenhouse gases (e.g., CO_2, NO_x, CO) are produced and have an influence on the formation of condensation trails (contrails). This may increase the cloudiness at high altitudes. All these perturb the upper atmosphere and have an influence on the climate.

13.9. HYBRID VEHICLES

The challenge for modern road transportation is to decrease oil consumption and CO_2 emissions. Improvements in cars and fuels can do that. Hybrid vehicles are a promising solution for the future. The idea is to use several sources of energy and regenerate lost energy when possible. A hybrid vehicle is propelled by at least two different energy sources and two different energy converters.

The simplest vehicle using a hybrid system is the electric bicycle. It uses two sources of energy. The first one is electricity provided by a battery and an electric motor to convert electricity into mechanical energy. The battery can be recharged on the grid. The second energy source is the human muscle. It can be argued that this is not too interesting because an ordinary bicycle does need any other energy source than muscles. However, in many situations the normal bicycle will not be used for practical purposes, such as going to work or going shopping, if the distances are too long and it is too tiring. An electric bicycle could provide an interesting alternative to a car for moderate distances.

There are two basic architectures which can be used: a parallel one in which both sources of energy are used to power the bicycle and a series architecture in which the user powers a generator which converts mechanical energy into electrical energy. Regenerative braking can be used to convert mechanical energy into electricity which can be stored for later use.

Hybrid technology is also used in heavy vehicles (e.g., railways locomotives, buses, trucks). No or little electricity storage has been used so far, but storage systems are now being developed. For trucks the extra cost of the hybrid system is often not compensated for by fuel savings. This is due to the fact that a relatively small number of these vehicles are manufactured. However, this will probably change in the future due to the large development of hybrid systems.

Cars account for the bulk of commercial hybrid vehicles. Hybrid car technology is expected to develop greatly in the short term. Hybrid vehicles may be used for many decades since as the technology improves these vehicles will require less and less fuel derived from crude oil. In the future a transition to other technologies which currently are not clearly defined will probably occur.

Commercial hybrid cars use an internal combustion engine and electric batteries to power electric motors. There are two basic configurations of hybrid power trains, as displayed in Figure 13.21.

The simplest one is the series hybrid. In this configuration, it is the electric motor which drives the wheels. The combustion engine is only used to charge the battery. The combustion engine can be small because it just charges the battery and does not provide power peaks. Since the engine can be used at the optimum rotation speed, the efficiency can be better than in a conventional use. The efficiency of an electric motor is also very good providing in the end a reasonable yield and lower gasoline consumption. The battery has to be more powerful than in the parallel hybrid since it is the only way to power the wheels.

In a parallel hybrid, both the combustion engine and the electric motor produce power to drive the wheels. A computer monitors the system, allowing both power sources to work together. The battery pack can be smaller than in

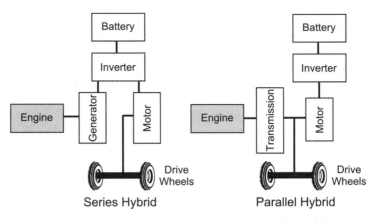

Figure 13.21. Schematic of principle of series and parallel hybrid power trains.

the series hybrid, but the combustion engine has to be more powerful. Such an architecture is especially efficient on highways.

The series/parallel hybrid combines both technologies. The combustion engine can directly drive the wheels or it can be disconnected from them. The combination is very efficient. This is the technology adopted for the Toyota Prius. More than 1 million of these vehicles have been sold since they were introduced to the market. The battery is never discharged and only about 10% of its capacity is used. The reason for that is to increase the lifetime of the battery. This lifetime can be more than a decade. The battery is an expensive component and completely discharging the battery would dramatically shorten its lifetime. The spectacular decrease of fuel consumption obtained is mainly due to good energy management. It is possible to have, in real use, a fuel consumption near 5 L/100 km in Paris, a heavy traffic area. This is about a factor of 2 better compared to a standard car similar in size and comfort. In the suburbs, where the speeds are higher, the gasoline consumption rises slightly. Small diesel engines can have similar fuel consumption, but they do not offer the same degree of comfort. Furthermore a hybrid gasoline emits fewer particles and nitrogen oxides.

In a hybrid vehicle regenerative braking allows the conversion of part of the kinetic energy of breaking into electricity to charge the battery or supercapacitors. Supercapacitors are interesting because they can be charged more quickly than batteries. The regenerated energy is used later as needed. Regenerative braking can provide about 25% of the car's energy in a city-driving mode. One of the advantages of regenerative braking is that the car requires less frequent brake maintenance.

The battery of a Toyota Prius is presently a NiMH battery. Changing to the Li ion technology would increase the driving range. Some manufacturers are moving in this direction. Apart from the cost, which is higher for the same quantity of electricity, the main issue for the Li ion technology is safety, as discussed in Chapter 12 devoted to energy storage. If millions of hybrid cars are on the road, the probability of a severe accident in a battery must be negligible. Any severe accident would jeopardize the hybrid technology.

The future of hybrid vehicles is plug-in hybrids for which the battery is recharged on the electrical grid during off-peak hours. In Europe, most people drive a daily distance less than 30–40 km. With batteries having a capacity allowing this distance to be driven, only electricity would be needed to power the car. The combustion engine would then be reserved for long trips only. Using this type of car would be clean as far as CO_2 emissions are concerned provided that the electricity is produced without CO_2 emissions, that is, by renewable energies or nuclear power, or if the CO_2 produced during generation of the electricity is captured and stored. The main requirements for the battery are that it have a reasonable weight, a long lifetime, and an affordable cost and be extremely safe.

13.10. ELECTRIC VEHICLES

Electricity was used to propel vehicles prior to gasoline. One of the advantages of electric motors is their large torque even at low rotation speeds. Electric cars were developed about a decade before the construction of the first gasoline car. An electric bus was operated in London in 1886. Vehicles were often involved in competitions at that time. In 1899 it was an electric car which set the world speed record of 105 km/h. Electric cars were so popular by the end of the nineteenth century that in 1900 more electric cars than gasoline cars were built in the United States. However, the development of electric cars was soon limited by the battery capabilities and gasoline cars, due to their greater autonomy, quickly became the vehicles of choice. With the modern progress in battery development and the demand of reduction of pollution, there is now a renewed interest in electric vehicles.

An electric vehicle uses one or more electric motors to propel the vehicle. The energy source comes from batteries, fuel cells, or a generator. Before electricity is used directly from the battery there are a number of efficiencies to be taken into account. First is the efficiency of electricity production, which depends upon the technology, as we have seen in the various chapters devoted to the different energy sources. There are also leaks during electricity transport. The order of magnitude of the losses is about 8%, but this depends upon the country and the grid. The losses in the inverter which transforms alternating current into direct current are around 10%, and in the electronics and electric motors losses are also about 10%. The efficiency in charging a battery is about 90%.

While batteries are normally put on board electric vehicles, it is possible, as is done for trains or trolley buses, to get the power by direct connection to conductors placed on the ground. In 1838 an electric locomotive reached a speed of almost 6.6 km/h and the idea of using conductor rails dates back to

1840. Hybrid vehicles, which we discussed above, also belong to the electric vehicle category, but part of the energy is supplemented by a conventional engine fueled with gasoline.

The main problem for electric vehicles is the limited range which can be covered with a fully charged battery. For a long time this limit was about 100 km, but with new storage technologies it is now possible to have car with a 200-km range. The range is of course reduced if the headlights of the car are used or if heating or air-conditioning systems are operating. Heating and air conditioning are real issues for electric vehicles since they decrease the range of the car by consuming extra electricity. Heating is more energy consuming than air conditioning. In a modern electric car a battery weighing about 200 kg allows a range of about 200 km at a speed of 120 km/h. Acceleration and slowing down can be partly managed using supercapacitors, which quickly store energy during slowing down and can give it back during acceleration. With about 20 kg of supercapacitors it is possible store around 30 Wh of electricity. Heating or cooling in the car can be done using heat pumps, which use electricity very efficiently. A power on the order of 1 or 2 kW is enough for a car.

In a modern electric car it is important to be able to regenerate the energy stored in the battery when the car brakes or decelerates. For that the battery should have the possibility to be charged quickly and often it is not possible to use all the energy available. Supercapacitors should help alleviate this problem. The energy density per unit of mass of batteries has greatly increased, going from about 30 Wh/kg for lead batteries to 160–170 Wh/kg for Li ion batteries. The cost has also greatly increased, going from about \$0.3/Wh for lead batteries to around \$2/Wh for Ni–MH or Li ion batteries. However, new manufacturing capabilities of Li ion batteries of 120 Wh/kg at a cost of \$0.3 are foreseen in China. The number of usable cycles is also an important parameter. For lead or Ni–MH batteries 1000 full-charge to full-discharge cycles are possible. For Li ion batteries only about 500 such cycles is possible. In the domain of the electric vehicle, 1000 cycles is a practical minimum. The realized number of cycles can be much larger if the battery is not fully discharged.

Safety is another important concern. Some Li ion technologies can be dangerous and lead to explosion. This has already occurred with cell phones and laptop computers. Because of the large amount of energy stored in the battery, this must not occur in a vehicle. This means that only safe Li ion technologies can be used. Indeed, if millions cars are sold (there will be in the near future one billion cars around the world), the probability of a serious battery accident should be much smaller than 1 in 10^9.

In France where 90% of the electricity is produced without CO_2 emissions electric cars provide an interesting way to reduce CO_2 emissions in the transportation domain. Electric vehicles emit no pollutants. In France about 15 nuclear reactors of 1 GW_e would be enough to provide electricity if all the cars would be electric vehicles. In a country where electricity is mainly produced by coal, the interest is not so large. The only advantage in electric car is to decrease local pollution and pollution is done by the power plant producing electricity.

13.11. CONCLUSION

Mobility is one of the hardest issues to solve as far as energy is concerned. At present our transportation systems depend almost entirely on petroleum products. Sea and rail transportation are the most economical ways to transport people and goods and they lead to small CO_2 emissions compared to other means of transportation. Planes demand a lot of energy and emit a lot of CO_2. But mobility is dominated by road transportation and the challenge is now to find a sustainable way to develop road mobility. The global number of cars is always increasing and despite the progress made in terms of energy consumption and pollution decreases, the energy demand for road transportation also increases from year to year. The challenge is to decrease energy consumption while keeping the freedom of travel.

Public transportation is a good way to ensure mobility in crowded areas, but it is difficult and energy consuming to develop this kind of transportation system in low-population-density areas. In many developed countries people often live great distances from their working place or other activities. When it is possible, the best way to use a car is to drive to the nearest railway or subway station, park it free, and use public transportation. This solution is often possible in Europe, but in the United States this is more difficult as the public ground transportation systems are generally not as well developed as in Europe. Energy demands of road transportation and housing are strongly correlated. Living in a low-energy-consumption house but far from the workplace may be worse than living close to the place of work in a less energy efficient house because less gasoline will be needed to drive to work.

In order to decrease fossil fuel consumption and CO_2 emissions, hybrid vehicles and, in some particular situations where only short distances are involved, electric vehicles are very likely to meet our needs in the near future. With plug-in hybrid vehicles, we may reach gasoline average annual gas consumption on the order of 1 or 2 L/100 km for distances smaller than 40–50 km/day. Gasoline will then be used for long driving distances only. Second-generation biofuels using clean hydrogen production (mostly by electrolysis) for their processing as well as marine biofuels may meet part of the fuel demand for road transportation. With these new technologies it is probably possible to reduce our road transportation petroleum demand by 50%.

As we shall discuss in Chapter 16, fuel cell vehicles are far from being used at a large scale and hydrogen is not a good fuel to power cars because the tank contains only a very small amount of hydrogen due to the physical properties of this element. What is important for transportation is to have a liquid fuel with a high energy density. This is the case of gasoline or diesel, for example. Nevertheless hydrogen will be useful in petrochemistry as well as to produce second-generation biofuels.

CHAPTER 14

Housing

A major part of the energy consumed in the world is used for space heating, cooling, or the production of hot water. In France, for example, 43% of the primary energy is consumed in the residential and service sectors. Housing is also responsible for 21% of greenhouse gas emissions. Thus, efficient use of energy in the housing sector is an important issue and significant energy savings can be realized in that domain. Unfortunately the timescale required to achieve this is long compared to the timescale needed to mitigate global warming due to anthropic greenhouse gas emissions. In France, there are about 30 million houses or apartments. They accommodate a population a little larger than 60 million. The number of new residential buildings built each year is 300,000–400,000. It would take about a century to replace all old buildings with new ones. A similar situation exists in the rest of Europe. This means that, while it is important to build new energy-efficient buildings, the improvement, refurbishment, and renovation of old buildings are even more important.

14.1. IMPORTANCE OF HOUSING

In the United States the population is about 300 million. In 2005 there were 111.1 million housing units. Some of these houses are old. The number of household units constructed in different time periods is shown in Figure 14.1. About half of the household units in the United States were built before 1970, a period in which energy efficiency was not a major concern.

The distribution of energy sources used to heat these U.S. households is displayed in Figure 14.2. The dominance of fossil fuels for space-heating applications is obvious, even more so when it is recognized that the electricity which is used is produced to a large extent with coal or natural gas.

The energy used for space heating and water heating accounts for most of the total energy consumption in residential units. For Great Britain this is illustrated in Figure 14.3, which shows the share of domestic energy

Our Energy Future: Resources, Alternatives, and the Environment
By Christian Ngô and Joseph B. Natowitz

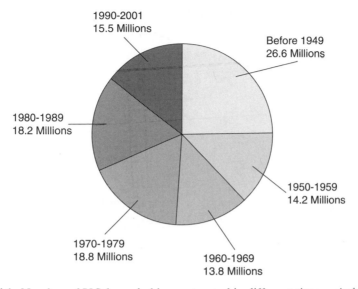

Figure 14.1. Number of U.S. households constructed in different time periods prior to 2001. *Source*: Data from www.eia.doe.gov/emeu/efficiency/recs_tables_list.htm.

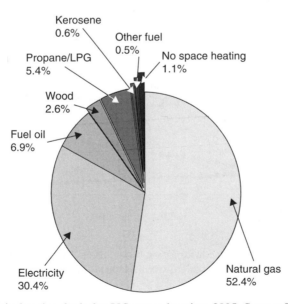

Figure 14.2. Main heating fuels for U.S. space heating, 2005. *Source*: Data from www.eia.doe.gov/emeu/recs/.

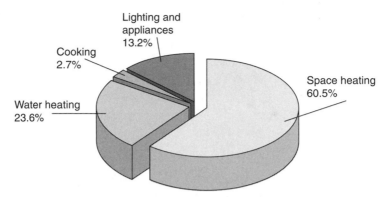

Figure 14.3. Share of domestic energy consumption in Great Britain, 2003. *Source*: Data from www.dti.gov.uk/files/file17822.xls.

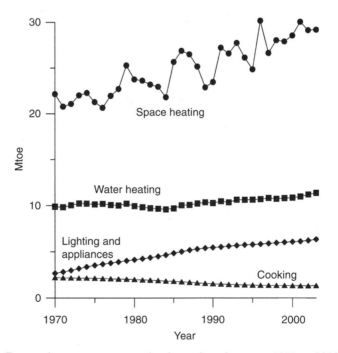

Figure 14.4. Domestic energy consumption by end use between 1970 and 2003 in Great Britain. *Source*: Data from www.dti.gov.uk/files/file17822.xls.

consumption by end use for 2003. Space heating and water heating account for more than 85% of the total domestic energy consumption.

The evolution of energy consumption as a function of time varies with the nature of the application. Figure 14.4 shows the evolution of domestic energy consumption by end use between 1970 and 2003 in Great Britain.

Space-heating energy consumption increased regularly over the years. This reflects the fact that houses have become larger and perhaps also that the temperature demanded to feel comfortable has changed.

The increase in size of residential living areas is a quite general phenomenon in developed countries. To some extent this counteracts gains made in energy efficiency. The energy demands for heating water remained quite stable over the years and energy used for cooking decreased. Apparently this reflects significant life-style changes. The energy consumption for lighting and appliances increased at about the same rate as space heating. There are more and more electronic devices in a house. These demand power to operate and often consume significant amounts of energy even when they are in standby mode.

The demand for an increasing level of comfort in the home also drives the energy demand. The level deemed comfortable depends very much on location and on the people concerned. For example, Americans typically find that a comfortable temperature is in the range of 20–26 °C while the English tend to prefer colder temperatures, in the range of 15 °C to 21 °C.

It should be noted that in the case of heating a decrease of 1 °C saves about 7% of the energy costs. Figure 14.5 depicts residential primary energy consumption in the United States over a 30-year period. After a short decrease, probably due to the oil shocks of the 1970s, the energy consumption increased smoothly from the early 1980s.

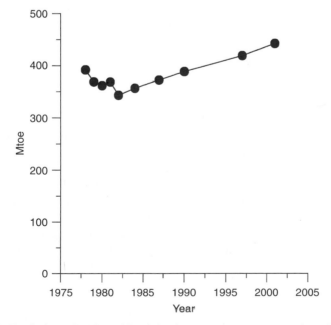

Figure 14.5. Evolution of U.S. residential primary energy consumption. *Source*: Data from www.eia.doe.gov/emeu/efficiency/recs_tables_list.htm.

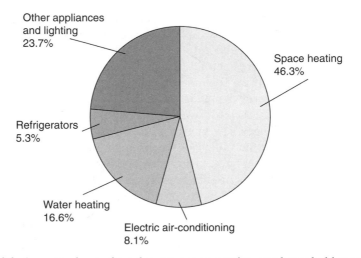

Figure 14.6. Average share of total energy consumption per household according to end use. *Source*: Data from www.eia.doe.gov/emeu/recs/recs2001/ce_pdf/enduse/ce1-2c_construction2001.pdf.

The end use of energy in residential units varies significantly with location. In the northern part of the United States, where very cold winters are common, houses can require a great deal of energy for space heating. In contrast, in the south, air conditioning may account for the bulk of energy usage. In Figure 14.6 the average U.S. share of total residential energy consumption for different end uses is depicted for 2001. Space heating is the largest single contributor to energy consumption.

14.2. TOWARDS MORE EFFICIENT HOUSING

A large part of energy consumed in the world is used to produce heat for residential space heating and hot-water production. Reduction of energy consumption in housing and the corresponding greenhouse gas emissions is becoming ever more important in developed countries. Houses are now built in such a way that they are heavily insulated and any exchange of air with the outside is controlled by ventilation systems. This is very different from earlier practices where no, or relatively primitive, insulation was the norm and air circulated easily between the inside and the outside in an uncontrolled way. Airtight houses allow control of energy losses which can decrease the energy demand while a high quality of comfort is maintained. However, for health considerations adequate ventilation is essential. Figure 14.7 shows the difference between older houses and newer energy-efficient houses.

In a standard European house with no special insulation, thermal energy losses are on average distributed as seen in Figure 14.8. Loss through the roof

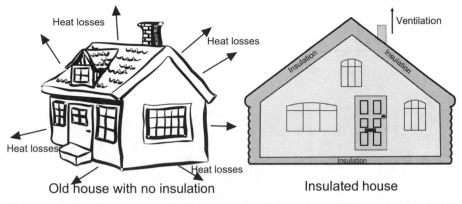

Figure 14.7. In older houses with little or no insulation air could flow relatively freely in or out the house. New houses are based on the concept of an airtight space with controlled ventilation allowing exchanges with the outside.

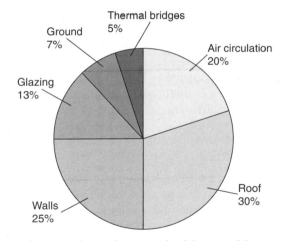

Figure 14.8. Thermal energy losses in a standard house with no special insulation. Percentages shoud be taken as estimates. *Source*: Data from J. Bonal and P. Rossetti, *Energies Alternatives*, Omniscience, 2007, www.eia.doe.gov/emeu/recs/recs2001/ce_pdf/enduse/ce1-2c_construction2001.pdf.

and the walls accounts for more than a half of the total loss. Thus additional insulation of the walls and the roof provides the most efficient means of obtaining significant energy loss reductions. The addition of more efficient glazing (double or triple glazing) could cut a bit more than 10% of the total thermal loss.

Heat losses from any hot object depend very much of the surface area which is in contact with the surroundings. All other parameters kept constant, the larger this area, the larger the losses. The same is true for houses. A more

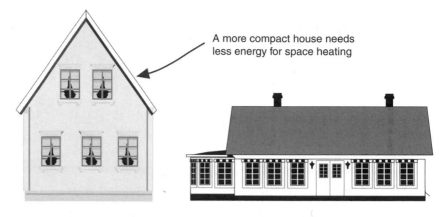

A more compact house needs less energy for space heating

Same living space and same volume space

Figure 14.9. Two comparable houses with different compactness.

compact house has a smaller contact area with the outside and lower energy losses. This is illustrated in Figure 14.9, which compares two houses assumed to have the same internal floor space and the same volume. The two-floor house on the left is more compact than the one-floor house on the right. The surface in contact with the outside is larger for the one-floor house. Even if they are built with the same materials, there may be a large difference in energy consumption. This difference may easily reach 50%.

A decrease of the internal temperature of a house from 20 to 19 °C in the winter can reduce the energy consumption by 7%. Regular maintenance of the furnace can save 8–12% and is also recommended for safety reasons. Replacing an old furnace with a new one could save ~15% and replacing it by a highly efficient one could decrease energy consumption by 30–40%. An intelligent active management system with electronic regulation allows a saving of about 10% of the energy. With existing technologies and responsible behavior it is possible to cut the energy consumption in a house almost in half without major changes in lifestyle.

Solar energy is freely accessible and its use in passive systems can significantly decrease the energy demand for space heating. If sunlight passes through normal glazing, the space behind the glass is warmed. This is nice in the winter but can be inconvenient in summer when it is hot. In the latter case it is better to avoid direct illumination from the sun. Since the height of the sun in the

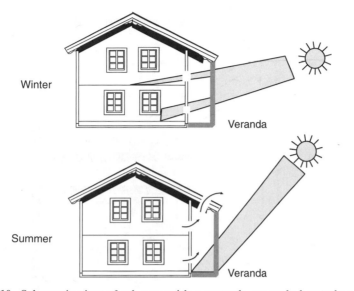

Figure 14.10. Schematic view of a house with a veranda properly located to provide additional heating in winter and shading in summer.

sky depends upon the season, simple construction techniques can be used to manage solar energy between the cold and the hot seasons. For example, the veranda sketched in Figure 14.10 is designed in such a way that it allows a greenhouse effect in winter when the sun is low in the sky but blocks the sunlight in the summer when the sun is high in the sky.

The above system is called *passive* because it does not require any source of energy to function. Actually, many of the systems dedicated to renewable energies require electricity to work. This is the case for most modern solar heaters. Their pumps need electricity to function. Even if the pumps have powers of only a few tens of watts ($\approx$30–40W), this can result in nonnegligible electricity consumption over the course of a year and has to be taken into account when energy costs are calculated.

There are many clever systems that allow the heating or cooling of houses without any use of fossil fuels, and it is not possible to go through all of them within such a short chapter. We briefly describe some of them to give an idea of what can be done.

For example, the veranda in Figure 14.10 could be replaced by a Trombe wall. This is a southerly facing wall built from material than can store thermal energy (e.g., stone, concrete, water) with a glazing on the outside and an enclosed air space. It is a passive system. This concept, patented in 1881 by Edward Morse, was developed by the French engineer Felix Trombe in 1964. A sketch is shown in Figure 14.11 for a particular kind of Trombe wall. The idea is that sunlight going through the glazing heats the air and the Trombe

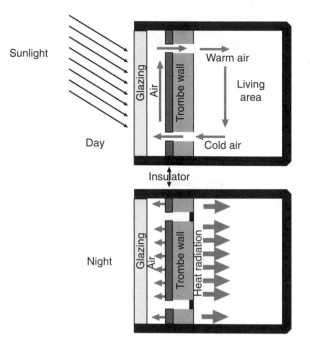

Figure 14.11. Principle of a Trombe wall arrangement (there are other possibilities).

wall. The warm air between the Trombe wall and the glazing can eventually be channeled by convection into the building or outside.

Figure 14.11 shows an extension of the basic Trombe wall. It is augmented with two vents and addition of insulation between the air space and the Trombe wall. During the day, sunlight passes through the glazing. The air between the glazing and the Trombe wall benefits from the greenhouse effect and warms up substantially. The Trombe wall also warms up. Hot air rises and goes through a vent at the top of the wall. Cool air from the house goes through a vent located at the bottom of the wall. This makes a natural convection phenomenon and an air circulation pattern sets in, warming the house. At night, the vents can be closed and the Trombe wall radiates thermal energy into the house. The insulation between the collector space (air) and the Trombe wall decreases radiation toward the outside.

It is possible to control a Trombe wall system such as that above by simple methods. For example, a movable curtain between the Trombe wall and the glazing can modulate the incoming flow of sunlight. Controlling the flow through the vent is another possibility to control heat transfer to the house.

Heat recovery ventilation provides an efficient system to reduce heating costs. Part of the heat energy in the exhaust air is transferred to the fresh air entering the house. Two different air flow circuits and an air heat exchanger are necessary. Therefore such systems are better suited for new buildings rather than for old houses. It is necessary that the incoming air at the level of

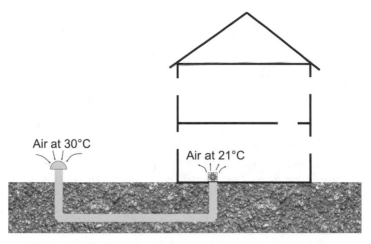

Figure 14.12. Sketch of a *puit provencal* or *puits canadien*.

the heat exchanger is at a temperature greater than $0\,^{\circ}\text{C}$; otherwise freezing of water and formation of ice could block the outgoing air. In case of temperatures below $0\,^{\circ}\text{C}$, the intake air should be warmed to a temperature higher than $0\,^{\circ}\text{C}$. The humidity level of the exhaust air should also be transferred to the intake air.

Subsoil, which can typically be at a temperature on the order of $18–22\,^{\circ}\text{C}$ at a depth on the order of $2\,\text{m}$, can also be used to control housing temperatures. A tube located underground can contribute to cooling in summer or heating in winter. It can work in a closed-loop circuit, but additional ventilation of the house is necessary, or it can be open to the outside and take air there. Of course some caution should be exercised in the installation to prevent problems (e.g., corrosion and dirtying of the tube). It can work as a passive system or an active one with an external venting system. This technology, called *puits Canadien* or *puits provencal* in France, is schematically presented in Figure 14.12.

> Two thousand years ago the Romans sometimes cooled their houses using *puits provencal*, geothermal heat exchangers which were made using buried clay tubes. Today, *puits provencal* coupled to heat recovery ventilation allows recovery of up to 95% of the heat losses in a house.

14.3. DIFFERENT REGIONS, DIFFERENT SOLUTIONS

There is no unique solution to meeting housing energy needs. Quite varied environmental and climate conditions can be found, even in a single country.

These conditions require somewhat different approaches to providing comfortable residences. The techniques to be preferred depend on the location and the available resources. Consider, for example, the United States, where there are five major climate areas.

- For the northern part of the United States, large changes of temperature between very cold winters and hot summers are common. Houses built in this region often have a slab, a crawlspace, or a basement, all of which have to be managed as far as heat and moisture is concerned. The main problem is to manage moisture. In summer, when air conditioning is necessary, humid air flows from outside the house to the inside. In winter water vapor has to flow to the outside but the formation of ice blocks must be avoided.
- In the southeastern United States, the climate is hot dry or medium dry. Air conditioning is required more than heating. Houses should employ architectural designs that minimize the needs for air conditioning. Heavy rains are common and moisture must be managed properly.
- A hot and humid climate zone is found in the southwestern United States. High humidity and periodic heavy rainfalls occur. Air conditioning is heavily used. Condensation of water from the air must be managed properly.
- A marine climate zone exists in the western part of the United States, in a narrow band between the Canadian border and Los Angeles. Temperatures are mild but the climate is fairly humid.
- Finally a mixed-humid climate zone exits in the eastern part of the United States. The weather is cold in winter, hot and humid in summer. Water vapor must be transported from the inside to outside in winter and from the outside to inside in summer.

While the needs vary significantly, many clever energy-efficient methods have now been developed to meet the needs. With new materials and new technologies and using a scientific approach great progress has been made. In principle it is now possible to build zero-energy and even positive-energy houses.

14.4. BIOCLIMATIC ARCHITECTURE

Bioclimatic architecture focuses upon energy saving and reducing negative impacts on the environment from the very beginning. Using well-known techniques can significantly decrease energy demand and any associated pollution.

Interestingly, some of the design solutions employed use techniques well known to our ancestors. Harnessing part of the incident solar energy using glazing, verandas, or thick walls allows decreasing the demand of heat in winter. Proper positioning of a house and its windows or breezeways can have a significant influence on comfort. A well-oriented window can replace a large part of heating during winter just by harnessing some of the solar energy. It is also possible to use solar light to provide part of the lighting of the house during the day. Thick-walled houses can remain relatively cool in warm climates, even without an air-conditioning system. Thick walls also allow storage of thermal energy and smoothing of temperature variations outside.

As already discussed, houses that are relatively compact minimize heat exchanges from the walls and the roof. In such cases natural or forced ventilation is necessary since efficient houses are airtight and air should be renewed regularly. Air circulation also prevents moisture and mold development in some parts of the house. A large part of any additional heat which is still necessary can be produced using renewable energies or new technologies such as heat pumps.

The improvement which bioclimatic design can make in the area of reducing home heating costs can be illustrated by the example of *La maison des négawatt* suggested by T. Salomon and S. Bedel, *La Terre Vivante*, 2002. Consider a conventional French house of $100\,m^2$ with a volume of $250\,m^3$. This house is taken to be one with $16\,m^2$ of glazing ($3.2\,m^2$ oriented south), 7-cm-thick insulation for the walls, and 14-cm-thick insulation for the roof. Window shutters remain always open. The temperature in the winter is to be maintained at $19\,°C$. The energy needed for heating is about 14,300 kWh/year.

Assume now that this house is properly oriented on its lot and has $11.2\,m^2$ of glazing facing south. The temperature in the house is maintained at $19\,°C$ during the day and $15\,°C$ during the night. Shutters are closed during the night and 85% closed during the summer. The energy necessary to maintain these conditions decreases to about 9400 kWh/year—that is, 34% less than in the previous case.

Finally, consider a bioclimatic house of the same size but with $28\,m^2$ glazing ($22\,m^2$ facing south), 10 cm for the walls, and 20-cm-thick insulation for the roof. As in the previous case, the shutters are again closed during the night and 85% closed during the summer. The space temperature is to be maintained at $19\,°C$ during the day and $15\,°C$ during the night. The total heating energy requirement drops to about 5100 kWh/year, which is 65% less than in the first case.

"Positive-energy houses," houses that produce more energy than they consume, can also be built. The excess of energy is usually in the form of excess electricity produced by the house's dedicated energy system. In many locations this excess may be sold for use on the grid. (This is practical only when the number of such houses is small compared to the total number of houses.) Because the excess of electricity generally occurs at off-peak hours when the total demand on the grid is low, it is desirable that there be a system for storage of excess electricity for use during peak demand. Employing modern storage technology for home use would add about 0.15 €/kWh to the overall costs. In practice, "zero-energy houses" would be sufficient in most cases, especially if they are coupled to cheap and efficient electricity storage systems.

14.5. INSULATION

The natural direction of heat exchange between two objects of different temperature will always be from the hotter object to the colder object. As soon as two bodies in contact have a different temperature, heat flows from the hot body to the cold one until the temperatures are equal. This heat exchange will take place spontaneously, but it can be slowed down dramatically if insulating materials are placed between the two bodies.

The transfer of thermal energy can be accomplished by three different physical phenomena—conduction, convection, and radiation:

- *Conduction* is a transfer of thermal energy without flow of matter. It is by conduction that a saucepan filled with water is heated when placed on an electric hotplate. Metals are the best thermal conductors. It is interesting to note in passing that diamond is also a good thermal conductor, even better than copper (but very expensive). However, it is an electric insulator while copper is also a good electric conductor.
- *Convection* is thermal energy transfer by the motion of matter. Particles in a hot region move to cooler regions carrying heat energy with them. When the saucepan containing water is heated on a hotplate, convection also occurs; the water located close to the bottom is warmer than the water on top of the saucepan. This warmer water moves upward. The colder water at the top sinks to the bottom and gets heated.
- *Radiation* uses electromagnetic waves to transfer thermal energy. This can occur even in a vacuum where conduction and convection are negligible. When our skin is exposed to sunlight, it is warmed by radiation.

Insulating materials are those which provide a high resistance to heat transfer. The resistance depends on the type of material, its density, and its thickness. Assuring adequate insulation is a major consideration in the construction of energy-efficient housing. In winter insulation slows down heat leaks toward the outside and in summer it slows down heat transfer into the house. There exist many materials with useful insulation capabilities. These can vary widely in effectiveness and price.

Fourier's law expresses ΔQ, the heat transferred per unit of time, as $\Delta Q = \lambda (S\,\Delta T)/e$, where λ is the thermal conductivity, S the heat transfer area, ΔT the temperature difference across the material, and e the thickness of the material. The thermal conductivity λ is the quantity of heat conducted in a unit of time through a material of 1 m thickness whose opposite faces of area, S, differ in temperature by 1 K. In the International System of Units (SI) λ is expressed in $W/(m \cdot K)$. In the United States it is typically expressed in $Btu \cdot ft/(h \cdot ft^2 \cdot °F)$. [The conversion between the two systems is $1\,Btu \cdot ft/(h \cdot ft^2 \cdot °F) = 1.73\,W/(m \cdot K)$.] The thermal resistance (R value), $R = e/\lambda$, is expressed in $m^2 \cdot K/W$. The convenience of using R values is that they are additive (Figure 14.13). The thermal conductance is the reciprocal of the thermal resistance R. The thermal transmittance, or U value, takes into account not only conduction but also convection and radiation:

$$U = \text{conduction} + \text{convection} + \text{radiation} \quad [W/(m^2 \cdot K)]$$

Vacuum, the absence of matter, is an almost ideal insulator. Thermal energy can pass through a vacuum only by radiation. If reflecting surfaces are used together with vacuum a very efficient insulating system can be constructed. Dewar flasks used to store cryogenic liquids are constructed based on this principle. For various engineering and safety considerations, vacuum is not normally used to insulate buildings. Instead, materials with low thermal conductivities λ are employed. The heat flow through a material depends on its thermal conductivity. If the thermal conductivity is large, the heat flow is significant.

Table 14.1 lists the thermal conductivities of some selected materials.

In the construction domain the thermal resistance is called the R value. One usually defines the thermal resistance R of a material of thickness e as $R = e/\lambda$. The larger the resistance, the smaller the energy flow through the material. If several materials are used in layers, the total resistance is the sum of the individual resistances (Figure 14.13).

TABLE 14.1. Average Values of Thermal Conductivity λ for Selected Materials

Material	Thermal Conductivity (W/m · K)
Copper	380
Concrete	1,5
Glass	1,15
Water	0,6
Fir	0,12
Motionless air	0,024

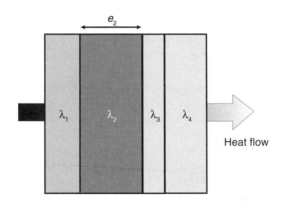

$$R = e/\lambda \qquad R = R_1 + R_2 + R_3 + R_4$$

Figure 14.13. If several materials with different values of thermal conductivity are layered, the thermal resistances are additive.

Air by itself is a good thermal insulator provided it is motionless; otherwise it allows thermal energy exchange through convection. A well-constructed insulator traps air in small volumes and keeps that air motionless. See the schematic representation in Figure 14.14. This is also the reason why a good insulator is not typically very thin because it is the volume of trapped air which does the job. Thin insulators such as aluminized foils are sometimes used for survival blankets, for example, or in the roof of a house to supplement conventional insulation material. However, these foils stop only radiative transfers.

Water has a thermal resistance about 20 times less than air. It is important to prevent moisture because materials containing water lose a large part of the thermal insulating properties. It turns out that a 10-cm-thick insulating panel has about the same thermal insulating properties as a 5-m-thick stone wall.

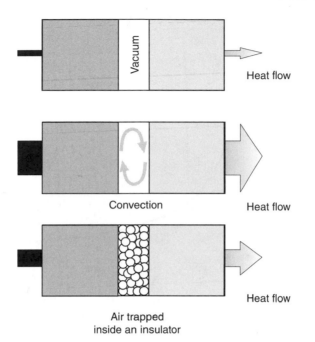

Figure 14.14. Sketch of different types of thermal insulating techniques. Vacuum is a very good insulator but not used in buildings. In order for air to be a good thermal insulator, it should remain motionless.

The first few centimeters of insulating material are the most efficient. Consider fiberglass, for example. Using 7 cm of fiberglass as an insulator decreases home heating energy requirements by about 40%. With 20 cm of fiberglass, approximately three times more, savings of 50% may be realized. A more complete example of this law of diminishing returns can be seen in Figure 14.15, which shows the thermal energy loss over the period of one year for a concrete wall covered by different thicknesses of vegetal wool insulation. Vegetal wool is a combination of latex, coco fiber, and wool. With only 2 cm of vegetal wool the energy loss is already reduced by a factor of 4. The improvement then becomes asymptotically slower as the thickness of the insulation increases.

A comparison of the insulation efficiencies for several different materials is presented in Figure 14.16. The thicknesses necessary to provide the same insulation as 100 cm of regular concrete are displayed. Of the materials considered, extruded polystyrene is the best because it is most effective at trapping and immobilizing air.

About $60 \times 10^6 \, m^3$ of concrete is used each year in France. This corresponds to about $1 \, m^3$ per inhabitant. Today's concrete contains less than 10% cement. The remaining part is made of gravel and sand.

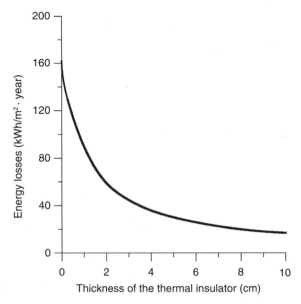

Figure 14.15. Thermal energy losses over a year from $1 \, cm^2$ of concrete wall having different thicknesses of vegetal wool for thermal insulation. *Source*: Data from J. P. Oliva, *L'isolation écologique*, La Terre Vivante, 2008.

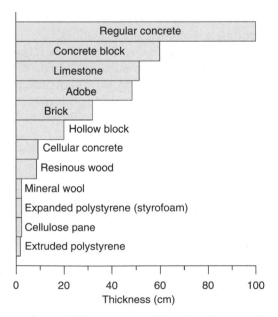

Figure 14.16. Comparison of different materials giving the same insulation as $100 \, cm$ of concrete. *Source*: Data from T. Salomon and S. Bedel, *La maison des negawatts*, La Terre Vivante, 2002.

14.6. GLAZING

The area covered by the windows of a house is a parameter that has a major impact on the energy necessary to heat or cool that house. In France, which has a temperate climate, the ratio of the area of the windows to that of the room normally depends on the orientation of the house on the lot. For southerly exposures this ratio should be 0.2–0.35. For easterly and westerly exposures it should be 0.1–0.25. Finally for a northerly exposure, 0.1 or less is recommended. The actual optimal values depend of course on the location. They will be different in the northern part of France or in the south.

In a standard house, energy losses through windows account for a little more than 10% of the total energy losses (Figure 14.8). Most older houses have single-pane windows which are not energy efficient. There are now new glazing technologies with several panes of glass. These are becoming standard for energy-efficient houses.

An insulated window is a set of two or three panes of glass spaced apart and sealed hermetically with air or another gas trapped between the two panes of glass. A sketch of a *double-glazed unit* is shown in Figure 14.17. These units are also called *insulating glass units* or *sealed insulated glass*.

The transmission coefficient of glass (thermal transmittance) is not very dependent on the thickness of the glass. On the order of $5.7 \, W/m^2 \cdot K$, it is basically about three times less insulating than a medium-quality concrete wall. For a single-pane window, including the frame, the transmission coefficient can decrease to about $5 \, W/m^2 \cdot K$, which is still high. With a typical double glazing the thermal transmittance can be reduced to about $3–3.5 \, W/m^2 \cdot K$. With low-emissivity glass one can go down to $1.4–1.6 \, W/m^2 \cdot K$ or even $1.1 \, W/m^2 \cdot K$ with argon double glazing. With triple glazing the values are a little smaller, $\approx 0.7–0.8 \, W/m^2 \cdot K$. In Figure 14.18, the thermal transmittance is given for some window frames with good efficiencies.

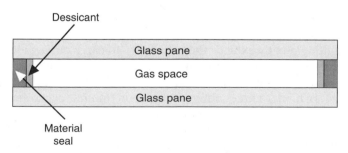

Figure 14.17. Sketch of double-glazed unit.

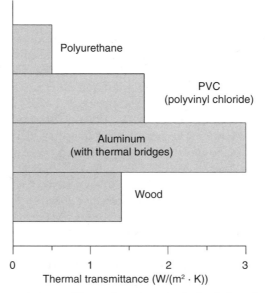

Figure 14.18. Thermal transmittance for some typical window frames of good efficiency. *Source*: Data from J. P. Oliva, *L'isolation écologique*, La Terre Vivante, 2008.

A triple-glazed window is about two times more efficient as far as thermal insulation is concerned than a double-glazed window. A good double-glazed window is about five times more efficient than a single-glazed window.

Shutters can also play an important role in saving energy. For example, for an annual input energy of about 210 kWh/m² in the Paris area the energy loss through double glazing in a south-facing wall equals 200 kWh/m² without shutters and 120 kWh/m² with shutters.

14.7. LIGHTING

Artificial lighting greatly extends the possibility to carry out a wide variety of human activities in a comfortable manner. For many centuries, before the phenomenon of incandescence was discovered, fire was the only source of artificial light. In modern times other phenomena (luminescence and phosphorescence) have provided new possibilities to produce artificial light at low cost.

A carbon arc lamp consists of two carbon electrodes separated by air. It produces light by establishing an electric arc between the two electrodes. Discovered by Sir Humphry Davy at the beginning of the nineteenth century it proved to be a seminal breakthrough in lightning. This technology was widely developed in the second part of the nineteenth century in parallel with the development of electrochemical batteries. The "electric candle," patented in 1876 by the Russian Paul Jablockhoff, was commercialized in France and Great Britain. Arc lamps were an important part of the lighting market until Thomas Edison invented the incandescent lamp in 1878. Modern arc lamps or arc lights have tungsten electrodes inside a bulb containing a gas: neon, argon, xenon, krypton, sodium, or mercury.

The international measure of light is the *candela*. The candela was initially defined as the light intensity of a "standard English candle." At night and in clear weather, a standard English candle could be seen at a distance of 27 km. The candela is 1/683 W/sr. Other derived units are the lumen and the lux. One lumen corresponds to the flux of light emitted within a solid angle of one steradian by a source of one candela. One lux is the intensity of light received uniformly on a surface of one square meter illuminated by one lumen. Thus $1 \, lx = 1 \, lm/m^2 = 1 \, cd \times sr/m^2$.

The first incandescent lamps did not produce much light—about 16 "candles"—while arc lamps were then producing intensities between 2000 and 4000 candles. Incandescent lamps established themselves as the new standard for artificial lighting, replacing arc lamps, mainly because they could be easily mass produced, their lifetimes are much longer, and they need little maintenance. These features spurred a rapid development of the incandescent lamp market and, concurrently, the establishment of widespread distribution grids for electricity. An incandescent lightbulb has a very thin filament of tungsten housed inside a glass container. The joule effect heats the filament at high temperature and light is emitted. The efficiency of light production is small and one gets of the order of 12–14 lm/W of input electricity. Throughout the world this technology is widely used in private homes because incandescent bulbs are cheaper than the alternatives. Their useful lifetime is about one year.

The global market for lighting is around $300 billion. There are close to 33 billion lamps in the world and half of that number need to be replaced each year. In 2005, the global consumption of electricity for lighting was about 2650 TWh, which is about 19% of the total electricity consumption (but less than 2% of the global energy consumption). Almost half of the global lighting energy is used to power incandescent lamps. Compact fluorescent lamps consume less energy and they are slowly replacing incandescent lights but are more expensive. Fluorescent lamps are however extensively used in the service sector and in industry.

In 1999 France used 41 TWh for lighting. In the year 2000, the United States used 659 TWh for lighting, an amount of electricity approximately equivalent to the total annual electricity production of France plus Italy. On the average lighting accounts for between 7 and 15% of the total electricity consumption of developed countries, but it is close to 20% in the United States. In developing countries, lighting represents an even larger share of the energy consumption. It is on the order of 30% of the total electricity consumption in Tunisia and 86% in Tanzania.

The distribution of lighting requirements between the different sectors of the market in France is indicated in Figure 14.19 for the year 1999. About 60% is in the service sector, 30% for home lighting, and 10% for public and road lighting. Over the past two decades there has been a large increase in home lighting consumption (from about 5 TWh in 1979 to 14 TWh in 1999).

The number of different types of lamps used across the world in 2007 is displayed in Figure 14.20. Of these fluorescent lamp usage dominates the energy and service sectors while incandescent lamps dominate the home usage sector.

A *fluorescent lamp* produces less heat and more light and is about four to six times more efficient than an incandescent bulb. Fluorescent bulbs consist of a tube with two electrodes and a gas typically containing argon and mercury vapor. In operation, there is flow of electrons between the two electrodes. The electrons collide with mercury atoms and excite them. Mercury atoms deexcite from their excited state to the ground state, emitting ultraviolet radiation. This invisible radiation is converted into visible light using a phosphor coating inside the tube. Typically these bulbs can produce 50–100 lm/W. The lifetime of a fluorescent bulb is about five years. However, these lamps contain toxic

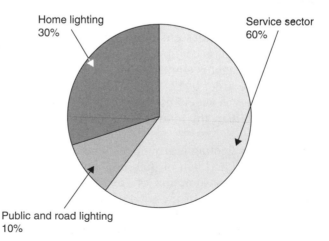

Figure 14.19. Distribution of lighting requirements between different sectors in France, 1999 (total lighting 41 TWh). *Source*: Data from *Nouvelles technologies de l'énergie*, Vol. 4, Lavoisier, 2007.

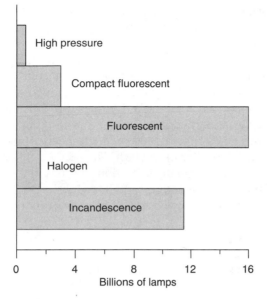

Figure 14.20 Number of lamps of different types in the world. There is a total of nearly 33 billion lamps. *Source*: Data from G. Zissis, in *Nouvelles technologies de l'énergie*, Vol. 4, Lavoisier, 2007.

materials such as mercury, lead, and cadmium as well as some radioactive materials (thorium, ^{98}Kr). In the United States 2.5 tons of mercury is used each year to produce about 750 million new fluorescent tubes.

Halogen lamps are incandescent lamps containing a halogenous compound of iodine or bromine. The halogenous compound increases the lifetime of the bulb because it decreases the rate of evaporation of the tungsten filament and prevents darkening of the inner surface of the bulb. Improvement in the evaporation of tungsten is used to increase either the lifetime of the lamp or light efficiency. Depending upon the application, the optimization may be different. For home applications, light output is chosen to be on the order of 20–25 lm/W. Halogen lamps have long operational lifetimes, typically 2000 h, and a color of light which is often appreciated by the consumer.

Compact fluorescent lamps work like the linear fluorescent lamps described just above. They have a good efficiency (40–80 lm/W) and their lifetimes are on the order of 6000–12,000 h if they are switched on only once a day. The large dispersion in the efficiency comes from the quality of manufacture. Low-cost compact fluorescent lamps are less efficient than those manufactured with greater care and with better quality electronic components.

High-discharge lamps work at high pressure (from 100 to several hundred bars). A plasma is created during operation. Most of the high-pressure discharge lamps contain mercury to control electronic mobility inside the plasma as well as other elements to define the wavelength of the emitted light.

Efficiencies between 50 and 110lm/W can be obtained. Heat losses are 25–50%, much smaller than those for incandescent lamps. The lifetimes are on the order of three years.

There is currently a strong focus on the development of *light-emitting diodes* (LEDs) for illumination purposes. An LED is a semiconductor device which converts electricity into light. The typical efficiency for commercial residential applications is on the order of 20lm/W, but it is foreseen that 50lm/W could be reached in a few year and perhaps 200lm/W within one or two decades. In the laboratory efficiencies of about 100lm/W are being obtained. The lifetime is expected to be on the order of 50,000h. They emit a monochromatic light, which means they are not normally used alone. Different methods are used to produce white light (e.g., three or more LEDs of different colors, LEDs emitting in the near-UV region coupled to a light-converting phosphor).

In France, installing a central system for monitoring lighting in a house with programmable switches to save electricity leads to a cost increase a little lower than 10%. Installed in a new house the payback occurs within a few years. More generally monitoring the whole house with a central electronic system can decrease the energy needs.

14.8. VENTILATION

Since new homes are airtight, ventilation is necessary to control air exchanges with the outside. Adequate ventilation should always be present for health reasons. However, it is often better to use controlled ventilation than to leave windows open because this allows the control of energy losses. Ventilation is required because it increases comfort and removes dilute pollutants emitted by many housing materials and manufactured objects: for example, paint, glue, and carpets. Ventilation can be passive, exploiting the density differences of cold and hot air, or active, using fans. Draft accelerates thermal energy exchange. However, in general too much draft is uncomfortable. The flow of air should be kept as slow as possible inside a house.

In France, about 2000kWh /year is lost in a five-room house using a simple ventilation system. With a heat and recovery ventilation system, 60–90% of this energy loss can be avoided. A simple mechanical ventilation system costs about 400€. A mechanical recovery ventilation system is more expensive, between 1500 and 3000€, but a quick payback can be obtained.

There are several kinds of active ventilation systems. The simplest is mechanical extracting ventilation, which continuously extracts air from kitchens and bathrooms (rooms having significant moisture). Ventilation can also be designed in such a way that part of the heat contained in the outgoing air is recovered. In single-room heat recovery ventilators the fan includes a heat exchanger and about 65% or more of the heat of the outgoing air is recovered. The system can be generalized to the whole house with a central unit. In this case the heat recovery can amount to about 85–90%.

14.9. WATER

Air typically contains water vapor. The concentration of water in air varies but cannot exceed a given value at a given temperature. If it does, water vapor condenses into liquid water and creates unwanted moisture that may lead to mold and can corrode various materials. The dew point temperature is the temperature threshold at which air is saturated (100% relative humidity). The dew point curve is shown in Figure 14.21. Below this curve air is unsaturated and the relative humidity can range from 0 to 100%. Above that curve the air is oversaturated and water condenses.

Condensation generally takes place at cold points. Condensation leads to increases in humidity which can foster growth of mold. Besides producing a musty odor, mold spores may lead to health problems, asthma for example. Good ventilation is necessary to prevent these problems. Indeed mold

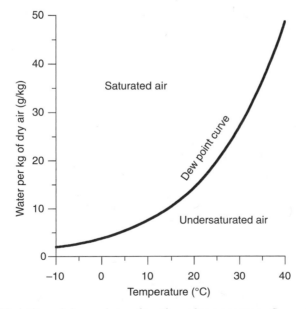

Figure 14.21. Variation of dew point as function of temperature. *Source*: Data from J. P. Oliva, *L'isolation écologique*, La Terre Vivante, 2008.

develops in areas where air flow is limited and water remains (e.g., corners, junctions between walls, ceilings, or floors). Some molds are toxic.

At 20 °C, 1 kg of dry air will become saturated when 14.4 g of water vapor is added. If we are in a room where 1 kg of dry air contains 10 g of water vapor, 69% of the maximum for that temperature, we feel comfortable. If it is winter and the temperature of the window pane is 10 °C, water vapor will condense because at that temperature the air is oversaturated (the dew point at 10 °C corresponds to 7.6 g of water vapor in 1 kg of air). As a result 2.4 g of water per kilogram of air will condense on the glass.

In Figure 14.22 the connection between the dew point temperature and the surrounding temperature is shown for different values of relative humidity. For example, if we consider a room at 22 °C with 70% relative humidity, the dew point is 16.3 °C. This means that there will be condensation on any part of the room which is at a temperature lower than 16.3 °C unless good ventilation exists to prevent this phenomenon. At two different temperatures the same value of relative humidity corresponds to different quantities of water contained in 1 kg of air. At high temperature the air can contain more water than at low temperature.

A major problem of condensation is that the water may be produced in the insulation. Water itself is a better thermal conductor than air. The thermal conductivity of water is given as $\lambda = 0.6 \, W/m \cdot K$ while for air $\lambda = 0.024 \, W/m \cdot K$. Thus condensation in the insulation can sharply reduce the insulating capabilities. Water vapor can be stopped by use of appropriate materials or the

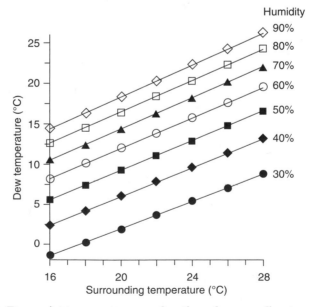

Figure 14.22. Dew point temperature as a function of surrounding temperature at different values of humidity. *Source*: Data from J. P. Oliva, *L'isolation écologique*, La Terre Vivante, 2008.

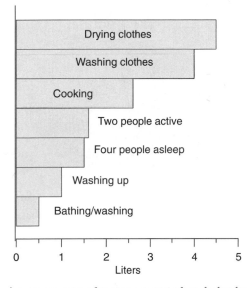

Figure 14.23. Approximate amount of water generated each day by an average household consisting of four people. *Source*: Data from www.buildingpreservation.com/ Condensation.htm.

insulator can be made permeable to water so that it escapes; this depends upon the strategy chosen to manage water vapor.

Most of the humidity created in a house comes from the activities of the inhabitants. The quantity of water emitted depends upon life-style and activity level. For a household of four, Figure 14.23 shows the approximate amount of water emitted per day in different household activities. Globally this amounts to about 15 liters per day. It turns out that the relative humidity indoors is usually larger than that outside.

14.10. ENERGY USE IN A HOUSEHOLD

Heating, air conditioning, and hot-water production represent a large part of the energy consumption of a household (Figure 14.3) but other energy-consuming activities exist. For example, electricity is used to power many appliances. The share of the energy used for such purposes in a typical French household is shown in Figure 14.24. Refrigerators and freezers account for almost one-third of the electricity consumption. This corresponds to a total energy consumption of about 17 TWh/year in France, more than two times the quantity of electricity needed by the country to power its electric trains.

The cost of different fuels used for space heating in France is shown in Figure 14.25 for the year 2004. It should be noted that the price of fossil fuels has strongly increased since then, which makes them currently less

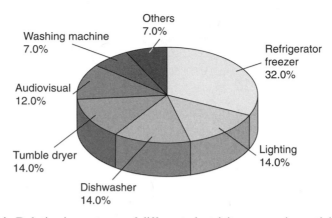

Figure 14.24. Relative importance of different electricity-consuming activities (assuming no electric heating or cooling) in a typical French household. *Source*: Data from www.ademe.fr.

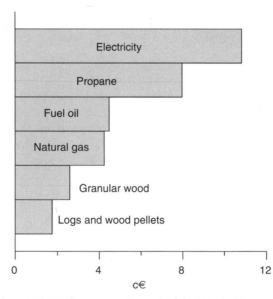

Figure 14.25. Price of 1 kWh (lower heating value) in 2004 in France, all taxes included. Data from www.ademe.fr and J. Bonal and P. Rossetti, *Energies alternatives*, Omniscience, 2007.

competitive. The cost for producing 1 kWh of heat with electricity is large. However, heating in this fashion turns out to be desirable because electric heating allows precise control of room temperature. The same level of control is difficult to achieve with gas-fired or fuel oil–fired systems. Thus, in the end the total energy consumption is lower. Furthermore, the use of a heat pump (see below) could reduce this cost by a factor of 3 or 4, making this technology extremely competitive.

14.11. HEAT PUMPS

Heat is the most degraded form of energy. Low-temperature heat cannot normally be exploited for purposes other than space heating and hot-water production. As indicated above, for an isolated system, the thermal flow between two subsystems of different temperatures put into contact is from the high-temperature subsystem to the low-temperature subsystem. Having a method to reverse this flow and take heat from a low-temperature source and transfer it to a high-temperature can provide a great deal of flexibility in energy management. For the isolated system this cannot be done spontaneously. However, this can be done for an open system if we supply the work necessary to accomplish the transfer. This is the principle of a heat pump.

A heat pump is a device which uses work to transfer energy from one source to another. Energy can be transferred from a low-temperature source to a high-temperature one or from a high-temperature source to a low-temperature one. In the end the low-temperature source is hotter in the first case and the hot source is cooler in the second case. Refrigerators, freezers, and air conditioners all work on this principle. Heat pumps operate in some sense in reverse to a heat engine. The principle is depicted in Figure 14.26 for a vapor condensation heat pump. Most heat pumps currently being used are of this type.

The advantage of heat pumps is that various low-quality heat sources can be used. Ambient air, soil (at a depth between about 0.6 and 1.2 m below the surface), and groundwater are mostly used as heat sources for small heat pump systems. Even waste heat sources (domestic, commercial, or industrial) can be exploited. Large heat pump systems can use rivers, lakes, oceans, geothermal resources (rocks at a depth greater than about 100 m), and wastewater.

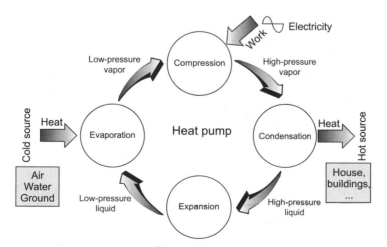

Figure 14.26. Principle of a heat pump extracting heat from a cold source.

An important parameter for a heat pump is its coefficient of performance (COP), which is the ratio of the amount of heat delivered by the heat pump to the energy absorbed by the compressor. For typical heat pumps the COP is on the order of 3–4. For heat pumps driven by an engine or by a thermal system, the PER coefficient, which is the ratio between the usable output energy and the primary input energy, is usually used. For an electricity-driven heat pump the PER is just the COP multiplied by the efficiency of power generation.

The use of 1 kWh of electricity in a conventional way, with convector heaters, for example, produces at most 1 kWh of heat if the efficiency is 100%. If this 1 kWh is used to power an air–air heat pump, the amount of heat delivered will be 3–4 kWh. For the cost of 1 kWh, 3–4 kWh of heat is made available. In this sense, heat pumps play the role of energy amplifiers because the heat pumps draw heat from the environment and make it available for the consumer.

There are two main families of heat pumps. The first is based on vapor compression and its principle is illustrated in Figure 14.26. The majority of heat pumps on the market are of this type. The second family is based on an absorption cycle.

In a vapor compression heat pump (Figure 14.26), the working fluid has a boiling point lower than the temperature of the cold source from which one gets energy. Placed in thermal contact with the cold source, the working fluid will evaporate and remove energy from the cold source. The vapor is then compressed to a higher pressure and temperature using external energy to power the compressor. The hot vapor flows into the condenser where it condenses and releases heat which will be used by the consumer. The high-pressure working fluid then passes through an expansion valve where it goes back to its original liquid state.

In absorption heat pumps thermal energy (heat) rather than mechanical energy (the compressor) is used to drive the cycle. Absorption systems are based on the fact that some liquids or salts can absorb the vapor of the working fluid. They are often used for space air-conditioning with gas as the fuel.

There is a large variation in heat pump installations. Some are used just for space heating or water heating. Others provide space heating, water heating and cooling, and so on. Commonly used heat sources for heat pumps are ambient air (in the temperature range $\approx$-10–15 °C), exhaust air (in the temperature range $\approx$-15–25 °C), groundwater (in the temperature range $\approx$-4–10 °C), ground (in the temperature range $\approx$0–10 °C), and rock (in the temperature range $\approx$0–5 °C). Wastewater at temperatures higher than 10 °C can also be used. Ambient air systems are simple but they are a little less efficient than ground or rock systems. They can only be used in regions which are not too cold in winter. Indeed, below about -15 °C an air–air pump becomes inefficient. As the temperature decreases below freezing, part of the energy produced is used to prevent freezing of the pump and that notably decreases the efficiency of the system.

The efficiency of a heat pump increases as the difference between the output temperature and the temperature of the heat source decreases. This is one of the reasons why geothermal heat pumps have a greater COP than air–air heat pumps. The theoretical maximum COP, COP_{max}, expected between a hot source at temperature T_{hot} and a cold source at temperature T_{cold} is

$$COP_{max} = \frac{T_{hot}}{T_{hot} - T_{cold}}$$

where the temperatures are given in kelvin. Cooling is less efficient than heating. For cooling the theoretical $COP_{cooling}$ is given as

$$COP_{cooling} = \frac{T_{cold}}{T_{hot} - T_{cold}} = COP_{heating} - 1$$

For residential heating, the theoretical maximum COP is about 15 but practical heat pumps presently have a COP of 3–4. The power of heat pumps for domestic purposes is on the order of 15–20 kW. Use of a heat pump is a particularly efficient way to warm swimming pools if the temperature difference between the water and the outdoors is low. For example, if the ambient air temperature is 18 °C and the water temperature is 21 °C, a COP between 5 and 7 can be obtained.

The choice of the heat pump's working fluid is a real issue. In the past most working fluids were chloroflurocarbon compounds (CFCs), which are greenhouse gases and have high ozone depletion potential. Even though they are used in closed-circuit systems leaks are always possible and when they are decommissioned releases of gas may occur. Chlorofluorocarbons are now prohibited because of their impact on the ozone layer. New working fluids, hydrochlorofluorocarbons (HCFCs) are now used. They have a much lower ozone depletion potential than CFCs because of their lower chemical stability in the atmosphere. Their global warming potential is also lower.

14.12. IMPACT ON ENVIRONMENT

Housing the world's population has a large impact on the environment. Building sites replace agricultural land. Their surroundings are usually waterproofed as concrete, roads, and so on, are placed around these sites. The soil which is covered is isolated and lost, usually for centuries. Meeting the energy demands for housing also generates significant CO_2 emissions, which increase the greenhouse effect. We focus on this subject in this section.

Housing and life-style are closely interrelated as far as energy consumption and CO_2 emissions are concerned. Many of the decisions taken by the consumer have a major impact on CO_2 emissions. Consider a typical French family living in a house of $160 \, m^2$ built in 1970 and heated by an oil furnace. Heating this house produces approximately 15 tons of CO_2 per year. Hot-water production provides ≈ 1.9 tons. The household appliances account for $\approx 880 \, kg$. The contribution from cooking is small, $\approx 180 \, kg$. The total emissions of CO_2 from this house are about 18 tons/year. (These data are taken from O. Sidler, www.enertech.fr.) The values are of course approximate but give a good indication of the relative importance of different uses. Decreasing the space temperature a little, improving the insulation, and using a more efficient fossil fuel heating system could reduce the amount of energy consumed for heating and the associated CO_2 emissions by a factor 2. Nevertheless emissions would still be ≈ 10.5 tons CO_2/year.

For the year 2004 the distribution of global greenhouse gas emissions from various human activities is shown in Figure 14.27. The direct contribution of residential and commercial buildings is small, around 8%. However, there

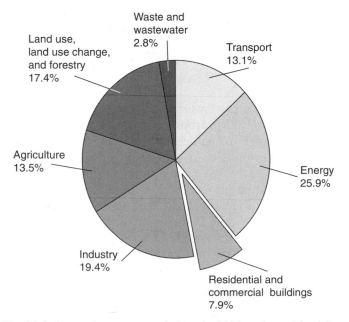

Figure 14.27. Global greenhouse gas emissions in 2004 evaluated in CO_2 equivalents (all greenhouse gases included). *Source*: Data from www.ipcc.ch and DGEMP, http://www.industrie.gouv.fr/energie/sommaire.htm.

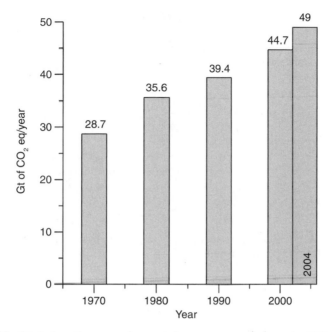

Figure 14.28. Global anthropogenic greenhouse gas emissions in CO_2 equivalent (includes CO_2, CH_4, N_2O, and F gases). *Source*: Data from www.ipcc.ch.

many indirect contributions resulting from transportation, electricity production and distribution, and so on, increase this amount.

Global greenhouse gas emissions have in fact increased greatly over the four last decades. Figure 14.28 illustrates this evolution. There has been a 70% increase of global emissions since 1970. At a global level, we are far from a stabilization of the emissions.

In the residential sector, the largest part of the energy consumed is used for heating. Figure 14.29 shows that almost 90% of the total residential sector energy consumption in France in 2003 was dedicated to heating, hot-water production, and cooking.

In the commercial sector, specific electricity requirements account for a larger part of the energy consumption than in the residential sector (see Figure 14.30). Heating and hot-water production represented only 68% of the total.

14.13. CONCLUSION

Housing energy demands are high. The level of energy consumption and greenhouse gas emissions depends strongly on the technologies employed to

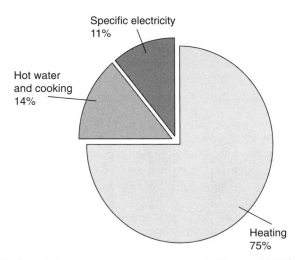

Figure 14.29. Residential sector energy consumption in France, 2003 (total 47.2 Mtep). *Source*: Data from DGEMP, http://www.industrie.gouv.fr/energie/sommaire.htm.

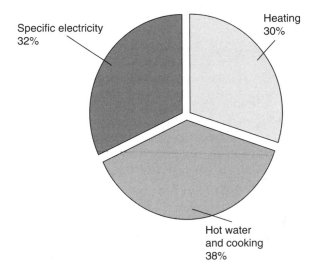

Figure 14.30. Commercial sector energy consumption in France, 2003 (total 21.7 Mtep). *Source*: Data from DGEMP, http://www.industrie.gouv.fr/energie/sommaire.htm.

meet these demands. There are now many well-known technologies which can be used to substantially decrease the energy consumption of buildings without loss of comfort. This can be done relatively easily for new buildings at manageable costs. It is even possible to construct zero-energy or positive-energy houses.

However, buildings built at a time when energy efficiency and pollution were not primary considerations may need major refurbishment or renovation to become energy efficient. This requires capital investment and many individuals may not be able to afford it even if the payback time is short. New financing modes may be needed to realize the societal benefits of increased residential energy efficiency.

The housing sector is an area in which it is possible to make great progress on energy efficiency at relatively low cost. But it is the consumer who makes the final decision concerning energy use. While government regulations and housing codes may be used to spur people to build more efficient homes and to improve old ones, education and information are necessary so that everybody understands why it is necessary to save energy and how that can be done.

CHAPTER 15

Smart Energy Consumption

Saving energy is a major concern in a world in which energy demand is rapidly increasing and in which we should strive to minimize the use of fossil fuels whose actual reserves are slowly but surely decreasing. The goal of efficient energy use is to provide a given level of energy service or even improve on that level while using less energy.

Energy can be saved or used more efficiently in all domains. In this chapter we shall briefly consider some of the ways to save energy. The question of energy efficiency has already been tackled in most of the other chapters and this chapter supplements that material. Our intent is to stir the interest of the reader in an area in which the consumer has an important role to play.

There are basically three agents involved in developing an energy-efficient economy:

- The first is *technology*. Laboratories and industries must be capable of developing energy-efficient technologies that can meet society's energy requirements at a competitive price.
- The second is the *authorities* (federals, state, and local governments, regulatory agencies, etc.). Using educational campaigns, regulatory procedures, and tax policies, government can orient people toward the use of one energy source or one technology in preference to another. For example, in some European countries, diesel fuel is cheaper than gasoline because taxes are lower. This has encouraged people to buy diesel cars rather than gasoline cars. Currently, about 70% of the new cars sold in France have a diesel engine. While the policy is obviously effective, one can question its wisdom since France now needs to import diesel fuel while exporting gasoline. Further, because a diesel engine consumes less fuel than a gasoline car for the same distance traveled, owners of diesel cars drive longer distances annually than the owners of gasoline cars. The net global energy savings which might have been realized from the use of diesel cars is sharply reduced by this trend.

Our Energy Future: Resources, Alternatives, and the Environment
By Christian Ngô and Joseph B. Natowitz
Copyright © 2009 John Wiley & Sons, Inc.

- The third agent is the *consumer*, who finally decides how the energy is to be consumed. The collective decisions of consumers determine actual energy demand. For example, if a consumer decides to take a vacation just a few hundred kilometers from home, energy consumption is far less than if the choice were to vacation several thousand kilometers from home.

Of course these three agents are interconnected and the culture of the society in which they exist is of great importance. Changing a culture is a question of education and takes a long time.

Rich countries have the ability to reduce their energy consumption. Done properly, this can be accomplished with relatively small impact. Living standards could even be improved because alternative ways of producing and using energy may provide better solutions to meet our needs. Unfortunately most of the people on the earth are below the energy use–poverty threshold (cf. Figure 1.5). To significantly increase their standard of living, they will need to increase energy consumption.

15.1. HOUSING

Humans have significantly modified the environment in which they live. Increasing numbers of people are now living in cities. The construction industry uses more than half of the raw materials consumed annually and produces more than half of the waste. Better building technologies and more efficient energy systems (e.g., thermal solar systems, heat pumps) can be used to greatly reduce energy demands for the housing sector. For a given house, the energy consumption depends very much on lifestyle and the way in which people use energy. Several of these factors are discussed in detail in Chapter 14 devoted to housing.

Different solutions can be used to reduce energy consumption in a house: New buildings can be built applying new standards and regulations that can assure the consumption of much less energy. However, these new buildings will represent only a small part of the total buildings. For example, in France about 300,000–400,000 new houses and/or apartments are built each year, but there are already around 30 million existing houses and apartments. Ignoring population increases, replacing all existing residences would require a century. Refurbishment and renovation of old housing to improve energy efficiency are therefore of great importance. Better insulation of the walls, the roof, and the floor could sharply decrease the amount of energy necessary to heat a house. Replacing old single-glazed windows by more efficient ones with double or triple grazing would also reduce energy leakage. Air tightness and a good ventilation strategy can allow energy management of the building and decrease the amount of energy necessary to maintain comfortable conditions inside the rooms. Where possible, efficient space-heating and hot-water systems using

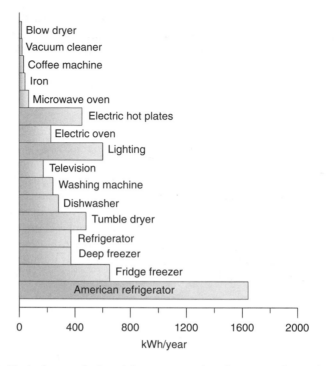

Figure 15.1. Typical annual electricity consumption for some domestic appliances. *Source*: Data from www.ademe.fr.

renewable energies and carbon-free technologies could be installed. Low-energy consumption lamps and efficient appliances can also significantly reduce the amount of electricity needed to power the house.

In Figure 15.1 we show the typical annual electricity consumption for some domestic appliances.

15.2. IMPROVING THE WAY WE CONSUME ENERGY

Producing and consuming energy to improve our quality of life are natural goals for humankind. However, wasting our natural resources for minimal gain is both shortsighted and foolhardy. In some situations the energy-efficient behavior is obvious and simple to implement. For example, leaving an incandescent light on while there is nobody in the room is a waste of energy. In other cases making the energy-efficient choice requires much more information and analysis. For example, electric heating is more expensive than heating with natural gas. However, with electric heaters, it is easier to modulate the temperature of a room than in the latter case. With electrical heating or heat pumps it is possible to do so very easily. This controllability can reduce overall energy consumption and therefore makes electric heating more competitive.

On a broader scale, net energy use has to be considered as a whole. For example, energy demands for housing and transportation are closely interconnected issues. Since it is often more expensive to build a house close to the workplace, people often buy lower cost real estate farther away. In such a case an energy saving in energy-efficient housing may be lost in transportation because the house is located great distances from work or shopping. More quantitatively, decreasing the energy necessary to heat a house by $80\,kWh/m^2/$ year is negated if the extra distance to drive to the working place exceeds $20\,km/day$.

As a second example let us compare living in an apartment downtown, close to daily shopping (baker, butcher, etc.), and living in a house located a few kilometers away from the shopping center. In the first case a deep freeze is not necessary while in the second it might cost less energy to keep things deep frozen than to drive to shopping as needed. In France, the average energy consumption of a deep freezer is about $600\,kWh/year$. This corresponds to the energy content of 60 liters of gasoline.

15.3. COGENERATION

Conventional power plants have electricity-generating efficiencies ranging from about 30 to 55%. The remaining part of the energy used is emitted into the environment and is lost. In dedicated installations heat can be produced with an efficiency near 80%. The principle of *cogeneration*, also called *combined heat and power*, is to simultaneously produce power and usable heat. The advantage of cogeneration is that more usable energy in the form of either heat or electricity is produced with the same amount of initial primary energy. When cogeneration is done using fossil fuels, CO_2 emissions for the same amount of useful energy are lower. Cogeneration is therefore an efficient way to exploit primary energy.

Combined heat and power generation is not a new technology. In the early 1900s, 58% of power used in industrial sites resulted from cogeneration facilities. By the 1950s about 15% of electricity in the United States was produced by cogeneration, but this value dropped to 5% a quarter a century later.

Figure 15.2 depicts a situation in which heat and electricity are produced separately with two different dedicated plants. About 40% of the initial energy is lost, although state-of-the-art efficiencies are assumed.

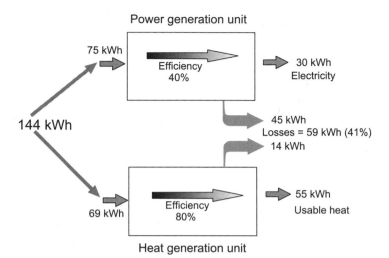

Figure 15.2. Electricity and heat production by two plant types with typical efficiencies. Large losses occur and the primary energy is not well exploited.

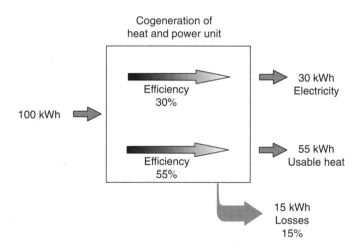

Figure 15.3. Cogeneration system. With an input of 100 kWh, 30% goes into electricity and 50% into usable heat. Only 15% is lost. The individual efficiencies of power generation and heat generation are smaller than in the case where electricity and heat are produced separately because in the latter case it is possible to optimize using more efficient technologies.

The advantage of cogeneration is illustrated in Figure 15.3 for a typical installation. The efficiency for producing each of these final energies is lower than for systems dedicated to just one of them, but the global efficiency is better. Because both the generated electricity and the generated heat are used, the total loss for this system is 15%. The main problem is that heat is most efficiently used on-site. This may limit the possibilities for cogeneration.

Comparing the systems in Figures 15.2 and 15.3 shows that the cogeneration plants have lower efficiencies for heat and power production than is realized in dedicated systems. However, the total efficiency of the combined heat and power system reaches 85% in this example compared to 59% in the case of separate heat and power generation.

The total number of cogeneration units in the European Union in 2003 was on the order of 12,600. The total capacity was about $92\,GW_e$ and the quantity of electricity produced was almost 300TWh.

The largest cogeneration system in the United States is in New York City. It is used to provide heat to about 100,000 buildings in Manhattan. Seven cogeneration plants produce steam at more than 500 °C which is dispatched throughout the buildings. Each year about $13.5 \times 10^9\,kg$ of steam is produced. On the average, 1 liter of water leads to about 0.95 kg of steam.

Three main cycles are possible for heat and power generation:

- *Topping cycles*, in which priority is placed upon power production and heat is harnessed. This is the case, for example, when a diesel engine generator produces primarily electricity and the heat generated is used to produce hot water.
- *Bottoming cycles*, which primarily produce heat but recover the unused heat to produce power. Heat rejected by industrial kilns may be used in this way.
- *Combined cycles* are a combination of topping and bottoming cycles. For example, a gas turbine produces power and the exhausted heat from the turbine is used to produce steam which then expands in a steam turbine to produce electricity or mechanical energy. Steam can also be extracted to be used in industrial processes which require it.

Distributed generation is the generation of heat and power at a small scale (between a few kilowatts to 25 MW). Distributed generation systems can be used locally. The primary fuel can be natural gas, liquefied petroleum gas, kerosene, or diesel fuel. Distributed generation units may be connected to the grid or not. The constraint imposed by the relative amounts of heat and electricity which can be produced may make it difficult to match such a system to consumer demand.

15.4. STANDBY CONSUMPTION

Many household appliances are not actually off when they are not in use. They remain in a standby mode. They consume energy while waiting to be activated. Standby power is the power used while the product is in this mode. It is often difficult for consumers to escape from this standby mode and they usually do not realize how much energy is consumed each year for no service. Standby consumption is therefore an important issue. Developing labels to keep the consumer informed on standby power requirements and enacting regulations to impose more efficient standby consumption standards on the manufacturers could improve the situation. It is possible to build appliances which consume less standby power, but this can entail an extra cost that the manufacturers and the retailers are not ready to pay. Reduction of standby power requirements might also require an increase in the complexity of the electronics of the appliance, which may increase the cost of maintenance.

Watching TV consumes energy (electricity). If the TV is switched off by remote control, many people think that it is not consuming energy. This is not true. The TV set is usually in a standby mode, waiting to be switched on again by the remote control. During that time the TV set consumes energy and this amount may be noticeable. Consider an 80-W TV with a standby mode power of 15W. If that TV is on for 3h a day, it consumes 240Wh of energy. If we leave the TV in standby mode during the remaining 21h it consumes 315Wh. The TV set energy consumption when the set is "off" is greater than when it is on.

A VCR may consume up to 19 times more energy in the standby mode than in the reading or recording mode.

In Figure 15.4 we show, for selected electric appliances, the range of standby mode powers going from the minimum to the maximum according to present levels of the technology. A black dot indicates the mean value of standby power for the electric appliance considered.

In member countries of the Organization for Economic Co-operation and Development (OECD), a power between 20 and 60W per household is devoted to the standby mode. This corresponds to an annual electricity consumption of between 175 and 525kWh. For comparison, the annual average electricity consumption of a French refrigerator is around 380kWh. Standby consumption is equivalent to the consumption of a typical domestic appliance in a household. The total standby power in the residential sector of OECD countries is estimated to be 15GW. This is considerable and is equivalent to the power of 15 typical nuclear reactors. At the global level 15GW power

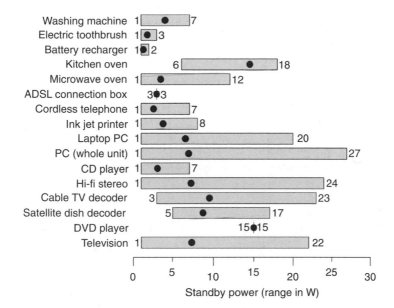

Figure 15.4. Standby power for selected electric appliances found in French households in the years 1998–1999. For a given appliance there is a range of standby power which is dependent upon the technology employed. The minimum and maximum figures are indicated. The black dots indicate the average standard powers. *Source*: Data from *Things that Go Blip in the Night, Standby Power and How to Limit It*, IEA-OECD, Paris, 2001.

represents 131 TWh of electricity per year. This is a huge value. For example, all the wind turbines in Germany in 2003 had an installed power of 14 GW and have produced 18.6 TWh of electricity. This was not sufficient to meet the standby power requirements of all electric appliances of the country. Even with a good efficiency ($\approx$30% while in Germany it was only $\approx$15% in 2003) it would require many years and a lot of money to offset the energy consumed in standby modes by building wind turbines. It is therefore far more effective to decrease the standby power requirements of these devices.

Figure 15.5 shows the lowest standby power consumption attainable with present technology for selected electric appliances.

Even in OECD countries, the average power per home differs from country to country, as is shown in Figure 15.6.

The total electricity consumption depends on whether the appliances are connected or not to the electrical grid. Figure 15.7 shows that the average power used can be notably different from the installed power.

The level of CO_2 emissions associated with standby power usage depends very much on the way the required electricity is produced. Figure 15.8 shows data for selected countries and Figure 15.9 indicates the percentage of CO_2 emissions in these countries attributable to standby modes. For countries

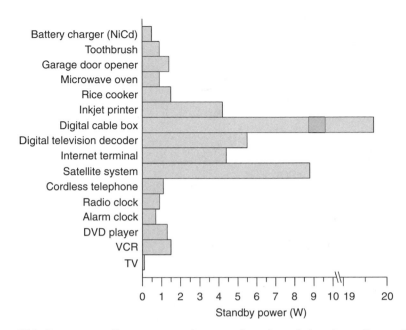

Figure 15.5. Lowest standby power requirements for selected electric appliances. Note the break in the axis. The digital cable box has a high-standby-power requirement. *Source*: Data from *Things that Go Blip in the Night, Standby Power and How to Limit It*, IEA-OECD, Paris, 2001.

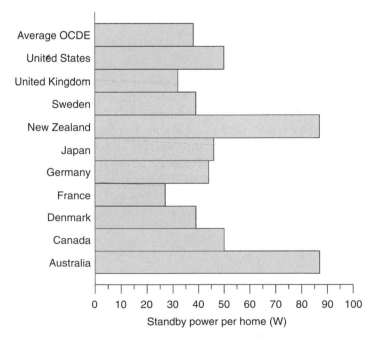

Figure 15.6. Average standby power requirements per home for selected OECD countries. *Source*: Data from *Things that Go Blip in the Night, Standby Power and How to Limit It*, IEA-OECD, Paris, 2001.

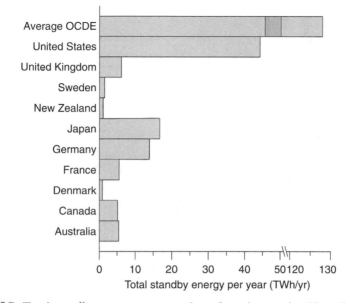

Figure 15.7. Total standby energy per year for selected countries. Note the break in the axis. *Source*: Data from *Things that Go Blip in the Night, Standby Power and How to Limit It*, IEA-OECD, Paris, 2001.

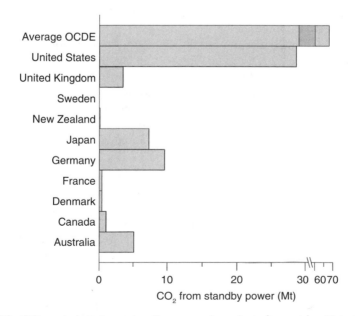

Figure 15.8. CO_2 emissions from standby power for selected countries. Note the break in the axis. *Source*: Data from *Things that Go Blip in the Night, Standby Power and How to Limit It*, IEA-OECD, Paris, 2001.

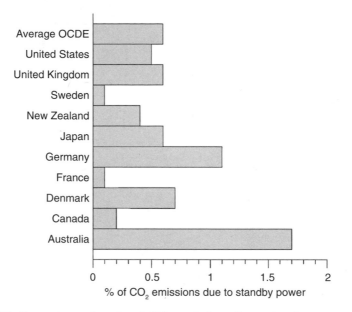

Figure 15.9. Percentage of national CO_2 emissions from standby power (Mt) for selected countries. *Source*: Data from *Things that Go Blip in the Night, Standby Power and How to Limit It*, IEA-OECD, Paris, 2001.

which produce electricity with little associated CO_2 emission, the impact of standby power is minimal.

Figure 15.10 shows the estimated fraction of total residential electricity used by equipment in standby modes. We see that it ranges between 5 and 12% depending upon the country.

In OECD countries, it is estimated that standby electricity in the residential sector represents about 1.5% of total electricity consumption and contributes 0.6% of the CO_2 emissions. This is not a small contribution. It represents the equivalent of emissions from 24 million European-style cars.

15.5. LIGHTING

Lighting in the residential sector represents about 30% of its global electricity consumption. Clearly the electricity consumption presently dedicated to lighting can be reduced with existing technologies or with those which are still at the research stage. Replacing incandescent lamps of 75 W by compact fluorescent lamps of 15 W producing the same quantity of light would reduce the annual electricity consumption of a typical household by about 150 kWh. At the European level that would represent savings of about 20–25 TWh, that is, a little more than 4 Mtoe. This change would also have an impact on CO_2 emissions, but the amount of reduction depends strongly upon the way the

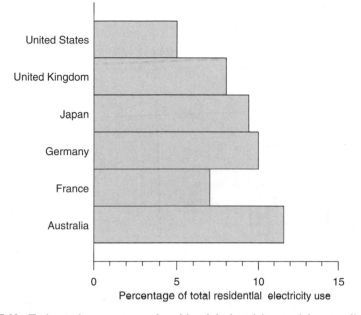

Figure 15.10. Estimated percentage of residential electricity used for standby power requirements in selected countries. *Source*: Data from *Things that Go Blip in the Night, Standby Power and How to Limit It*, IEA-OECD, Paris, 2001.

required electricity is produced. In countries like France or Sweden, where most of the electricity is CO_2 emission free, the reduction is small. In other countries, for example, Germany or Denmark, which rely heavily on coal-fired power plants, the reduction of emissions is much more significant.

15.6. TRANSPORTATION

Modern transportation accounts for a very large fraction of global energy consumption and depends very heavily on fuels derived from oil. The distribution of transportation modes available in a country and the frequency of usage of these different modes are major determiners of that country's rates of energy consumption and CO_2 emissions.

Figure 15.11 shows an evaluation of fuel consumption for different transportation modes in the urban area of Paris, France. The values of fuel consumption are expressed in grams of oil equivalent (goe) per passenger and per kilometer. The subway is the most economical way to travel in the city. Private commercial vehicles are the least economical, but they are necessary to transport freight. These numbers are calculated for an assumed average occupancy rate. For a car with twice as many passengers as the average, fuel consumption per passenger would drop by a factor of 2.

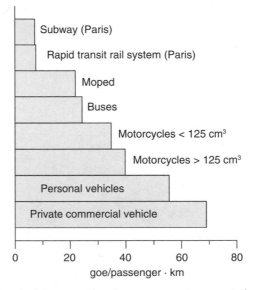

Figure 15.11. Urban fuel consumption for passenger transportation in Paris for differ-ent means of transportation, 2000. *Source*: Data from J. Bonal and P. Rossetti, *Energies alternatives*, Omniscience, 2007.

Figure 15.12 presents a comparison of fuel consumption in France for dif-ferent modes of interurban transportation. Again these evaluations assume average occupancy rates. The most fuel efficient means of transportation is the fast train (TGV) and the least fuel efficient is the airplane.

Finally, in Figure 15.13 the fuel consumption is presented in grams of oil equivalent per ton of freight and per kilometer. The electric train is again the most economic mode of transportation. This is especially the case in France where 90% of the electricity is produced without CO_2 emissions. The values would be a little different in a country, like the United States, where coal plays an important role in power generation.

As mentioned in Chapter 13 on transportation, about two-thirds of the annual oil production of the world is used for transportation. Of this, more than three-quarters is used for road transportation. There have been continu-ous improvements in energy consumption over the years leading to vehicles which need significantly less fuel than before. However, since about three-fourths of the cars purchased each year are used cars, the car fleet is renewed only slowly and the improvements which decrease fuel consumption are felt only slowly. In France in 1980 the average fuel consumption of a car was 9.3 L/100 km. This dropped to 7.4 L/100 km in 2000. At the same time, the average distance traveled in a year rose from 12,800 km in 1980 to 13,800 km in 2000.

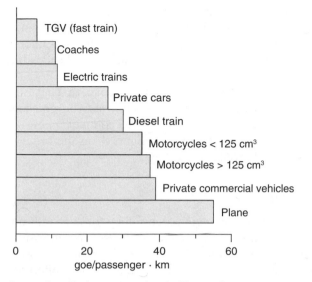

Figure 15.12. Interurban fuel consumption in France for passenger transportation for different means of transportation, 2000. *Source*: Data from J. Bonal and P. Rossetti, *Energies alternatives*, Omniscience, 2007.

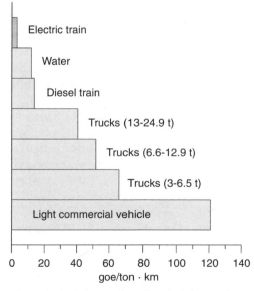

Figure 15.13. Interurban fuel consumption for freight transportation for different means of transportation, 2000. *Source*: Data from J. Bonal and P. Rossetti, *Energies alternatives*, Omniscience, 2007.

In the quest to improve vehicle energy efficiency, properties of both the engine and the fuel must be taken into account. Improvements in both have been made in recent years. However, the role of the consumer is essential to realizing real gains because it is the consumer who decides how to use the vehicle.

For airlines, fuel consumption is a major concern because of the rapid increase in the price of oil. New planes consume less fuel and some companies are now replacing their old planes with more economical new ones. Currently fuel consumption in a plane full of passengers is on the order of $5 L/100 km$, but new planes can achieve $3.5 L/100 km$.

15.6.1. Technology

There are two basic types of engine technology for cars: internal combustion engines using spark ignition and diesel engines using compression ignition. The fuel for the first type is gasoline and the fuel for the second type is diesel. There are more diesel engines in Europe than in the United States or Japan because diesel fuel is subject to lower taxes than gasoline.

Great progress has been made in diesel engine technology and the power has doubled during the last 15 years to $65 kW/L$, making diesel-powered cars quite pleasant to drive. Compared to similar gasoline vehicles, diesel vehicles have lower fuel consumption and higher torques. Consequently they emit less CO_2 per unit of distance than the equivalent gasoline vehicle. If only CO_2 emissions are considered, diesel engines appear to be the better solution for our transportation needs. However, diesel engines have other serious drawbacks. The first one is that their NO_x emissions are larger than those of gasoline engines. Furthermore, diesel engines emit particulates. The smallest ones are probably dangerous, especially for humans. Indeed they can travel into the lungs, penetrate deeply inside, and stay there. These particulates can also trap incompletely burned hydrocarbons and other dangerous chemical products produced during the combustion on their surfaces. This could result in serious health problems for many people. Fortunately, by using new particulate filter technologies and better catalytic converters, some progress has been made in reducing both particulate emissions and toxic gaseous emissions.

A person breathes about $15 m^3$ of air per day. Because of this, air pollutants resulting from road vehicle emissions have a major impact on human health. Among the gases which are emitted by cars, motorcycles, and heavy-duty vehicles, CO is toxic for blood; SO_2, NO_x, and O_3 have a negative impact on lungs; and volatile organic compounds may be carcinogenic. Particulates which are emitted in large quantities by diesel engines have an impact that varies according to their size. Particulates with a size larger than $10 \mu m$ are stopped in the upper part of the respiratory tract. Particulates with a size between about 3.5 and 10 μm are stopped in the middle part of the respiratory tract. Most dangerous are small-size particulates (<3.5 μm) because they go deeply into the lungs and are potentially very dangerous.

In 2004, the French agency AFSSE (Agence française de sécurité sanitaire de l'environnement et du travail, www.afsse.fr) published an analysis of health effects of the long-term exposure of people to particulates of size less than 2.5 μm. It considered 76 urban areas in France and about 15 million inhabitants more than 30 years old. The analysis resulted in the following estimates. There are between 6000 and 9000 deaths of people older than 30 years due to all causes. Among them between 600 and 1100 were due to lung cancer (6–11% of the total deaths by lung cancer) and between 3000 and 5000 due to cardiorespiratory troubles.

There have also been important improvements in fuels which make them significantly cleaner. They now have low sulfur and aromatic content. New technologies such as hybrid vehicles and perhaps in the longer term fuel cells promise even greater benefits in this area.

A variety of technologies are being applied to make road vehicles more energy efficient. For example, new cars are designed to reduce aerodynamic drag. Improvements in lubrication systems and transmissions reduce energy losses due to internal friction. Electronic monitoring of fuel consumption helps to decrease the fuel demand. New tire technologies using silica instead of black carbon, "green tires," allow a decrease of fuel consumption by 3–8% depending upon the type of trip. Car manufacturers are also promoting *downsizing*, developing smaller engines that can provide power equivalent to existing ones. FIAT has developed an engine which is 20% lighter and 25% smaller than a conventional engine but which develops 105 HP and emits only 69 g of CO_2/km.

15.6.2. Individuals

There are several ways for individuals to reduce automobile fuel consumption. Good maintenance and a properly adjusted engine can reduce fuel consumption up to 25%. Moderate driving techniques can also reduce fuel consumption. This can be understood by reference to Figure 15.14, which shows the typical ratio between the power and the maximum power needed at constant speed as a function of the ratio between the velocity and the maximum velocity. At 25% of the maximal velocity of a car only 5% of the maximal power is needed. At half the maximal velocity a power a little below 20% of the maximal power is required to ensure a constant velocity for the car. Gunning the engine is wasteful of fuel and is not needed to maintain a reasonable velocity.

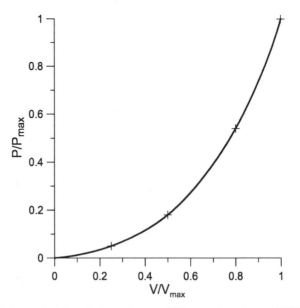

Figure 15.14. Schematic of evolution of ratio P/P_{max} as function of V/V_{max}, where V is the speed of the car, V_{max} its maximum speed, P the power necessary to ensure a constant speed V, and P_{max} the maximum power of the car.

Driving fast costs energy and fuel consumption is closely related to the speed of the car. For a midsize European car, driving at 120 km/h rather than at 130 km/h decreases fuel consumption by about 1 L/100 km (Figure 15.15). Air conditioning is also fuel consuming, increasing fuel consumption by about 10% on the highway and around 25% in urban areas where there are traffic jams. Heating a car requires even more energy and this is a major concern for hybrid vehicles.

Tire pressure is another important consideration in fuel economy as well as in safety. An underpressurization of the tires by 0.5 bar leads to an increase of fuel consumption of about 2.5%. One hundred kilograms extra weight in the car increases the fuel consumption by about 5% or more. Roof racks should only be used when necessary and removed otherwise. Indeed, they increase the fuel consumption by about 10%.

Urban driving is usually very fuel consuming because of traffic jams. Often there are also parking difficulties which make public transportation more appropriate. In urban driving, the amount of energy transmitted to the wheels for a midsize passenger car is about 13% of the energy taken from the tank. For comparison, it reaches 20% for highway driving.

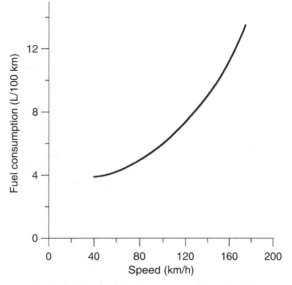

Figure 15.15. Plot typical fuel consumption of European car.

There are several ways to save energy and reduce automobile fuel consumption. The most obvious is not to drive a car unnecessarily. For short distances walking or biking is an alternative. Using public transportation when it is available may also be a solution. However, it should be noted that public transportation is not always the best solution for the broader community. In areas where the population density is small, public transportation is quite inefficient, especially at off-peak hours. Clearly one person in a bus consumes more energy that one person in a car. Public transportation is most efficient in high-density areas, towns, or cities. Car sharing is also a way to reduce fuel consumption since several passengers share the same car for the trip. This of course imposes constraints on the trip and on the time schedule.

15.7. CONCLUSION

Energy efficiency is a primary concern for modern society. It should be done without losing anything in terms of standard of living or the population will not accept it. Technology can improve energy efficiency. Government authorities can use laws, standards, regulations, and taxation to foster the adoption of energy-efficient solutions. But the consumers are the real key to energy efficiency. They decide how to spend their energy dollars. To make the best choice they must be informed. Development of an energy-efficient society requires that everyone be involved. Small efforts by a large number of people are better than a big effort by a few. As a rule it is usually less expensive to save 1 kWh than to produce 1 kWh. Thus the cheapest kilowatt-hour is the one which is not consumed.

At the same time the impact on the environment may be lower. This of course depends very much of the nature of the energy used. For example, replacing an incandescent lamp by a fluorescent lamp saves energy for the same amount of operation time. If the electricity is produced by fossil fuels, this replacement also reduces the amount of emitted CO_2. However, if the electricity is produced by renewable or nuclear sources, energy is saved but there is no gain in CO_2 emissions.

There may be several ways to save energy for a given application, and it is sometimes even possible to avoid consuming energy. Since energy is becoming increasingly expensive, a new mindset based on smart energy consumption should emerge.

We are rapidly reaching a situation in which we will have to change the way we use energy. This does not mean we will be less comfortable. In many cases we may be more comfortable. For example, changing from central heating using natural gas to the use of heat pumps to heat a house is capable of providing greater comfort to the consumer since the heat pump also provides air conditioning in summer while natural gas does not. Hybrid vehicles can provide the same service as conventional ones but consume less energy and are quieter.

CHAPTER 16

Hydrogen

More than a hundred years ago, Jules Verne, in his book *The Mysterious Island*, predicted that hydrogen would be an important future source of energy. In that book he said : "I believe that water will one day be employed as fuel, that hydrogen and oxygen will furnish an inexhaustible source of heat and light." Many people today think that this will soon be the case. One problem, however, is that hydrogen is not an energy source but an energy vector in the same way as is electricity. Energy is needed to produce the precious molecule of hydrogen by extracting the hydrogen atoms trapped in other molecules.

Electricity can be used to produce hydrogen from water. Compared to electricity, hydrogen has the advantage of being able to be stored more easily. It can be carried in suitable containers or in pipes which transport it.

The first part of this chapter considers the issues of hydrogen production, storage, transport, and distribution. As this is an emerging technology, many of the figures which are quoted have to be taken to be estimates because there are still large uncertainties in determining efficiencies and costs for certain production strategies. The use of hydrogen, primarily in fuel cells, to meet our energy needs will be treated in the second part of the next chapter.

16.1. FROM PRODUCTION TO DISTRIBUTION

16.1.1. Properties

The hydrogen atom, H, is the smallest and lightest of the atoms. The most plentiful isotope of hydrogen is ^1H, which accounts for 99.985% of the atoms in natural hydrogen. It is composed of only one proton and a single electron. Hydrogen is the most abundant element of the universe, comprising 75% of the visible mass. It forms 0.15% of the crust of the earth and is a major constituent of water (11.11% in mass). It is found in a wide range of materials: water, hydrocarbons, biomass, for example. Hydrogen accounts for 14% of the weight of biomass.

Our Energy Future: Resources, Alternatives, and the Environment
By Christian Ngô and Joseph B. Natowitz
Copyright © 2009 John Wiley & Sons, Inc.

Although the hydrogen atom is present in great quantities on the earth, the hydrogen molecule, H_2, formed by two hydrogen atoms is scarce. There is only 0.5 ppm of hydrogen, H_2, present in our atmosphere. The reason is that this molecule is too light to be retained by the earth's gravitation.

In 1766, the British chemist Henry Cavendish used pig bladders to collect a mysterious gas resulting from the action of vitriol (sulfuric acid) on metals. He noted that this gas burned in the air and produced water. He baptized it the "flammable air."[1] In 1781, the French chemist Antoine-Laurent de Lavoisier, assisted by Jean-Baptiste Miller of Laplace, carried out the first chemical synthesis of water. He understood that it was a composite material, formed of flammable air (hydrogen) and "vital air" (oxygen). Lavoisier later named this strange flammable air *hydrogen*, which means "forming water."

The burning of hydrogen releases energy. However, hydrogen is not a primary energy source since energy is needed to remove it from molecules containing hydrogen atoms. It is a secondary source of energy which requires a primary energy source in order to be produced. More precisely, it is an energy vector similar to electricity, which also requires energy to be produced. It facilitates the storage and transport of energy from the place where it is produced to the place where it is used.

In spite of its wide use in industry, the image of hydrogen in the public mind is that of a dangerous gas and its degree of acceptability is still very low. This perception of hydrogen probably has been strongly influenced by its association with the accident of the German zeppelin *Hindenburg*, which caught fire in 1937 when landing at Lakehurst, New Jersey, in the United States. This dramatic accident marked the psyche and rang the death knell for transport by airship. Even though experts have since shown that it was not hydrogen which was responsible for the fire but rather the extremely flammable varnish covering the envelope of the airship, public fear and distrust of hydrogen-containing vessels remain.

In many places it is probably also not forgotten that "town gas," mostly a mixture of carbon monoxide, hydrogen, and methane, was responsible for numerous accidents and deaths in the early part of the twentieth century. The toxicity of town gas does not come from hydrogen but rather from carbon monoxide, which is a blood poison. This mixture was generally replaced by the end of the 1960s by natural gas, which is essentially methane (CH_4).

The activation energy of hydrogen is smaller than that of natural gas, and thus, a simple spark is enough to cause an explosion. But, except in situations of confinement, hydrogen burns like natural gas. Its flame is blue, nearly invisible to a naked eye, and no smoke is emitted. This can prove to be a

TABLE 16.1. Properties of Hydrogen (H^2)

Properties	Values
Gas density (at 273.15 K)	0.08988 kg/Nm^3
Liquid density [at 20.3 K (−252.85 °C)]	70.79 kg/m^3
Boiling point (10^5 Pa)	−252.85 °C (20.3 K)
Freezing/melting point	−259.2 °C
Lower heating value	3 kWh/Nm^3 or 33.33 kWh/kg
Higher heating value	3.55 kWh/Nm^3 or 39.41 kWh/kg
Specific heat at constant pressure	3.96 Wh/kg · K
Specific heat at constant volume	2.86 Wh/kg · K
Specific volume	11.99 m^3/kg
Specific density (air = 1)	0.0696
Flammable range	4–74% by volume of air
Flame temperature	2318 °C
Liquefaction energy	3.92 kWh/kg

Note: Nm^3 means normal cubic meters. It is the volume of the gas at normal temperature and pressure conditions.

disadvantage because one does not see the flame. It has however the advantage that the weak radiation from the flame limits the fire hazard. The large fires which devastate petrochemical sites generally propagate by radiation.

In an unconfined situation H$_2$ tends to diffuse very quickly into the air (four times more quickly than natural gas) and can escape through very small interstices. It is odorless and colorless. Therefore, to detect possible leaks, it is necessary to add an odorous product. In balance, hydrogen is hardly more dangerous than natural gas. The main properties of hydrogen are given in Table 16.1.

16.1.2. Production

At present the total world production of hydrogen is approximately 50 million tons per year. In energy terms, this corresponds to 130 million tons oil equivalent (130 Mtoe), which is barely 1.5% of the world production of primary energy. By way of comparison, the energy content of the totality of the food consumed annually on our planet is 500–600 Mtoe. Converted into a volume of gas under normal conditions of temperature and pressure, the production of hydrogen is approximately 550 × 10^9 Nm^3/year.

Today, hydrogen is primarily used in very specific applications, such as the synthesis of ammonia or the desulfurization of gasoline (the refiners often use hydrogen, which they produce during oil refining). Its use as an energy vector is, on the other hand, not very widespread, except in very particular and marginal cases, such as the propulsion of spacecraft. The different world uses of hydrogen are shown in Figure 16.1. Those in the European Union are shown in Figure 16.2.

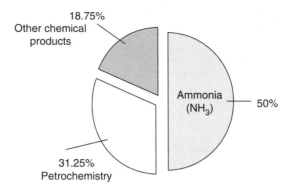

Figure 16.1. Uses of hydrogen in the world. *Source*: From www.afh2.org.

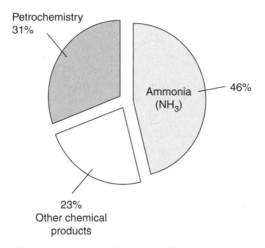

Figure 16.2. Uses of hydrogen in the European Union. *Source*: From www.afh2.org.

Producing hydrogen from fossil fuels proceeds through the manufacture of syngas, which is a mixture of hydrogen and carbon monoxide (CO and H_2). There are several ways to obtain this gas, but industry prefers those methods which make it possible to optimize the total process, allowing produced hydrogen to be used to manufacture new compounds, for example, ammonia. The three most common ways to produce H_2 are the vaporeforming of natural gas, partial oxidation of oil residues, and gasification of coal. Unfortunately these techniques all have the disadvantage of producing CO_2. Therefore they contribute to the increase in the greenhouse effect if this CO_2 is released into the atmosphere. The formed gas is purified by various techniques to obtain more or less pure hydrogen. The distribution of the origin of hydrogen production is shown in Figure 16.3.

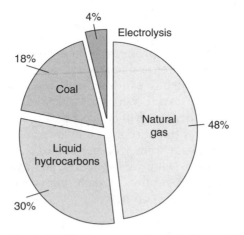

Figure 16.3. Share of origin of hydrogen production. *Source*: From www.afh2.org.

Vaporeforming

Vaporeforming consists of using water to produce hydrogen from hydrocarbons. It is the process most widely used by industry. It is used mainly with light hydrocarbons like methane, naphta, or liquefied petroleum gas. The chemical vaporeforming of natural gas, containing mainly CH_4, may be written as

$$CH_4 + H_2O \rightarrow CO + 3H_2$$

This reaction proceeds at moderate pressure (from 20 to 30 bars) and at a temperature of about $900\,°C$ in the presence of a nickel catalyst. The natural gas must first be desulfurized because sulfur is a poison for catalysts. The syngas obtained is a mixture of hydrogen (H_2), water, methane (CH_4), carbon monoxide (CO), water (H_2O), and carbon dioxide (CO_2) obtained by oxidation of carbon monoxide.

The carbon monoxide is then reacted with water vapor by an exothermic conversion reaction called water gas shift:

$$H_2O + CO \rightarrow CO_2 + H_2$$

This reaction produces one extra hydrogen molecule compared to the preceding reaction and thus, overall,

$$CH_4 + 2H_2O \rightarrow CO_2 + 4H_2$$

During this process parasitic reactions leading to the production of soot (pure carbon) occur. By adjusting the amount of water vapor, one can eliminate most of them using a reaction which is also used to produce syngas from coal:

$$C + H_2O \rightarrow CO + H_2$$

By vaporeforming and conversion, one can obtain approximately 75% hydrogen and 25% CO_2. The initial reforming reaction is endothermic: It consumes 206 kJ/mol. The second is slightly exothermic: It releases 41 kJ/mol. On the whole, the balance of these two reactions is largely endothermic and 165 kJ/mol must be supplied. The yield of this process is approximately 65% and produces 11 tons of CO_2 per ton of produced hydrogen. Afterward it is necessary to separate carbon dioxide from hydrogen and to eliminate the impurities, which again requires energy. There are two industrial ways to purify hydrogen: methanation and selective adsorption on molecular sieves, referred to as the PSA (pressure swing adsorption) process.

Partial Oxidation

Residues of hydrocarbons can be converted, in the presence of oxygen, into hydrogen and carbon monoxide. This method is more constraining than vaporeforming in terms of investment and operational processes. However, costs are often reduced because less expensive hydrocarbons can be used. The process is exothermic. An initial reaction uses oxygen in insufficient quantities so that oxidation is only partial. Oxygen is used at high temperature (between 1200 and 1500 °C) and under high pressure (from 20 to 90 bars) in the presence of water vapor to slow down the reaction speed. This reaction can also be performed at lower temperature (approximately 600 °C) in the presence of catalyst. The syngas which is produced is then treated by reforming in the same way as in vaporeforming. The resulting gas often contains sulfur and it is necessary to desulfurize it.

The two main processes of partial oxidation are those used by Shell and Texaco. Shell carries out the desulfurization stage before the conversion of the carbon monoxide whereas Texaco carries it out afterward.

The process of partial oxidation can be applied to natural gas or to oil residues. In the latter case, one produces about 15 tons of CO_2 per ton of produced hydrogen. The installation capacities go from 50,000 to 100,000 m^3/h of hydrogen. The cost of the hydrogen produced by this technique is approximately twice than in vaporeforming.

Autothermal Reforming

The production of syngas is endothermic, whereas partial oxidation is exothermic. It is possible to simultaneously carry out both reactions in a proportion such that the reaction is autothermal, that is, without release or absorption of heat. One thus compensates for the heat absorption in the reactions of vaporeforming by that released in the partial oxidation reactions. Autothermal reforming is still at an experimental stage but could be an interesting technique for large manufacturing units, in particular in the gas-to-liquid (GTL) process, in which one synthesizes fuels starting from natural gas and using the

Fischer–Tropsch reaction. By adjustment of the relative proportions of hydrogen and carbon monoxide in the syngas, this process would allow optimization of the use of oxygen and thus minimization of the costs.

The Fischer–Tropsch process is a chemical reaction allowing conversion of a mixture of carbon monoxide and hydrogen into synthetic hydrocarbons according to the equation

$$(2n+1)\text{-}H_2 + n\text{-}CO \rightarrow C_nH_{2n+2} + n\text{-}H_2O$$

This reaction needs a catalyst. The catalysts most currently used are iron and cobalt. This process has a good yield but requires large investments. It was first developed in 1925 and was used by Germany and Japan during the World War II. By the beginning of 1944 Germany was able to produce about 124,000 barrels of synthetic oil per day using this technique.

Coal Gasification

Gasification is a process similar to the partial oxidation described above. The first stage is a process in which coal is gasified in the presence of water and oxygen in order to produce syngas. This is then converted into hydrogen and CO_2. There are three main technologies to accomplish this:

- In the fixed-bed technology, the gases circulate through coal particles whose dimensions vary from 3 to 30 mm. The temperature is between 800 and 1000 °C and the pressure between 10 and 100 bars.
- In the fluidized-bed technology, the particles of coal, from 1 to 5 mm, are in a suspension in a gas current; the range of temperatures is of the same order as the fixed-bed method.
- The forced-flow technology requires higher temperature (between 1500 and 1900 °C) and higher pressure (from 25 to 40 bars). It employs very fine particles (0.1 mm) which flow at high speed.

Producing Hydrogen In Situ

Hydrogen can be produced in a centralized way before it is conveyed to the places of use or it can be produced in a semidecentralized way in stations located throughout a country. It can also be produced on demand using reformers.

In the latter case, reformers allow the use of carbon compounds to produce hydrogen in situ. In a reformer, one breaks the fuel molecules (ethanol, methanol, gasoline, etc.) using air or water to get a gas mixture containing hydrogen (from 30 to 35%), carbon monoxide (from 10 to 15%), and CO_2 (from 10 to 15%). As the carbon monoxide is a poison for a fuel cell, it is necessary that its concentration be reduced below 10 ppm. That is done through a series of chemical reactions carried out at high temperature in the presence of a catalyst.

Electrolysis of Water

Today, 96% of the hydrogen produced and used in the world (primarily by the chemical industries) is derived from fossil fuels in processes which also produce carbon dioxide. The latter should be captured and stored if we want to prevent CO_2 emissions into the atmosphere. However, a sustainable economy based on hydrogen should produce the hydrogen using decarbonated energies such as nuclear or renewable energies. The electrolysis of water using such energies produces only hydrogen and oxygen. It would be a sustainable solution for the future, although it is not yet economically competitive.

An electrolysis cell is composed of two electrodes (Figure 16.4), an anode maintained at a positive potential and a cathode at a negative potential, connected to a current generator. Both electrodes are in an electrolyte, a medium in which electric charges are transported by ions. The electrolyte is generally an acidic or basic solution, but one can also use a polymeric exchange membrane of ions or a conducting ceramic membrane.

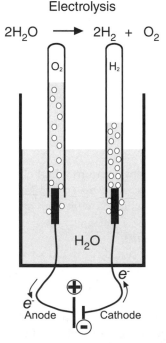

Figure 16.4. Principle of electrolysis of water.

Industrial electrolysis generally uses an aqueous caustic potash solution as the electrolyte at temperatures from 80 to 160 °C. The system can be monopolar, as displayed in Figure 16.4, or bipolar, where one plate plays the role of an anode on one side and of a cathode on the other side. Bipolar electrolysis systems are more efficient and commonly used at the industrial level. Alkaline electrolysis is done typically in modules producing between 0.5 and 800 Nm^3 of hydrogen per hour.

Electrolysis in an acid medium uses a solid electrolyte (conducting polymeric membrane of protons) in more compact modules. Better yields are obtained than with alkaline electrolysis but at a higher cost. The modules can operate at atmospheric pressure or at much higher pressures. This technology is expensive because of the cost of the membranes and the noble metals used as catalysts. There are nevertheless high hopes that this technology will benefit from experience gained in the development of the fuel cells which reverse the water electrolysis process (see Section 16.2).

Anthony Carlisle and William Nicholson discovered the phenomenon of electrolysis in 1800. They noticed that when they plunged two metal wires connected to the poles of a Volta battery into water, gas bubbles formed on the surface of the wires. The gases formed were oxygen at the positive pole and hydrogen (with a volume twice that of the oxygen) at the negative pole. This phenomenon would be explained three years later by the English Chemist Humphry Davy as the dissociation of water molecules into oxygen and hydrogen induced by the circulation of an electrical current.

Electrolysis at high temperature (900–1000 °C) is a technology under development. Derived from research on solid oxide fuel cells, this technique employs a conducting ceramic membrane and oxygen ions. In principle, such a technique is capable of reaching yields higher than 80%. The difficulties associated with realizing such efficiencies are associated with development of materials useful for sustained high-temperature operation. High-temperature electrolysis is interesting because less input energy is needed than at lower temperatures, approximately 3 kWh/Nm^3 at high temperature compared to 4 kWh/Nm^3 for conventional electrolysis. Since it is possible to use higher current densities, the device is about twice as compact. Furthermore, using a significant part of the energy in the form of heat instead of electricity makes it possible to reduce the cost of exploitation. Heat is less expensive than electricity. The use of an external heat source loses some of its interest if it produces CO_2.

The cost of hydrogen produced by electrolysis is strongly correlated with the cost of electricity. The cost is estimated to be between 12 and 19 €/GJ and can sometimes exceed this value. One needs about 1 kg of hydrogen, perhaps more depending upon system optimization, to drive 100 km in a car using a fuel cell. The cost of hydrogen produced by electrolysis would be about 2 €/kg if it is produced at large scale or 3 €/kg in smaller size stations. It is more expensive than the hydrogen which comes from fossil fuels, which costs below 1.3 €/kg. Using biomass to produce hydrogen would lead to prices between 2.6 and 2.9 €/kg.

Electrolysis is particularly interesting if the electricity is produced without emitting CO_2 into the atmosphere, that is, by using renewable energies or nuclear power. This production mode is not yet economically competitive compared to hydrogen production from natural gas. However, progress in electrolysis technologies and the possibilities of taxes on carbon dioxide emissions could reverse this situation.

Electrolysis of water needs a lot of energy. It is presently used to obtain very pure hydrogen or when cheap electricity is available because it is not being used for other purposes. The current contribution of electrolysis to the production of hydrogen is small: approximately 4% of the total production. It is however an interesting process because production of hydrogen provides a means for long-term storage of a surplus of electrical energy. That could be interesting, for example, in the case of a wind turbine which produces electricity (because there is wind) but for which no immediate demand exists. Evaluations of costs according to different production methods are displayed in Figure 16.5.

The theoretical potential of decomposition of a water molecule is 1.481 V at 298 K (the enthalpy of dissociation is 285 kJ/mol). However, one needs a higher potential than this to compensate for the phenomena of polarization of the electrodes and the losses by the Joule effect. The voltages applied are in general between 1.7 and 2.1 V and the yields between 70 and 85%. To produce 1 m^3 of hydrogen at normal conditions (1 Nm^3), one needs approximately 1 liter of water and between 4 and 6 kWh of electricity.

Thermochemical Cycles

The combustion of hydrogen with oxygen releases a lot of energy. The "higher heating value," which takes into account the energy for condensing water vapor, is equal to 286 kJ/mol or 39.4 kWh/kg. This means that it is extremely difficult to make the inverse reaction occur, that is, water decomposition into hydrogen and oxygen. Temperatures higher than about 3000 °C would be

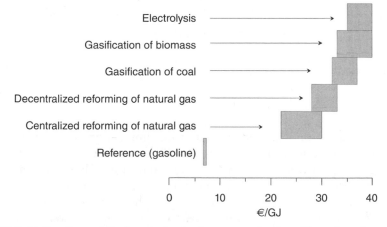

Figure 16.5. Estimation of final costs for hydrogen production. The price of gasoline is the reference. The cost includes production, storage, transport, and distribution. *Source*: Data from *Clefs CEA*, No. 50/51, 2004–2005, p. 30.

needed. One way to circumvent this difficulty is to use thermochemical cycles. These consist of a series of chemical reactions allowing the dissociation of water to be realized at a much lower temperature.

The first work on this subject dates to 1964. Since then, a large number of thermochemical cycles have been imagined. All are unfortunately not practical because of technical constraints, and only some are potentially possible. They depend on the nature of the heat source available. One possibility would be to use nuclear reactors functioning at very high temperature (approximately 1000 °C). Other ways would be to use solar energy, which provides even higher temperatures.

The iodine–sulfur cycle illustrates this technique. This possibility has been studied by the French Atomic Energy Commission (CEA) and the U.S. Department of Energy (DOE) within the framework of a program (Généra-tion IV) to develop nuclear energy for the future. It is based on the high-temperature decomposition of sulfuric acid and hydrogen iodide. The sulfuric acid is broken up under these conditions according to the cascade of following reactions:

$$H_2SO_4 \rightarrow H_2O + SO_3 \quad \text{at 400–600 °C}$$

$$SO_3 \rightarrow SO_2 + \tfrac{1}{2}O_2 \quad \text{at 800–900 °C}$$

It is this latter chemical reaction which specifies the required characteristics of the heat source. The hydrogen iodide breaks up at lower temperature and the formed iodine reacts with sulfur dioxide and water to form sulfuric acid and hydrogen iodide, thus regenerating the products of the cycle:

$$2HI \rightarrow H_2 + I_2 \qquad \text{at } 200\text{--}400\,°C$$

$$SO_2 + 2H_2O + I_2 \rightarrow H_2SO_4 + 2HI \quad \text{at } 25\text{--}120\,°C$$

The sum of the four chemical reactions in this cycle is

$$H_2O \rightarrow H_2 + \tfrac{1}{2}O_2$$

Thus, instead of carrying out the water decomposition reaction in only one stage, at a temperature larger than 3000 °C, one accomplishes it at less than 1000 °C. It is however necessary to make use of other chemicals and to separate the species formed during the reaction. With respect to the direct decomposition of water, the efficiency of a thermochemical cycle varies according to whether or not one recovers a part of work or heat released in intermediate stages (cogeneration). A theoretical estimate of the efficiency gives a value a little smaller than 50% without cogeneration and close to 60% with cogeneration.

Beyond the technological difficulties, using thermochemical cycles in nuclear reactors functioning at high temperature as a heat source requires use of significant quantities of chemicals (H_2SO_4, HI, I_2), some of which are corrosive and dangerous. Their localization close to a nuclear reactor raises questions of safety and thus of acceptability by society: Any accident of a chemical nature occurring at a reactor might indeed be perceived as a nuclear incident.

There are many other possible thermochemical cycles using other reagents to carry out the decomposition of water into hydrogen and oxygen. Hybrid cycles combining thermochemical and electrolytic reactions could avoid certain difficulties specific to each process. The thermochemical cycles are interesting in theory, but it will always be necessary to compare the price of a given cycle with that of normal electrolysis. The objective is to reach a cost of 9–13 €/GJ. This remains higher than the cost of reforming of natural gas and using it as a source of energy to produce heat (5–10 €/GJ).

From Biomass

One can produce hydrogen from biomass by either thermochemical or biological processes. We have seen, in Chapter 6, how gasification of the biomass is possible. This provides a way to produce hydrogen (mixed with CO, CO_2, and CH_4 in the syngas), but this use of biomass would put it in competition with biofuel production. As we have noted above, the cost is much higher than for production of hydrogen using fossil natural gas. The problem in biomass usage is the low energy density per unit of volume compared with that of fossil fuels. Collection, transport, and storage problems can also increase the cost.

Using Enzymes

Microbes can produce gases and finding new methods to produce hydrogen using microbial techniques is an open field of investigation. Experiments have shown that it is possible to produce hydrogen in aerobic and anaerobic biotopes with photosynthetic microorganisms. The mechanism relies on specific enzymes. The principal enzyme is hydrogenase, which catalyzes the decomposition of water.

The biological production of hydrogen may be carried out in three ways:

1. In the photosynthetic process, photosynthetic organisms directly produce hydrogen from solar energy. The process is limited by a poor conversion yield and by the fact that the production of hydrogen also produces oxygen, chemically very reactive element and one that is toxic for certain hydrogenases.
2. Anaerobic (without oxygen) digestion uses a specific mechanism to produce hydrogen with the hydrogenases present in the microorganisms. The maximum yields obtained in laboratory have difficulty exceeding 50% of the theoretical yield.
3. Photofermentation associates a stage of anaerobic digestion leading to acetates which will be transformed into CO_2 and hydrogen by photosynthesis. One application is to produce hydrogen from the fermentable share of household refuse.

These processes have some drawbacks which will be difficult to surmount. In all the cases, the principal process limiting the yield of hydrogen production during fermentation is its transformation into methane and there is a close correlation between the production of hydrogen and that of methane.

Photolysis

Photolysis of water consists of directly using sunlight to dissociate water molecules into oxygen and hydrogen. Illuminated large-gap photocatalyst semiconductors, such as titanium dioxide (TiO_2), or gallium arsenide (GaAs), are used for this purpose. Improvement can be made by modifying the structure of the semiconductor or by coupling it with photosensitive structures containing dye molecules which make it possible to better absorb sunlight. The photolysis of water is a technology still at the research stage. The yields, which depend upon the specific technology, can reach are about 15% but, in practice, are often lower than 10%. Such a technology may lead to inexpensive arrays which could be integrated into the roofs of houses or buildings.

16.1.3. Storage

Hydrogen is a very reactive chemical element. It diffuses easily into cracks and, when transported in metallic tubes, may make them brittle. Under normal conditions, 20 °C and 1 bar, small masses of hydrogen will occupy large volumes. Energy must be expended to compress or liquefy it. While hydrogen storage can be difficult, a number of technologies offer viable solutions to this problem.

Taking into account hydrogen's physical and chemical properties there are three main types of storage systems which may be utilized: gas storage, liquid storage, and sequestered storage in solids.

Hydrogen is light and occupies a great volume for a small mass. One gram of hydrogen under normal conditions of temperature and pressure fills 11.2 liters. One gram of methane under the same conditions occupies a volume eight times lower.

At $-253\,°C$ and under 1 bar of pressure, the density of liquid hydrogen is $71\,kg/m^3$. At $20\,°C$ and 350 bars, it is $23.7\,kg/m^3$. It is only $0.09\,kg/m^3$ at atmospheric pressure (1 bar).

Storage of Compressed Hydrogen Gas

Hydrogen can be stored on board vehicles or in stationary tanks depending upon the application envisaged. For onboard storage the tank must be compact, light, reliable, inexpensive, and able to be filled rapidly. Some of these constraints apply to the fixed tanks intended for stationary applications. For the majority of cases where a few tens of kilograms of hydrogen fuel is needed, compressed gas storage is probably the best solution, particularly in the case of automotive applications where hydrogen is used as a fuel for a fuel cell. Around 1 kg of H_2 is needed to drive 100 km, which means that 5 kg onboard gives a range of about 500 km. High pressures are needed to maintain this fuel within a reasonable volume and one has to be sure that there is no leak.

Traditional steel bottles have been routinely used at pressures around 200 bars. Tanks which can withstand pressures near 350 bars are now common, and it is also possible to construct composite material tanks capable of storing hydrogen at pressures up to 700 bars. Safety is a major concern and this puts strong constraints on the properties of these tanks. Special materials are needed because hydrogen can embrittle and weaken certain materials in particular metals. Compressing hydrogen to high pressure requires a lot of energy. The estimated energy required to compress this gas to 700 bars is equivalent to about 10% of the energy contained in the hydrogen.

The energy density of storage in a bottle is about 0.45 kWh/kg. In this case the hydrogen contributes only about 1.1% of the total weight of the bottle. Depending on the design of the bottle and the choice of the material, it is possible to reach 1.5–2.6% by weight. Using composite materials is much more efficient and higher pressures can be used and can make it possible to reach 5 kWh/kg ($\approx11.3\%$ by weight).

For very large scale storage it is possible to use large natural or artificial underground cavities: old mines, aquifers, salt caverns, or natural caves. For example, in Kiel, Germany, a gas cavern is being used at a depth of 1330 m for storage of town gas (containing 65% H_2) since 1971. Typical pressures in caves are between 80 and 160 bars, giving energy densities of 250–465 kWh/m^3. The energy densities reached in aquifers are much smaller. Losses due to leaks represent about 1–3% per year.

Storage of Liquid Hydrogen

Liquid hydrogen occupies less room than gaseous hydrogen. A given volume stores more than twice as much energy in the liquid state than does the gas compressed to 350 bars. Because liquid H_2 is light and less dangerous than compressed H_2, it has been used for a long time as a fuel in space applications. It is however difficult and energy consuming to liquefy hydrogen. The energy needed to liquefy hydrogen is about 11 kWh/kg. This corresponds to 28% of the energy content of the H_2. Well-insulated vessels are required to store it.

The first liquefaction of hydrogen was achieved independently by the chemists S. Wroblewski and J. Dewar at the end of the nineteenth century. At the beginning of the twentieth century, the method was strongly improved by George Claude, a French chemical engineer and cofounder of the company Liquid Air who improved the "machine of refrigeration by compression" originally developed by the German Carl Von Linde and created the process known as the Claude cycle. In this cycle, hydrogen is initially compressed in a closed container. This compression requires work (energy). Once the gas is compressed, heat is evacuated through the walls. Hydrogen is then at ambient temperature in compressed form. The gas is then allowed to decompress, which cools the gas as it returns to the external pressure. This cycle is repeated several times in order to reach the liquefaction temperature, 253 °C. The last cycle, phase transition of the hydrogen gas to liquid hydrogen, is more difficult and requires a special expansion technique. An alternative method is to cool the gas using a cooling agent: very cold gas helium, which has a temperature of liquefaction lower than that of hydrogen.

Achievable energy densities, including the mass of the storage container, are around 13.8 kWh/kg (25.9% hydrogen by weight) or 2760 kWh/m^3. Boil-off losses depend on the size and insulation of the vessel and range from 3% per day for small containers to 0.06% per day for large ones. The cost to liquefy H_2 approaches 6 €/GJ (to be compared with about 2 €/GJ for the gas compression). The storage capacities for the various possibilities we shall consider are summarized in Table 16.2.

TABLE 16.2. Hydrogen Capacities of Selected Storage Methods

Storage Method	Energy Capacity (kWh/kg)	Hydrogen Capacity (%weight)
Gaseous H_2	5	11.3
Liquid H_2	13.8	25.9
Metal hydride	0.8–2.3	2–5.5
Nanotubes	1.7–3	4.2–7
Activated carbon	2.2	5.2
Glass spheres	2.5	6
Fullerenes	2.5	6
Zeolites	0.3	0.8
Chemical	3.8–7	8.9–15.1

Source: From Tero Hottinen, www.tkk.fi/Units/AES/studies/dis/hottinen.pdf.

Storage in Metal Hydrides

It is difficult to solidify hydrogen because of the large energies required and the very low temperatures which must be reached. One can nevertheless store it in a "solid-state" form at a range of temperatures close to normal by exploiting the adsorption properties of other elements. Some metals such as palladium, vanadium, or certain alloys containing magnesium are able to adsorb large quantities of hydrogen by forming metal hydrides. Metal hydrides are composed of a host lattice of metal atoms and hydrogen atoms which are trapped in interstitial sites.

The hydriding of a metal can be done either by direct dissociative chemisorption or by electrochemical splitting of water. In the latter case a Pd catalyst is needed. For example, Mg_2Ni forms $Mg_2NiH_{3.9}$ and Mg_2NiH_4. The mechanism is very tricky. The essential point of this combination is its reversibility: If one changes the conditions of temperature and pressure, the metal can desorb stored hydrogen, thus making storage possible. To empty a tank employing hydride storage, it is generally necessary to heat it or lower the external pressure. In contrast, to fill it, it is necessary to cool it or increase the pressure. The interesting metal hydrides are those with useful ranges of temperatures and pressures that are compatible with the intended application: typically from 0 to 300 °C and 0.1 to 10 bars. It is noteworthy that some prototype cars directly use the heat released by a fuel cell to carry out the hydrogen desorption which precedes its use as fuel; however, starting these cars can be problematic, especially if the weather is cold.

Although the volume capacities of storage of metal hydride can be equal to or even higher than those of liquid hydrogen tanks, they have a low mass capacity relative to the mass of the metal. The metal atoms are much heavier and, indeed, account for most of the weight of the filled tank. In addition, the high price of hydrides (larger than 20 €/kg) is a limiting factor. This technique also requires the use of very pure hydrogen so that the adsorption capacity is not degraded after many filling cycles. The problem of the availability of sufficient raw material resources is also an important issue. It is not clear whether there is enough raw material resources to synthesize the hydrides necessary for ground transportation on a large scale. These considerations are likely to limit the use of hydrides. They could nevertheless be interesting in niche applications.

Carbon Nanotubes

Hydrogen can be stored in carbon nanotubes by chemisorption or physisorption. Theoretically, single-walled nanotubes can absorb around 14% by weight of H_2 and multiwalled nanotubes about 7.7% by weight. The multiwalled nanotubes can absorb hydrogen in the spaces between the concentric single-walled nanotubes from which they are constructed. One current drawback in the use of nanotubes for this purpose is their high price, which may reach several hundred euros per gram. There is also the problem of safety since nanotubes can be dangerous if inhaled.

Carbon nanotubes are particular geometric structures of carbon, tubes whose interior dimensions are on the order of 1 mm (10^{-9} m). Their length can be typically several millimeters. Some have been obtained that are almost 2 cm long. In addition to single-walled nanotubes there are also multiwalled ones in which single-walled tubes are nested inside each other. Carbon nanotubes are 100 times stronger than steel and weigh six times less. A carbon nanotube conducts heat better than diamond (another extended carbon structure) which itself is a better thermal conductor than copper (between two and five times better). Nanotubes are also better conductors of electricity than copper.

Carbon nanotubes were discovered accidentally in 1991, although they have always been present in soot. Their discovery followed that, in 1985, of other previously unknown carbon species, the fullerenes. The best known molecule of the fullerene family, C_{60}, consists of pentagonal and hexagonal subunits much like those of a soccer ball or a geodesic dome. Because of its similarity to the latter the molecule is called Buckminsterfullerene.

Other Possibilities

Other possibilities for hydrogen storage exist. Among these are additional forms of carbon such as graphite nanofibers, fullerenes, and porous carbon with high surface area (also called activated carbon) which can be used to store hydrogen. Zeolites, which are microporous inorganic compounds with a pore size between 0.3 and 1 nm, are also a possibility, although the capacity of storage is poor ($\approx$0.1–0.8% by weight). Glass spheres, which are empty glass

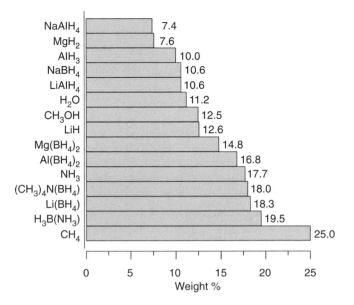

Figure 16.6. Hydrogen storage weight in percent for selected materials containing hydrogen. *Source*: Data from J. Wang, Sandia National Laboratory, http://cohesion.rice. edu/CentersAndInst/CNST/emplibrary/Wang.ppt.ppt.

microballoons with a diameter ranging between 25 and 500 µm, may store up to 5–6% by weight at 200–490 bars.

Organic compounds containing hydrogen atoms can also be considered a means of storing hydrogen. For example, Methanol (CH_3OH) and ammonia (NH_3) have high capacities of storage: 8.9% by weight for methanol and 15.1% for ammonia. Values are given in Figure 16.6 for selected chemicals.

16.1.4. Hydrogen Transport and Distribution

For hydrogen to become an energy vector, it is necessary that it is available at the time and place where it is needed. A supply infrastructure is necessary to transport hydrogen from the place of production to the place of consumption.

Transport
Already during antiquity, networks of pipes were used to transport water from one place to another (aquaducts). In the last century this technique was extended to the transport of other liquids (e.g., oil) or gases (e.g., natural gas). The first hydrogen pipeline was built in the Ruhr Valley (in Germany) in 1938. More than 200 km long this pipeline is still used today by the company Air Liquide and supplies 14 chemical and petrochemical sites. The gas most frequently transported by gas pipelines today is natural gas. In France there is

more than 30,000 km of pipes in the distribution backbone and 165,000 km if the gas pipes connecting homes to the general gas pipeline network are counted. Compared to that, the hydrogen network in Europe and in the United States is jointly very modest with a total of only 2400 km of pipes. Nevertheless, they demonstrate that hydrogen transport is feasible, even though the associated cost is approximately 50% more expensive than that of transporting natural gas and a transported volume of hydrogen contains three times less energy than the same volume of natural gas. Hydrogen is transported in gaseous form, at ambient temperature, and under pressures ranging from 10 to 100 bars. The diameter of a pipeline typically varies between 10 and 30 cm. The cost of the gas pipeline itself is about one million euros per kilometer for a standard diameter of 30 cm. The expenditure related to this form of transport is of the same order of magnitude as that encountered in the transmission of electricity by cable, approximately 2 €/GJ. A large fraction of the cost is that of compressing the hydrogen in the pipes.

The investment necessary to build a hydrogen gas pipeline network in the United States so that cars can freely access a hydrogen fuel supply is estimated to be $260 billion. This is a significant cost but of the same order of magnitude as the investment necessary for a natural gas network. An important difference, however, is that natural gas is normally directed to stationary locations, power plants, homes, buildings, industries, and so on. In that situation, a progressive installation of the network can be pursued. In contrast, assuring the autonomy of hydrogen-fueled vehicles throughout a large territory would require installation of a complete network. Therefore, hydrogen transport systems are likely to first be directed toward sites with stationary applications.

An advantage to storing and transporting hydrogen in liquid form is that the same quantity of hydrogen in liquid form occupies a volume one thousand times smaller than in gas form at normal temperature and pressure. Today, hydrogen is transported in liquid form for the glass and electronic industries. In these cases, trucks of 40 tons typically carry 3.5 tons of liquid hydrogen at a cost which ranges between 2 and 4 €/GJ. The fuel energy necessary for driving a truck of 40 tons over a distance of 150 km corresponds to about 20% of the energy contained in transported hydrogen. The advantage of road

TABLE 16.3. Comparison of Different Hydrogen Transports

	Gas Pipeline	Liquid Hydrogen
Investment cost	$1 million/km	Truck: $250,000
		Boat: $400 million
Running cost for 100 km	$2.5/GJ	$10/GJ
Yield over distance of 1000 km	92%	Truck: 45%

Source: From *Hydrogène, énergie de demain?* ECRIN, Omniscience, 2007.

transportation is of course the lower capital cost (about 250,000€ for a truck compared to a million euros per kilometer of a gas pipeline). This means of transport is not appropriate on a large scale and it will remain confined to very specific needs. If the needs to transport liquid hydrogen were to increase dramatically, railroad transport would be less expensive, with a cost between 0.5 and 1.25€/GJ. Table 16.3 compares costs for different means of hydrogen transport.

In terms of mass, hydrogen is a very good fuel: It contains approximately 3 times more energy per unit of mass than gasoline and more than two times that of methane. In terms of volume, which is the significant parameter for road transportation, the situation is much different. In a gas pipeline, compressed hydrogen at standard pressures transports 15 times less energy than oil. Compared to natural gas, on a volume basis, hydrogen contains 3.3 times less energy under the standard conditions of temperature and pressure and 4.6 times less if the comparison is made at 200 bars. Liquid hydrogen contains, by unit of volume, 2.5 times less energy than liquefied natural gas.

Boat transport is also a possibility to connect overseas zones of strong H_2 production to locations of strong H_2 consumption. This would be similar to the transport of natural gas in methane tankers. Two prototype cargo liners capable of transporting 14,000 tons of hydrogen in a volume of 200,000 m^3 have been developed.

Distribution

Two methods by which hydrogen fuel may be provided for a car are by filling up a tank on demand or by exchanging an empty tank for a full one. The first solution requires widely distributed hydrogen pumping facilities. To be acceptable for the consumer, it is necessary to be able to fill up the tank of a car in a reasonable time, say less than 5 min. This would require a minimum flow rate of 16 L/mn for an 80-liter tank. This is not so easy to achieve as it requires a pressure difference (about several tens of bars) between the two tanks and maintenance of this pressure difference throughout the exchange. In addition, during the transfer, the pressure in the tank of the vehicle will increase. This will make an increase in the temperature of gas of a few tens of degrees. The problem is that when the tank cools down and the pressure decreases the tank will no longer be full. The pressure change can be about 20%, which means it is not possible to completely fill up a tank. The exchange of tanks is easier but requires a standardization of tanks and connectors. Certain test vehicles, such as the Partner taxi from Peugeot, already use this type of exchangeable tank. An important advantage is that the filling of the tanks may be deferred until periods of low electricity demand, nighttime, for example.

In 2007, about 40 hydrogen-supplying service stations existed in the world, most in the United States, Japan, Germany, and Iceland. In any country, development of a complete network of hydrogen stations will take a long time. The European Commission estimates that equipping only one-third of the current service stations in Europe with hydrogen-pumping capabilities would cost between 100 and 200 billion euros. Further, many current service stations will not be suitable for hydrogen distribution if they cannot increase their areas to accommodate underground storage of hydrogen. In Munich, where a hydrogen station is operated by the Total Company, the area is approximately 10 times larger than that of a standard service station. The cost of the storage and pumping equipment is also still very expensive. For example, the cost of a distribution pump is about 150,000 €.

16.1.5. Conclusion

Although hydrogen offers many interesting advantages as a fuel and as a means of storing electricity produced by intermittent renewable energy sources, we are still far from large-scale use of hydrogen as a fuel. There are many storage, transport, and distribution problems to solve. Switching to a hydrogen economy will require large initial investments and take a long time. In addition, continued large increases in the price of fossil fuels would have to occur before hydrogen would become economically competitive. Given environmental concerns, it will probably be necessary to produce the hydrogen without emission of CO_2, which means that renewable energies and nuclear power will be preferred over fossil fuels at the production stage. Applications of hydrogen as a fuel will mainly be done using fuel cells. These cells are the subject of the next section.

16.2. HYDROGEN: ENERGETIC APPLICATIONS

As we have already discussed, there are several ways to use hydrogen as an energy vector. Fuel cells, which have the capability to convert the chemical energy of hydrogen directly into electricity, appear to offer the most practical and efficient way to do this. This section is mainly devoted to a description of fuel cell technology. A brief consideration of other energy applications of hydrogen is also included.

16.2.1. Fundamentals of Fuel Cells

Consumers want electrical energy on demand. In recent years, this has spurred the development of significantly improved batteries which are convenient portable power sources. However, batteries have limited running times. They can be expensive and have to be replaced when their usable energy content is exhausted. Rechargeable batteries can be used a large number of times but require an available recharging source and a waiting period while the battery is recharged. Electric vehicles, which have relatively small driving ranges and significant recharge times, have met with limited acceptance. They have difficulty competing with gasoline (or other similar) fueled vehicles because the energy density of gasoline is much higher than that of a typical battery. To store of 1 kWh of energy about 25 kg of lead battery is needed but only 70 g of oil. In addition, gasoline tanks can be refilled in a few minutes.

To have a longer lasting portable source of chemical energy, it is necessary to have reserves of the fuel which can replenish the supply at the electrodes as needed. In "redox batteries" the reductant and oxidant are liquid stored in containers and can be refilled as they are consumed. Another possibility is to use hydrogen and oxygen gases in fuel cells. In such a device, hydrogen is the fuel and serves as the reductant. Oxygen is the oxidant. The reaction which generates energy, $2H_2 + O_2 \rightarrow 2H_2O$, is the inverse of that which occurs in the electrolysis of water.

In 1839, Christian Friedrich Schönbein published the basic principles of fuel cell operation in the *Philosophical Magazine*. Schönbein was a gifted scientist who also discovered the ozone molecule, O_3, in 1840 and "guncotton" (nitrocellulose), a replacement for gunpowder, in 1845. In 1842, his friend, Sir William Robert Grove, constructed the first fuel cell (which he called the *gas voltaic battery*). Grove is considered the father of this technology. Although the basic idea of a fuel cell is simple, its achievement at an industrial scale proved to be quite difficult and it took more than a century before fuel cells became really usable in applications.

Interestingly, space travel was the first well-known domain where fuel cells were extensively employed. Starting in 1932 Francis Bacon, building upon the earlier work of Grove, resumed research on fuel cells. He developed the first 1-kW hydrogen fuel cell prototype in 1953 and a 5-kW cell in 1959. These would prove to be the starting point for development of more powerful fuel cells that would used in the Apollo space missions.

In a fuel cell there is no storage of electricity. The term "fuel cell" implies a continuous feeding of a chemical fuel that reacts with the oxygen contained in the air. It was first expected that a wide variety of fuel–oxygen couples could be employed in fuel cells. In fact, various practical difficulties have limited the use to a relatively small number of fuels, primarily hydrogen, alcohols, and natural gas. In the hydrogen fuel cell, the hydrogen is referred to as the "combustible" and the oxygen as the "combustive." The system will function as long as one brings hydrogen and air to it. Except for ageing, the electrodes of a fuel cell are almost unmodified during the process. A fuel cell is therefore a sustainable transformer of energy that is nearly ideal for producing electricity, heat, and water. Fuel cells have many outstanding properties. They have high efficiencies compared to internal combustion engines and there are no moving parts inside the cell. Consequently they are not noisy and have no vibration.

To produce electricity, a fuel cell must be designed to support the controlled reaction between the fuel and the combustive matter. There are three key components in a fuel cell: two electrodes (the anode and cathode) and the electrolyte. In a gas-fed cell the electrolyte section of the cell must be gastight. At the anode there is an oxidation process supported by a catalyst in which the hydrogen molecules are transformed into two H^+ ions by releasing two electrons (Figure 16.7). The H^+ ions move from the anode region to the cathode region through the electrolyte, whose principal function is to transport the ions but not the electrons. Thus, positive electric charge flows inside the electrolyte. At the same time, electrons carrying the same magnitude of negative charge circulate through an external circuit to the cathode. When the H^+ ions arrive at the cathode they combine with oxygen ions which have been produced at the cathode from the oxygen gas by combination with the electrons from the external circuit. From the H^+ and O^{2-} ions, water is formed. The circulation of the electrons in the external circuit produces an electrical current which can be used to power devices such as electrical motors, for example. The choice of reactants determines the "potential" at which each electrode carries out the electrochemical transformation. The potential difference between hydrogen and oxygen is 1.23 V, but when the current circulates inside the fuel cell there are losses and the usable voltage is decreased to about 0.6–0.7 V. This means that for practical applications several cells have to be connected in

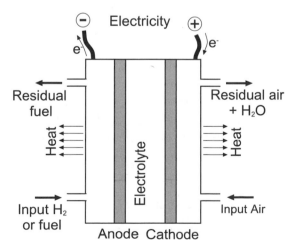

Figure 16.7. Principle of a fuel cell. Hydrogen and oxygen are combined to form water. This generates electricity and heat.

series. The core of a commercial fuel cell consists of multiple individual cells usually called the stack.

The choice of electrolyte plays a major role in fuel cell design because the property of ion conduction is not commonplace. For that one needs a material which is itself naturally composed of ions. This is the case, for example, for acids which contain H^+ ions or bases which contain OH^- ions. These ions have to move very rapidly in the electrolyte. The ion mobility depends in particular on the temperature of the medium. In the earlier stages of development of fuel cells, most electrolytes were liquid because solids capable of conducting ions were primarily high-temperature (1000 °C) ceramics. These materials were difficult to produce. The electrolytes commonly in use today are:

- Concentrated aqueous solutions of potassium hydroxide (potash) functioning between −30 and +80 °C. The mobile ion is OH^-.
- Very specific solid polymers which conduct H^+ ions functioning between 0 °C and 90 °C. For that purpose, PEMs (proton exchange membranes) or SPEs (solid polymer membranes) are often used.
- Hot phosphoric acid (at a temperature greater than 200 °C). The conducting ion is H^+.
- Molten salts (melted carbonates at a temperature above 650 °C). The conducting ion is CO_3^{2-}.
- Special ceramics at very high temperature (currently at temperatures reaching more than 1000 °C). The conducting ion is O^{2-}.

Electrodes of the fuel cell must perform several functions. The most difficult one is to ensure the transformation of hydrogen or oxygen gases to the ionic state, H^+ or OH^-. This transformation is not as easy as expected and sometimes a catalyst is required to increase the rate of the reaction. This catalyst, all the more necessary at low operating temperatures, is usually finely dispersed in the electrode. It is often an expensive material such as platinum.

16.2.2. Different Types of Fuel Cells

Many research studies have been and continue to be devoted to the improvement of fuel cells. There currently exist several families of fuel cells designed for different applications.

Alkaline Fuel Cell (AFC)

The oldest type of fuel cell and also the least expensive is the AFC. Such cells were used by NASA in space vehicles for the Apollo missions starting in the mid-1960s. This technology is still in use for manned flights. It employs an alkaline electrolyte, potassium hydroxide (KOH) or sodium hydroxide (NaOH). The chemical reactions involved are

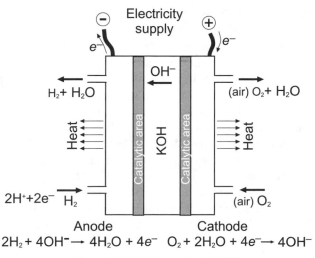

Figure 16.8. Schematic of AFC.

$$2H_2 + 4OH^- \rightarrow 4H_2O + 4e^- \quad \text{at the anode}$$

$$O_2 + 4e^- + 2H_2O \rightarrow 4OH^- \quad \text{at the cathode}$$

The operating principle of such cells is shown in Figure 16.8. The technology is simple and inexpensive but the air used must be free from CO_2 because this gas is a pollutant for the electrolyte. A broad range of catalysts can be used (Pt, Ni, certain alloys, etc).

These fuel cells function at around 80–90 °C at atmospheric pressure but can also be used at higher temperature (up to 260 °C) if the electrolyte concentration is higher and under high pressure (3–4 bars), as is the case in space applications. Their typical useful life spans are 15,000 h. In space one of these fuel cells functioned more than 80,000 h. The power supplied ranges from 1 to about 10 kW. For example, a prototype used in 1998 to recharge the batteries in a hybrid propelled London taxi provided 5 kW.

The fuel cell used in manned Apollo missions was developed by the Pratt and Whitney Company. Its weight was 110 kg. The electrolyte was KOH at a weight concentration of 85% by mass. Operating at a temperature of 260 °C and at a total pressure of 4.1 bars, it employed pure hydrogen and oxygen as fuel and combustive material, respectively. The nominal power of the unit was 1.42 kW at 27–31 V. The fuel cell developed for the space shuttle Orbiter and still in use delivers a power of 12 kW at 27.5 V and weighs only 91 kg.

Proton Electrolyte Membrane Fuel Cell (PEMFC)

The key element of a PEMFC is an electrolyte membrane with some very specific properties. The membrane must conduct protons (H^+) but not electrons, otherwise the fuel cell would short circuit and cease to function. Furthermore, the gases (H_2, O_2) must not pass through the membrane. The most commonly used material for these membranes is Nafion, a product developed by the E. I. Du Pont de Nemours Company. The degree of hydration of the Nafion membrane directly affects its ion conductivity. This material is however expensive ($\approx$\$520–650 m^2). More recently developed membranes, for example, those based on polybenzimidazole, also have good properties and can be used at higher temperatures. This type of fuel cell has attracted the major portion of recent investments in the field of fuel cells because it offers some particular advantages, especially for use in road vehicles. Among these is its ability to function at low temperature, which enables quick starting of the fuel cell. A schematic of a PEMFC fuel cell is shown in Figure 16.9.

The elementary reactions at the electrodes are as follows:

$$H_2 \rightarrow 2H^+ + 2e^- \quad \text{at the anode}$$

$$\tfrac{1}{2}O_2 + 2H^+ + 2e^- \rightarrow H_2O \qquad \text{at the cathode}$$

For a PEMFC, the energy produced by the fuel cell is about half electricity and half in the form of thermal energy. The PEMFC functions at temperatures

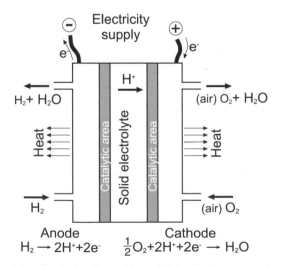

Figure 16.9. Principle of a fuel cell working with hydrogen gas and oxygen (provided by air).

of 80–100 °C but is very sensitive to contamination by carbon monoxide (CO), which is a poison for the platinum catalyst. The catalyst is essential for operation of the fuel cell. At a higher temperature, the fuel cell becomes more resistant to CO poisoning.

Since a PEMFC runs at rather low temperature, it is more difficult to dissipate heat produced during operation of the fuel cell than for an internal combustion engine which operates at higher temperature and is coupled to an exhaust pipe which removes part of the generated heat (about one-third). For heat dissipation and resistance to CO poisoning, it is very desirable to develop a PEMFC which functions efficiently at higher temperatures and there is currently much interest in increasing the operating temperatures of these fuel cell to the 130–150 °C temperature range.

Several applications are envisaged for such cells, for example, transportation (automobile, trucks, buses, trains, ships, submarines) with powers ranging from 50 to than 200 kW_e; cogeneration (combined heat and power generation) for large public, industrial, multifamily buildings, with powers around 100–300 kW_e; or even utilization in individual homes, with powers lower than 10 kW_e.

With the best technologies, outputs on the order of 2.9 kW/L and 1.4 kW/kg have been obtained. That is close to those obtained with internal combustion engines. However, for a complete energy system, including all components necessary for full operation (fluid and gas supply, heat exchanger, DC–AC converter, control elements of the system, exhaust system for removal of produced water, etc.), such performances must be divided by a factor close to 3. There remain many problems to solve to ensure reliability and to reduce the costs to make the PEMFC economically competitive. It is very important that the solutions to these problems involve a global systems approach in which not just the core of the fuel cell but the whole energy system is optimized.

Phosphoric Acid Fuel Cell (PAFC)

The PAFC is a fuel cell which allows large power outputs but has the disadvantage of using a corrosive liquid electrolyte: phosphoric acid. In addition to hydrogen, the fuel for this cell can also be natural gas, propane, or biogas. These fuel cells were the first on the market and more than 200 are operating around the world with powers ranging from 100 kW to 1 MW. Commercial fuel cells of this design have life expectancies of more than 40,000 h. An electricity yield of about 40% and a thermal yield around 50% are quite common. The cost is about $5200–6500 kW_e. More than 80% of the PAFCs used in the world have been produced by the ONSI Corporation, a joint American–Japanese venture between the UTC Fuel Cell Company and Toshiba. They produce the PC25 fuel cell, which has a power of 200 kW. Cumulatively, PAFCs of this type have logged more than 2 million operating hours at the time of this writing.

William Grove used sulfuric acid as the electrolyte to make the first fuel cell in 1842. Phosphoric acid, with its poorer ionic conductivity, was not employed until 1961 when G. V. Elmore and H. A. Tanner built a PAFC which functioned directly in air rather than requiring pure oxygen. Boosted by the military demand, the first commercial PAFCs appeared in 1965.

Schematically, the operating principle of the PAFC is similar to that of the PEMFC (displayed in Figure 16.9). The reactions at the electrodes of the PAFC are the same as those of the PEMFC.

For a trial test, a PAFC was installed in December 1999 in Chelles, France, by Electricité de France—Gaz de France. Having a maximum power of $200\,kW_e$ or of $220\,kW_{th}$, it functioned in the cogeneration mode (combined heat and power output). The total efficiency was 75%. It met the needs of 200 households for heat and electricity. This experiment stopped at the end of 2005. The investment cost for that fuel cell was ~$1 million or about $5000 per installed kW_e.

A PAFC must be used in a narrow range of temperature ranging from 190 to 210 °C to avoid degradation. The advantages of a PAFC are that phosphoric acid is inexpensive and the cell is not sensitive to carbon dioxide. However, as indicated previously, platinum catalyst, which is needed at the electrodes, is expensive.

Molten Carbonate Fuel Cell (MCFC)

This technology uses an electrolyte which is a mixture of molten carbonates. The reaction processes taking place at the electrodes are the following (Figure 16.10):

$$H_2 + CO_3^{2-} \rightarrow H_2O + CO_2 + 2e^- \quad \text{at the anode}$$

$$O_2 + 2CO_2 + 4e^- \rightarrow 2CO_3^{2-} \qquad\qquad \text{at the cathode}$$

High power outputs can be obtained. The MCFC can be used in the cogeneration mode with yields on the order of 45% electricity and 35% heat. The management of molten carbonates raises some problems and development of this technology on a commercial basis would be very difficult.

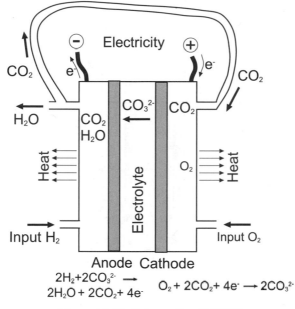

Figure 16.10. Schematic of MCFC.

The first MCFC prototype, developed in 1960 by G. Broers and J. Ketelaar, functioned for six months. This fuel cell used a mixture of lithium and sodium carbonates impregnating a porous structure of magnesium oxide. In the middle of the 1960s, the U.S. army tested several MCFCs manufactured by Texas Instruments. The power outputs ranged between 100 and 1000 W.

These cells function at high temperature ($\approx 650\,°C$), which allows one to use the thermal energy produced by the fuel cell and couple it to a gas turbine. Furthermore, the high temperature allows a direct reforming of hydrocarbons at the anode. However, in addition to the drawbacks mentioned above is the necessity to inject the CO_2 formed at the anode at the cathode.

Solid-Oxide Fuel Cell (SOFC)

The SOFC uses a ceramic made of solid oxide, usually zirconium dioxide doped with yttrium, as the electrolyte. It functions at high temperature ($800–1000\,°C$), which makes it less sensitive to pollutants or impurities present in the fuel. The fuel can be hydrogen, natural gas, or other molecules containing hydrogen atoms. The reactions at the electrodes are the following (Figure 16.11):

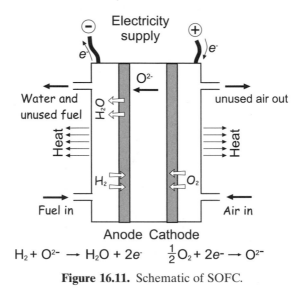

Figure 16.11. Schematic of SOFC.

$$H_2 + O^{2-} \to \quad H_2O + 2e^- \quad \text{at the anode}$$

$$\tfrac{1}{2}O_2 + 2e^- \to O^{2-} \qquad\qquad \text{at the cathode}$$

Although studies of this type of cell began in the 1930s, it was not until the period 1983–1989 that real progress was made. Even with this progress the SOFC technology remains the least advanced among the fuel cells we have discussed but is expected to make a significant contribution in the future, in particular for stationary cogeneration equipment. Potentially its ultimate performance should be better than that of the PAFC. In time, it will probably replace the PAFC. Powers up to 100 MW can be reached.

Since this fuel cell operates at high temperature, it is possible to exploit part of the heat released using a gas turbine which can possibly be followed by a vapor turbine. The total yield could then reach 80%. The high temperature of the anode also makes it possible to directly reform fuel and SOFCs do not require a scarce and expensive catalyst. For this reason one can imagine a stationary SOFC functioning initially with natural gas, gradually shifting to natural gas mixed with hydrogen, and eventually using pure hydrogen in the long run.

A cell takes several hours to reach the 800–1000 °C operating temperature, which is crippling for some applications. Moreover, it is not wise to start and stop the fuel cell often because thermal stresses can degrade the cell materials. It would be interesting to reduce the operating temperature of this fuel cell, in particular to extend the lifetime of the cell materials to lower the cost. The fuel cell would also reach the operating temperature quicker.

The technologies so far developed employ tubular or planar geometries. In the planar technology the electrodes are flat sheets while in tubular one they are cylinders. The tubular technology, developed by Westinghouse and Mitsubishi, has a long life span. Tests have shown that such cells can reach nearly 100,000 operating hours. Many companies are now working on the planar technology, which utilizes thin ceramic sheets that are more conducive to operation at lower temperature. Very high efficiencies can be obtained with planar technologies and cheaper materials can be used. Compared to the MCFC technology, the advantage of the SOFC is its use of a solid electrolyte, which means that no fluid circulation system is needed. Small and compact prototypes of a power around 1 kW could soon be available on the market.

Direct Methanol Fuel Cell (DMFC)

Fueled by methanol, the DMFC uses a solid electrolyte. It works at 120 °C or at lower temperatures (50–100 °C). Alcohol is not reformed but is directly used as fuel. It is quite promising for powering portable equipment (laptop computers, cellular phones, etc.). The electrolyte can be a polymer or an alkaline liquid.

Methanol is easy to obtain and can easily be mixed with water but is quite toxic. Ethanol can also be used. In this case the fuel cell is called a DEFC (direct ethanol fuel cell). At present, the yield with ethanol as a fuel is smaller by a factor of 2 than that for methanol.

Methanol (CH_3OH) is also known under the name wood alcohol. It is the simplest molecule in the family of organic alcohols. The next member of the family is ethanol (C_2H_5OH), the alcohol which is present in beer, wine, or bourbon, for example. The great drawback of methanol is that it is extremely toxic to humans: Its ingestion can cause neurological problems, blindness, and death if the concentration is sufficient. Without suitable precaution alcohol obtained by distillation from some biocompounds can contain methanol and be toxic. The potential danger of methanol comes from its ready miscibility with water. Leaks can lead to pollution of groundwater.

Open Issues

A summary of the different characteristics of the fuel cells we have discussed is given in Table 16.4.

These fuel cells have not yet reached large-scale usage and often remain confined to laboratories or pilot applications. Each type of fuel cell has both advantages and disadvantages. Some important problems are still unsolved and require concentrated research and development activities. Among these problems are:

TABLE 16.4. Summary of Characteristics of Fuel Cells

Type of Fuel Cell	Electrolyte	Temperature Range (°C)	Power Range	Possible Applications
Alkaline (AFC)	Potassium hydroxide (liquid)	50–200	1–100 kW	Space, portable devices
Phosphoric acid (PAFC)	Phosphoric acid (liquid)	160–220	200 kW-10 MW	Stationary units
Molten carbonate (MCFC)	Potassium lithium carbonate (liquid)	650–1050	200 kW–10 MW	Stationary units
Proton exchange membrane (PEMFC)	Polymer membrane (solid)	60–120	1 W-250 kW	Transportation, stationary units, portable devices
Solid oxide (SOFC)	Zirconium dioxide ceramic (solid)	700–1000	1 kW-10 MW	Stationary units, transportation
Direct methanol (DMFC)	Polymer membrane (solid)	60–90	1 kW-250 kW	Portable devices, transportation, stationary units

Note: Numbers should be taken as indicative of typical ranges.

Alkaline Fuel Cells The potash solutions of the AFC require that both the hydrogen and air be free of CO_2 because CO_2 causes an immediate irreversible formation of potassium carbonate, which crystallizes in the fuel cell. Furthermore, these solutions, which are rather corrosive, are difficult to confine by the electrodes, which they eventually breach.

Phosphoric Acid Fuel Cells At a temperature close to 200 °C, the PAFC poses problems for start-up, maintenance of temperature stability, and safety of operation. This technology is probably not the best one for extensive future deployment.

Proton Exchange Membrane Fuel Cells The polymeric membrane of a PEMFC is very expensive. Researchers are trying to find suitable substitutes.

Molten Carbonate Fuel Cells The MCFC, using molten carbonate at a temperature of almost 650 °C, has problems similar to those of the PAFC. Carbonates are "creeping" salts, which are difficult to confine.

Solid Oxide Fuel Cells In spite of requiring a high temperature, SOFCs have the advantage of being insensitive to CO which is oxidized into CO_2. One of the goals of research is to decrease as much as possible the temperature at which ceramics become ion conducting.

Each fuel cell technology works in a specific range of temperature. Practical cells should be capable of quickly reaching and maintaining their working temperatures. This leads to design requirements for coping with the very different ambient temperatures which might be encountered. For automotive applications, for example, cells should be capable of functioning in ambient temperatures below 0 °C or greater than 40 °C. If the operation is not continuous, the fuel cell system should also be able to endure a very large number of temperature changes encountered during its operational cycle. For this reason as well as for better management of CO, more research is needed. In the PEMFC field, a polymeric membrane functioning at temperatures greater than 120 °C is required. Efforts are also being made to develop ceramics for the SOFC that become conductors of ions in a temperature range of 700–800 °C.

For various and complex reasons, electrodes deteriorate with time, as do catalysts. Repair or regeneration of the working surfaces in situ is not presently possible. It is thus necessary that operational units have adequate diagnostics for monitoring the level of deterioration and be constructed so as to facilitate the replacement of defective components. These requirements place important constraints on the design of a unit that must remain hydrogen tight.

The catalysts that are normally used often have the drawback of being both expensive and sensitive to CO poisoning. Thus it becomes imperative to supply the fuel cells with gases which are free of CO impurities.

For an internal combustion engine power is increased when the volume of the combustion chamber is increased. For a fuel cell the available power could be increased by increasing the surface area where the reaction occurs. However, doubling the area requires an increase in the linear dimensions by an average of 44%. Increasing the surface power density is a more desirable solution. This is both technologically challenging and expensive. Another possibility is to increase the pressure. This would allow the fuel cell to deliver more power and is analogous to use of a turbocharger in an internal combustion engine to increase the pressure of the injected fuel. Research on compressors for fuel cells is ongoing. A disadvantage of this solution is that such compressors can prove to be noisy.

16.2.3. Transportation

Currently, developed countries rely almost entirely on oil as the energy source for their transportation needs. However, oil is becoming more and more expensive and cheap oil will not be so common in the future. It is time to prepare for the post-oil era. Used in fuel cells, hydrogen offers an appealing solution to the problem of providing portable sources for road transportation.

Use of hydrogen in transportation is not new. Lighter than air, its first recorded use in transportation was to provide buoyancy for a 4-m-diameter balloon designed by the physicist Jacques Alexandre Charles and constructed by the Roberts brothers. This balloon, named *le Globe*, rose from the Champ de Mars in Paris on August 27, 1783, and reached the city of Gonesse, approximately 25 km away. On December 1 of the same year, 400,000 spectators, among them Benjamin Franklin, America's first ambassador to France and himself a scientist, watched as Jacques Charles and Nicolas Robert departed on the first flight in a hydrogen balloon. They traveled over 15 miles. Robert got out there, and Charles went up once more in the balloon, reaching an altitude of over 1 mile.

Somewhat later, in 1805, François Isaac de Rivaz, a Swiss surveyor and notary, developed the first hydrogen-fueled vehicle. He employed a spark ignition engine supplied with a hydrogen and oxygen mixture. This was successfully patented in 1807. In 1813, Isaac de Rivaz drove a second vehicle measuring 6 m long with wheels 2 m in diameter at a mean velocity of 3 km/h.

Since a fuel cell produces only water, a switch to fuel cell–powered vehicles could meet our transportation needs while allowing a dramatic reduction of air pollutants. Even if the hydrogen were produced using fossil fuel sources in large centralized plants, the shift from delocalized CO_2 emissions to localized emissions would allow the adoption of efficient techniques to sequester the CO_2 gas.

Modern aviation uses planes heavier than air. Such vehicles rely upon aerodynamic lift, which requires the motion of some part of the plane in order to fly. An aerostat is a lighter than air ship, a balloon or blimp, for example. These are lifted by buoyancy, that is, aerostatic lift. Ballooning began initially with balloons inflated with hot air or with hydrogen. For a period early in the last century, airships inflated with hydrogen competed with the airplanes. Deutsche Luftschiffahrt AG, a company created in 1909 by Count Ferdinand von Zeppelin, transported 40,000 passengers in 1600 flights. In 1929, the airship *Graf Zeppelin* made a trip around the world. During its career it flew on 590 journeys covering 1.7×10^6 km. Airships offered passengers luxury comparable to the largest ocean liners. The use of hydrogen-filled airships continued until the tragic *Hindenburg* accident in 1937. The few existing modern airships employ helium gas.

Hydrogen and Fuel Cells

Hydrogen fuel cell stacks can reach approximately 50% efficiencies but only about 35% delivered power. This compares very favorably with the efficiencies of steam engines (~5%), internal combustion engines based on the Otto cycle

(~20%), and diesel engines (~25%). Fuel cell–powered vehicles would produce essentially no air pollution and the noise levels of such vehicles would be much lower than those of vehicles powered by internal combustion engines. All of these justify the use of hydrogen as a fuel. However, it should be kept in mind that production of hydrogen has its own efficiency limitations.

Internal combustion engines based on the Otto cycle, using gasoline, or those based upon the Diesel cycle, burning fuel oil, have benefited from more than a century of experience and are now very reliable. The useful life spans of cars can reach 400,000 km, which corresponds to approximately to 10,000 h of driving. Engines in buses, trucks, ships, or diesel locomotives can reach 10^6 km and about 10,000 h of functioning. Great progress has also been made in terms of pollution control and this is continually improving. For example, Toyota has recently developed a prototype vehicle whose exhaust, except for CO_2, which is a primary product of fuel combustion, is five times cleaner than the typical air of Los Angeles.

First used in spacecraft in the 1960s, the fuel cell was a technology which seemed very well suited for automotive applications. Two attempts to employ this technology took place toward the end of that decade and at the beginning of the 1970s. The first, by General Motors, used an alkaline fuel cell of 5 kW; the second was by Austrian Scientist Karl Kordesh, who equipped an Austin A-40 with an AFC stack of 6 kW. Hydrogen under pressure was stored in bottles put on the roof of the car. The second prototype functioned for three years and was driven 16,000 km.

Following these two attempts, a period of nearly 20 years elapsed before a new fuel cell–powered vehicle, the LaserCell, developed by the American Academy of Sciences, was presented in 1991. It was built on a chassis of a Ford Fiesta and was equipped with a PEMFC of 14 kW. Hydrogen was stored in a hydride form. In 1993, a PEMFC of 20 kW supplied with hydrogen compressed to 200 bars was used in the "Green Car" developed by the Energy Partners Company. In both cases, the vehicles were of the hybrid type, including a lead–acid battery, which recovers the energy of braking and later provides power during acceleration. Daimler Chrysler then developed the Necar I. Other manufacturers have followed. Today many operational prototypes exist.

In a fuel cell–powered vehicle, the cell is the central component of a complex system. It must be coupled with many subsystems performing a series of complementary functions. These include the compressor and the humidifier of air for the oxygen supply, the hydrogen tank (or the reformer if one uses another fuel to produce hydrogen), a circulator for supplying hydrogen to the fuel cell, the separator used to evacuate the water formed during combustion, and the circuit for electric power management. Without these additional components it would not be possible to have a vehicle with driving characteristics comparable to those of the current generation of vehicles powered by internal combustion engines. Rapid accelerations require peaks of power which are difficult to provide with a fuel cell alone. This would require the ability to very quickly introduce the necessary air and hydrogen and evacuate the water produced. This requirement can be moderated by use of a hybrid system in which the peaks of power are provided by a battery (of the nickel–metal–hydride or Li ion type) or by supercapacitors—high-energy-density electrochemical capacitors used as supplementary storage.

Although hybrid vehicles employing fuel cells are not yet being produced commercially, hybrid configurations combining internal combustion engines with batteries and an electric motor have made their appearance (e.g., Prius from Toyota). Internal combustion engines produce mechanical energy directly. Such energy is readily coupled to the drive train of a vehicle. In contrast, fuel cells produce electricity. Fortunately, electric motors, which convert electricity into mechanical energy with a good yield, are readily available.

Two architectures to combine fuel cells with an electrical storage capability (batteries and/or supercapacitors) are common. In the most frequently used approach, the fuel cell and the battery are coupled in parallel. The battery is used to meet peak power demands and it recovers part of the energy used in braking. The battery is sometimes replaced or complemented by supercapacitors. In this second architecture, the battery plays the main role by providing electricity to the electric motor. The fuel cell has a low power and is used only to recharge the battery as necessary.

There are several possible ways to share power between the fuel cell and the electrical storage device. The optimum choice depends on the application: urban vehicle, road vehicle, and so on. In all cases, a hybrid architecture is the best choice because it makes it possible to recover part of the energy of braking.

In many ways, buses seem particularly well suited for the application of fuel cells in the field of transport. There is more room for hydrogen storage in the vehicles and they generally belong to large fleets for which management of hydrogen supply and maintenance is easier than for vehicles of private individuals.

About 1 kg of hydrogen is needed to drive 100 km using a fuel cell–powered private vehicle. Under standard conditions of temperature and pressure that corresponds to approximately 11,200 liters or $11.2\,m^3$ of hydrogen (a cube 2.24 m on a side). To meet the requirement of a 500-km range it would be necessary to embark with an initial supply of 5 kg of hydrogen. The pressures used in the current prototypes are presently around 250–350 bars. At a pressure of 350 bars, one would need a storage volume capacity of about 160 liters. The tank must have a geometric form suitable for high-pressure storage. This is not always convenient to place in a car. An alternative would be to store liquid hydrogen. With a density of $70.8\,kg/m^3$, $5\,kg$ of liquid hydrogen occupies a volume of 72 liters. The problem is the constant evaporation of liquid H_2 (something like 3%/day), which is a problem if the car is not used often. Storage in the form of hydrides could provide a solution but only for market niches. Their use in a vehicle poses constraints and requires particular characteristics: One would need a host material with a hydrogen storage capacity of about 5% of the total mass of the host, the capability to desorb hydrogen at low temperatures, and an ability to employ the host in approximately 5000 load–discharge cycles. It is not yet possible to achieve all these goals simultaneously.

Another approach is to produce hydrogen on demand starting from liquid or solid fuels. Most convenient is the reforming of hydrocarbons or alcohols. Another possibility is to use an alcohol directly in a DMFC or DEFC. An advantage of such fuels would be that they require minimal changes in the current fuel supply system and corresponding consumer practices. Reforming is, however, a complex and expensive process to implement. Platinum catalysts are required. Carbon monoxide produced during the process is a poison. The output efficiency is variable according to the technology employed and the fuel used. It does not exceed 80% and can even fall to 40%. Of the alcohols, methanol is the one which is potentially the most interesting, but it is a water-soluble toxic compound. This last property is very constraining because leakage is always possible and could result in pollution of the groundwater. One can also use sodium borohydride, which, with water and in the presence of a catalyst, releases hydrogen.

For quick starting, fuel cells working at low to moderate temperatures are more convenient. This is why the manufacturers choose to develop the alkaline or PEMFC technologies. The latter are the most promising for road vehicles and the great majority of existing fuel cell vehicles are based on this technology. Although their yield is higher than that of the internal combustion engine used in today's vehicles, approximately 50% of the energy supplied is thermal energy which must be evacuated. This is more difficult to accomplish than in the current vehicles, where the temperature of the thermal engine is higher. With a vehicle of $10\,kW$ effective power, one must evacuate $5\,kW$ of heat at a temperature of approximately $75\,°C$, that is, at about $30\,°C$ lower than for an internal combustion engine. The size of a heat exchanger increases according to the quantity of heat to evacuate and as the difference in temperature between the hot source and the cold source decreases. The size of the exchanger is thus significantly larger in the case of a fuel cell vehicle and can reach twice that required for current vehicles. It is a serious problem which is more easily solved for large vehicles, buses, for example. This is one of the reasons that it would be interesting to have fuel cells which function at higher temperatures, around $120–130\,°C$. Although the operation of a fuel cell is quiet, the vehicle can sometimes be very noisy because the dissipation of heat requires cooling fans and because of the need for an air compressor.

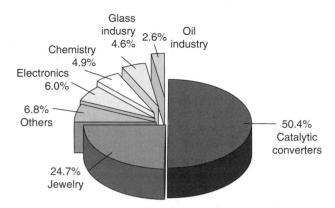

Figure 16.12. Share of platinum demand, 2006. *Source*: Data from CNUCED according to statistical data from Johnson Matthey, www.unctad.org/infocomm/anglais/platinum/uses.html.

Although enormous progress has been made in reducing the quantities of catalyst needed in a fuel cell, significant future reductions will be difficult. If, at this time, all of the vehicles existing in the world were equipped with a fuel cell, each requiring the same quantity of platinum catalyst as currently employed, the total platinum requirement would be 300 times the present annual world production of platinum. One can anticipate even larger needs in the future. At present, large-scale development of road transportation using the current technology of PEMFCs with platinum catalysts does not seem feasible. Research is necessary to develop other catalysts which are both readily available and not too expensive.

This demand for platinum for use in catalytic converters started to grow rapidly in the 1970s (Figure 16.12). In 2005 this application accounted for 45.6% of the demand for platinum. For comparison, in the same year the use of platinum for jewelry represented 30.1% of the demand. Although new techniques of coating very thin layers of catalyst on the polymer electrolyte have been designed to reduce the cost, the cost of platinum still amounts to about $150 per private vehicle. Any strong increase in the use of PEMFCs is likely to introduce significant tension in the platinum market.

Ships and Submarines
Fuel cells can also be used for maritime transportation. The systems used can be of the hybrid type, that is, associated with a battery. The type of fuel cells most commonly used are PEMFCs or MCFCs with power output ranging from

100W to several megawatts. The fuel used can be diesel fuel, natural gas, methanol, or pure hydrogen. Diesel fuel cell systems for commercial marine applications (which represent 98% of the market) are in the range 200kW– 1MW. Military applications demand more powerful systems, in the range 500kW–2.5MW.

For ships, fuel cells have the advantage of significantly reducing local pollutant emissions, and if the hydrogen fuel is produced using renewable or nuclear energies, greenhouse gas emissions are also reduced. Using natural gas in a SOFC, for example, or hydrogen produced from natural gas would reduce greenhouse gas emissions by about 20–40% relative to current technologies, depending upon the nature of the ship and its type of use. Large ships can benefit from technologies derived from stationary applications (e.g., SOFCs) while small ships can benefit from development made in the automotive domain (e.g., PEMFCs).

In 2003 an experimental fuel cell–powered catamaran, Hydroxy 3000, was developed in Switzerland. Using two electric motors supplied with electricity produced by a PEMFC of 3kW, this boat reaches a speed of 12–18km/h. Seven meters long and weighting 1500kg, it can transport six passengers.

The U.S. military is quite interested in the use of fuel cells for propelling ships or submarines. The technologies currently used for this purpose, aside from a small number of nuclear-powered submarines, are gas turbine generator systems and diesel generators. Gas turbine generators consume about twice as much fuel as diesel engines. Fuel cells would reduce the fuel consumption by about 30% compared to diesel engines and cause considerably less pollution. Another military benefit would be the reduction of the radar cross section and infrared signature as well as the noise level. The last is particularly interesting for submarines. In this respect they compete favorably with nuclear energy propulsion. Hydrogen and liquid oxygen can be stored onboard and the submarine is therefore air independent. For example, the U212 submarine manufactured by the German company HDW uses nine 34-kW Siemens PEMFCs. It can stay submerged for up to three weeks. The U212 combines these fuel cells and a conventional diesel generator with a lead battery.

SOFC and Transportation

SOFCs function at high temperatures. They need a long time to reach these temperatures and are hard to use when a vehicle is often started and stopped. In addition, the materials of the fuel cell do not respond well to abrupt changes of temperature. Nevertheless, one of their great advantages is the ability to reform hydrocarbons or alcohols without a catalyst and they could prove useful for some specific transport applications. One could imagine using an SOFC to power heavy trucks which, by using different drivers, could run essentially uninterrupted for long periods of time. When the truck is not in motion, the energy produced might be harnessed to power ancillary handling equipment or to provide an alternative supply of electricity complementary to that normally employed by the trucking company. Small SOFCs might be used to replace the batteries of vehicles, providing power for air conditioning and electronics.

Although hydrogen-powered fuel cells offer some very enticing advantages in the field of transportation, a considerable period of time will pass before we see their application on a large scale. Other technologies have been mastered for a long time and have a cost and reliability that the fuel cells have not yet reached. The use of fuel cells for surface transportation on a large scale will require two revolutions. First it will be necessary to replace the well-known technology of the internal combustion engine by that of fuel cells with all the attendant difficulties that can arise during market penetration of this technology. Second, it will be necessary to create an extensive and convenient hydrogen production and distribution network.

16.2.4. Direct Use of Hydrogen

For some applications hydrogen can already be used directly as a fuel. This has the advantage of using already existing technologies and avoiding the use of fuel cells which are not presently reliable and ready for a large-scale market.

Road Transport

Ford, BMW, and Mazda have developed experimental cars in which hydrogen is used directly as a fuel in an internal combustion engine. These cars provide test vehicles for the investigation of potential problems associated with hydrogen storage and transport. Compared to fuel cell technology, vehicles using hydrogen in an internal combustion engine emit a small quantity of nitrogen oxides. These result from the presence of nitrogen in the air and the fact that combustion takes place at high temperature. The efficiencies of such cars are not as high as with those using fuel cells, but the technology is simpler and less expensive to implement. Hybrid solutions using both gasoline and hydrogen or electric motors and batteries together with an internal combustion engine running on hydrogen show considerable promise. For example, the Ultim Car is an experimental vehicle built by H2-Développement by modifying a Peugeot Expert. It functions with a 2-L^3-capacity gasoline engine which runs on gasoline or pure hydrogen. Tests are being carried out to optimize the system in order to decrease the quantity of nitrogen oxides emitted while playing on the enrichment of the combustion mixture, by overfeeding and by decreasing the temperature of combustion. Hydrogen is stored in the back of the vehicle in four 20-liter standard bottles. The range of the latest version is more than

200 km. When it functions with hydrogen, this car emits water vapor. The nitrogen oxide emissions decrease by 50% and those of CO_2 by 99% compared to the use of the vehicle in the gasoline mode. The intention is that hydrogen be used for in-town driving and gasoline be used only outside the cities.

An internal combustion engine can also use a mixture of fuels such as gasoline–hydrogen or natural gas–hydrogen. For example, the goal of the Althytude project, in the French towns of Dunkerque and Toulouse, is to operate buses functioning with a mixture of natural gas and hydrogen in proportions ranging from 15 to 30%. This mixed-fuel solution is seen as a precursor to the transition from natural gas to pure hydrogen.

Planes

For airplanes, hydrogen can be used directly to power an internal combustion engine or in a turbine. A big advantage of hydrogen compared to kerosene is that it does not emit CO_2. Emitting CO_2 at high altitude is more harmful to the environment than emission at ground level. In addition, the use of hydrogen in place of kerosene avoids emission of CO, sulfur dioxide, and unburned hydrocarbons. Hydrogen-powered planes will produce a small quantity of oxides of nitrogen because there is nitrogen in the air and the engines function at high temperature. For airplanes, the best physical state of hydrogen to use is probably liquid hydrogen because flights do not last long enough for small leaks or boil-off to be important. The combustion of hydrogen produces water vapor 2.6 times more than in the case of a plane burning kerosene. Questions remain regarding the consequences of an increase in the water vapor on the greenhouse effect. Indeed, while water vapor remains a few days in the atmosphere at low altitude, it can stay between 6 and 12 months at high altitude.

The first plane using liquid hydrogen as fuel, a B57 Bomber, flew on February 13, 1956. This was the first of three successful flights. Since that time a number of other tests have occurred. The Tu-155 built by Tupolev, the Russian aircraft manufacturer, made its first flight in 1989. Recently, a California Company, AeroVironment, has flown an ultralight airplane, Helios, powered by liquid hydrogen fuel cells. Currently there are joint projects between Airbus and Tupolev to develop a hydrogen-fueled airplane for commercial use.

Air traffic and airport activities account for 13% of CO_2 emission associated with transportation and 2% of the total world emission. In 2005, the global airline kerosene bill was about 100 billion euros, which is twice as large as it was in 2003 but still less than today given the increasing price of oil. Among the alternative fuels which might be used to power airplanes in the future are synthetic kerosene produced from coal or biomass, cryogenic methane obtained from hydrocarbons or biomass, and hydrogen. There are however still many technological difficulties to be solved before hydrogen can be commonly used as airplane fuel.

Ships

Hydrogen can also be directly used to run ships. Several interesting concepts exist. For example, in 2006 Ivo Veldhuis, a young Dutch naval engineering researcher at the University of Southampton, the United Kingdom, and his collaborators proposed a liquid hydrogen–powered transport catamaran. Named H_2Oceanjet, It would embark with 1000 tons of hydrogen contained in 10 tanks. This hydrogen would be used to supply gas turbines associated with engines coupled to water propellers. This ship would have a range of almost 10,000 km and could carry 3000 tons of freight. A speed of 118 km/h, twice the average speed of currently operating container ships, could be reached.

16.2.5. Direct Combined Heat and Power

Electricity is often produced by large power stations located large distances away from the end users. Such centralization offers many advantages of scale. Given the pollution potential of the fuels currently employed, this advantage of scale also extends to pollution control in the sense that it is possible to invest in pollution-reducing technologies without severe increases in the cost of the electricity produced. This would be more difficult in small installations. At the same time, the distribution of the electricity produced requires long and expensive power lines to connect the consumers to the place of production. This is not a serious problem in highly populated areas. However, in cases of low population density, the cost per inhabitant can be very high. Centralization also has the disadvantage that many consumers are dependent upon a single production facility. In a number of countries, accidents, inclement weather, or peaking demands on the supply of electricity can, and often have, led to "black-outs" or "brown-outs" over very large areas.

The possibility that a less centralized system based on nonpolluting fuels could be developed to meet the electricity needs of modern society is certainly an intriguing prospect and well worth consideration. One advantage to producing electricity close to where it is used is that it would also become possible to use part of the heat energy produced during the process. Cogeneration, or combined heat and power generation, is then an additional way to decrease the cost of energy. By simultaneously generating mechanical or electrical energy and thermal energy from a primary fuel, extremely high efficiencies of fuel use can be achieved, that is, 80–90%. Such technologies inherently consume less fuel and pollute less. Hydrogen can be used in this way. Hydrogen fuel cells employed for cogeneration would allow independence from large electrical supply networks. They are clean and quiet and produce only water as waste.

For cogeneration in static facilities the size and weight of the fuel cell are not of great importance compared to its use in other applications such as transportation. The PAFC has been the type most tested for this type of application.

In the 1990s, cogeneration modules of 200 kW produced by the UTC Cell Fuel company were in service in Japan, the United States, and Europe (some are still working). These projects provided important operating experience in integration of fuel cells for the production of electricity and heat. Other types of fuel cells, MCFCs, SOFCs, and PEMFCs, have been studied and could have an increasing share in the stationary fuel cell market. The big difference between SOFCs (and MCFCs) and PEMFCs is that the first can directly use a variety of fuels like natural gas, liquid hydrocarbons, biomass, and gasified coal. The fuel reforming can take place in the cell. The PEMFC has a more restricted choice of fuels and reforming must be made outside the cell in a separate apparatus. In the first stages of application, natural gas will probably be the fuel of choice. It has the strong advantage of being available through an existing supply network. Methanol is currently being tested in some isolated sites. Market demand will make the selection among the different solutions. While the PAFC and PEMFC appear to have the lead in the short term, the SOFC (and MCFC) offers a greater potential for the future.

The principal disadvantage is currently the cost of these installations, estimated between 3500 at 10,000 € per installed kilowatt, compared to approximately 800 €/kW for traditional power stations. But this cost will probably decrease toward 1000 €/kW within the next decade.

16.2.6. Hydrogen and Portable Devices

There is demand to have increased self-sufficiency for portable devices. Currently this is done in two ways: development of low-consumption electronics and batteries with higher energy densities. In applications for which the delivered service and not the cost is the major priority, fuel cells can be of great interest. Among the different types of fuel cells, only two are likely to have portable applications: the PEMFC and the DMFC. The reason is that they work at low temperatures, between 60 and 80 °C. This allows them to function at ambient temperatures without releasing too much thermal energy, reducing thermal heat management problems.

The main challenge in providing fuel cells for such applications is the miniaturization of the cells. Either one miniaturizes current PEMFC or DMFC devices or one develops a completely different architecture. The lowest power of PEMFCs and DMFCs used in other applications is about 1 kW. Manufacturing smaller fuel cells requires the miniaturization of the different components (cell, pump, exchangers, and current converter) and management of the water and heat flows, which are traditionally the major difficulties in this type of system. The problem is that for the smaller dimensions surface effects may become important compared to volume effects. Consequently, at some scale, the operational characteristics of the microdevices may be different from those seen with larger cells.

It is also necessary to adapt to the different levels of power required. This in turn demands management of the fuel and the injected air to modulate the power. The choice of fuel is also a real issue. Pure hydrogen is not so convenient. Methanol or hydrocarbons seem to be more appropriate. Methanol has the problem of its toxicity. For hydrocarbons, it would be necessary to have a microreformer, a difficult device to build. Another possibility is the production of hydrogen from sodium borohydride ($NaBH_4$). Some laboratories, among them those of the CEA, in France are trying to adapt methods borrowed from microelectronics and materials science to the development of micro fuel cells.

16.2.7. Hydrogen Safety

Like any other fuel, hydrogen can be dangerous. Care must be taken to decrease the risks of accidents. This is true at all associated stages of production: storage, transport, and handling of hydrogen. Compared to other fuels, hydrogen burns faster (e.g., 10 times more quickly than oil for a 100-liter tank) but radiates 10 times less energy than gasoline. This latter point reduces the consequences of a fire because it is less likely to be ignited by the materials in the vicinity. In the case of an airplane accident hydrogen can be less dangerous than the kerosene currently used. If there is leaking, it will diffuse and escape quickly due to the low mass of the molecules. While it takes little energy to induce an explosion in a mixture of hydrogen with a little air, hydrogen mixed with large amounts of air detonates with difficulty.

The annual number of incidents of vehicle engines catching fire is significant. In the United States alone, 15,000 vehicles burn accidentally each year, causing approximately 500 deaths and 7000 burn injuries.

16.2.8. Conclusion

Hydrogen and fuel cells offer a very appealing way for meeting our energy needs. As oil and gas become scarce and/or very expensive, greater effort will be placed on development of hydrogen technologies. At that point the social acceptance of this new technology will also become an important consideration. Addressing the safety aspects is of utmost importance because any dramatic accident resulting from the use of hydrogen could stop or severely slow down the introduction of this energy vector. Still we are far from a large-scale development of hydrogen in the domain of transportation. There are still many difficult problems to solve, requiring extensive research and development efforts. Costs must drop considerably or hydrogen and fuel cell technologies will remain limited to niche markets.

We believe the first fuel cells to be introduced at a large scale will be those supplying portable devices. Consumers are probably ready to pay a premium for mini or micro fuel cells, which offer the advantages of self-sufficiency and the possibility of easy refueling. The SOFC technology, useful for large-scale stationary installations, is probably the one which will appear next. It can initially employ natural gas as a fuel and need not use a catalyst. It would allow a smooth transition to hydrogen fuel as it becomes available. The fuel cells used for transportation applications will probably be those which will be the last to appear. The automotive field needs devices which are inexpensive and reliable and can work in a large range of external temperatures and external atmospheric conditions. Hybrid vehicles using synthetic fuels from coal, gas, or biomass will be much more competitive in the near future.

Producing hydrogen from fossil fuels, currently the cheapest way to get hydrogen, does not remove the CO_2 generation problem but does move the place where CO_2 emissions are made. From decentralized pollution sources, that is, the vehicles which use oil derivatives as a fuel, one moves to centralized pollution sources, large manufacturing units producing hydrogen from fossil fuels.

A study made by the IFP (French Petroleum Institute) has shown that if the total production and usage chain is analyzed (from well to wheel), the quantity of CO_2 emitted per kilometer is minimized if one uses natural gas to produce hydrogen as a fuel for the vehicles. In this case CO_2 emissions are 108 g/km, that is, the same order of magnitude as that obtained with the Prius, Toyota's hybrid car (104 g/km). If one uses liquid hydrogen instead of compressed hydrogen, the emissions go up to 171 g/km. Lastly, if the hydrogen gas is produced by coal gasification, one reaches 230 g/km, equivalent to the emissions of a traditional vehicle of large engine size. It is thus very desirable to produce hydrogen using renewable energies or nuclear power or to capture and store the CO_2 emitted during the production of H_2 from fossil fuels. In this way CO_2 emissions associated with personal transportation could be decreased significantly.

As a summary of future fuel cell possibilities we note that the following can be expected:

- Fuel cells working at high temperature (SOFC and MCFC) which can produce heat at a maximal temperature from 800 to 1000 °C. The applications (for the SOFC) go from residential to industry and from the cogeneration to the centralized power production.
- The PEMFC, whose temperature will not exceed 80–100 °C, will be better used for domestic applications (hot water, heating).
- The PAFC, limited to 200 °C, will have other uses requiring higher temperatures.

In terms of power delivered we can expect:

- Small fuel cells of the PEMFC and SOFC type with a few kilowatts of power dedicated to residential applications.
- Medium-power systems (100, 200, 250 kW) with the PEMFC, SOFC, and PAFC.
- Installations of about 1 MW combining a high-temperature fuel cell (SOFC or MCFC) with a turbine to produce electricity.

Large (higher than 1-MW) cogeneration installations are generally based on the SOFC and MCFC.

CHAPTER 17

Conclusion

Energy is necessary at all the stages of daily life. It allows humankind to do things which would be difficult to do with only the energy of the human body. The three main domains where energy is used are the production of electricity, the production of thermal energy (heat or cold), and transportation. During the past two centuries fossil fuels (coal, oil, and gas) were extensively used to meet our energy needs. They currently sustain about 80% of the global primary energy consumption. These fossil fuels have been relatively cheap. For example, in France, producing 1 kWh of electricity costs on the average about 0.03 € and it is delivered at the consumer at a cost on the order of 0.1 €. The availability of cheap energy has created the wealth of modern civilization.

Oil, gas, and coal were formed by nature for free and we have just had to pay for retrieving these resources and using them. Fortunately for us the fossil fuels turned out to be very concentrated energy sources (about 10 kWh per liter of gasoline). This is not generally the case for renewable energies, which have low energy densities (37 liters of water falling from 10 m corresponds to an energy of only 1 Wh). A further problem is that some renewable energies are intermittent.

We are now faced with a major energy challenge consisting of two imperatives. The first is the need to progressively replace fossil fuels by other sources of energy. The reason is that fossil fuels exist in finite quantities in the earth. One day they will become scarce because we harness them at a much larger rate than they are produced by nature. Humankind has used fossil fuels to satisfy its need without paying much attention to the question of reserves. For that reason we may soon be faced with a shortage of oil. Oil can be synthesized from natural gas or coal and extracted from unconventional oil resources. Consequently oil will be available until the end of the century but very likely not at today's price. After oil, the same problem will arise with natural gas and, one or two centuries later, with coal. If we are more careful in exploiting these fossil resources, the problem will occur later, but it will occur since the resources

Our Energy Future: Resources, Alternatives, and the Environment
By Christian Ngô and Joseph B. Natowitz
Copyright © 2009 John Wiley & Sons, Inc.

are finite. There is currently no single alternative energy source able to quantitatively replace fossil fuels.

The second imperative of the energy challenge is to deal with climate changes which may be induced by an increase of the greenhouse effect due to human activities. A large part of this increase is due to CO_2 and use of fossil fuels is responsible for that. The timescale associated with possible negative climate change is shorter than that associated with the disappearance of fossil fuels.

Making a simple extrapolation of today's rate of energy consumption for the planet indicates that continuing "business as usual" will be unsustainable before the end of this century. This means that we have to start changing the way in which we consume energy and progressively replace fossil fuels by other energy sources. It will take decades to do that efficiently and we need to start now.

Meeting the energy challenge relies on a two-target strategy:

1. The first target is smart energy consumption, which means saving energy when it is possible and using energy in the most efficient way. A large amount of energy can be saved in developed countries where there is usually an overconsumption of energy beyond that required to meet the real needs of the population. In poor and developing countries this is more difficult. Many of the people already have difficulties obtaining energy because it is scarce and expensive. It will be quite difficult for them to save energy since they already do not have enough of it.
2. The second target is to use decarbonated energies, energy sources that do not produce CO_2 in operation. Renewable energies and nuclear energy are decarbonated energies. So far they provide less than 20% of the world's primary energy consumption.

Electricity can be produced with decarbonated sources. Nuclear energy is very efficient in this respect, allowing high powers and large quantities of electricity at inexpensive price. This is also the case for hydropower. Wind energy is rapidly being developed. Its main drawback is its intermittency. Solar energy can be converted into electricity using photovoltaic panels or concentrated solar power plants. Solar energy has a great future and inexpensive thin photovoltaic films will probably be available soon, making solar energy economically competitive. This will be helped by the increase in the price of oil and other fossil fuels. High-temperature geothermal energy can also be used to produce CO_2-free electricity. Biomass can be used to produce electricity, but this is not always a good idea since biomass can have more interesting uses.

In transportation a lot of progress has been made in manufacturing cars which require less fuel than previous models. Biofuels have been developed and are meeting some of our transportation needs. Second-generation biofuels should be even more useful because they will be produced from lignocellulosic biomass and there is no competition with food production. Furthermore no fertilizers are needed. Biofuels allow a decrease in oil demand to a certain extent but cannot completely replace oil.

Hybrid vehicles are now being made and offer a good opportunity to reduce gasoline consumption. In the near future, plug-in hybrids will be able to recharge their batteries on the grid and this will decrease gasoline consumption even more. All electric vehicles will also be useful in cities or for short trips. This means that electricity will play a greater and greater role in meeting the energy needs of the future.

Housing is a large energy-consuming sector because a lot of thermal energy is required for heating or cooling and producing hot water. In the future it will be possible to replace fossil fuel with technologies based on renewable energy sources: thermal solar energy, heat pumps using air, water or the soil as a source of energy, and so on. A good choice of architecture and the materials to build houses is of primary importance. Old housing will need refurbishing to meet new insulation standards and reduce energy consumption. Compared to buildings built before the 1970s, it is now possible with new housing to consume more than five times less energy for the same comfort level. If a very large number of buildings are improved, the global effect will be sizable.

The weak point in the energy supply chain remains energy storage. This is an important issue because appropriate storage techniques allow the smoothing of electricity production over time. This in turn allows decreases in installed power capacity. Electricity storage is particularly important for intermittent renewable energies, wind energy, for example.

Humankind has overcome several crises over the ages and has always adapted to new living conditions. However, in any crisis there are winners and losers. It is worth preparing for the future in order to assure that we are among the winners. In the energy domain, making significant changes usually requires decades. It will probably take between 30 and 50 years before we can replace a large part of our fossil fuels by other energy sources. We must meet this challenge. Meeting it also provides us with the opportunity to protect the environment.

Three main domains will have to be involved if we are to realize this evolution and revolution: research and industry, to provide efficient energy technologies; governmental entities, to adopt appropriate strategies for energy supply and consumption; and finally consumers, who will decide which energies to use and how to use it. We sincerely hope that this volume provides information which makes the task of the last two easier.

CHAPTER 1

1.1 A man weighting 80 kg walks up six floors on a staircase which has a total height $h = 20$ m. It takes him $t = 120$ s to do it. What is the energy necessary and the power developed during this effort? If the same person goes up these stairs in $t = 30$ s (and arrives at the top out of breath), what are the energy and power needed in this case?

1.2 Heating 10 liters of water from 0 to 100 °C requires an amount of energy of 1.16 kWh. Calculate the power needed to do this in 1 h in 5 mn. Assuming no energy loss while heating, what is the time necessary to do this with a 2-kW hotplate?

1.3 A 5-kg flowerpot falls from the third floor (≈ 10 m) of a building to the sidewalk below. Assuming no air resistance, calculate the speed of the flowerpot when it hits the sidewalk. Calculate its kinetic energy. If the flowerpot hits somebody, that person will die if the kinetic energy is more than about 700 J. What is the likely outcome of such an accident?

1.4 The mass of the earth, considered as a sphere, is $M = 5.97 \times 10^{24}$ kg and the radius is 6371 km. Calculate the rotational energy of the earth.

CHAPTER 2

2.1 One barrel of oil contains 159 liters, or about 127 kg, of oil. If the heat of combustion is 47 MJ/kg, calculate how much energy there is in a barrel of oil. In mid-1990 the price of petroleum was $12/bbl. It has reached $140/bbl at the beginning of summer 2008. Calculate the price per gigajoule at each time.

2.2 The higher heat of combustion of hydrogen is 143 MJ/kg. The heat of combustion of carbon is 32.8 MJ/kg. Calculate approximately the heat of combustion of a hydrocarbon C_nH_m in megajoule per kilogram.

Our Energy Future: Resources, Alternatives, and the Environment
By Christian Ngô and Joseph B. Natowitz

2.3 A steam turbine of a power generation plant works at 650 °C and ejects heat at a temperature of 30 °C. Calculate the maximum yield of this turbine.

CHAPTER 3

3.1 In the United States the annual electricity production from coal is about 2200 TWh/year. In 2007, the amount of CO_2 emitted by the United States with fossil fuel–fired plants was 2.3 Gt. Of that, 1.9 Gt came from coal-fired plants. Estimate the quantity of electricity produced. Comment on the result.

3.2 In 2007, the reserves of anthracite and bituminous coal are about 430 Gt. Those of subbituminous and lignite are equal to 416.5 Gt. Assuming a coal consumption of 3.2 Gt/year, calculate the useful life of coal reserves for high-quality coal and for all types of coal.

Assuming that coal is the only energy source, calculate the useful life in this case. The total primary energy consumption is equal to 11 Gtoe.

3.3 Each year a coal-fired plant consumes 1 Mt of coal and produces 3 TWh of electricity. The energy content of 1 ton of coal is about 2.9×10^{10} J. Calculate the yield of the coal plant. How much CO_2 is emitted per year?

CHAPTER 4

4.1 The power P emitted by a black body at absolute temperature T is given by the Stefan–Boltzmann law:

$$P = \left(\frac{T}{64.5}\right)^4 \ W/m^2$$

Calculate P for an average temperature of −18 °C on the earth (which would be the case if no greenhouse effect exists) and for a temperature of 15 °C (that with the natural greenhouse effect). Evaluate the power absorbed by the greenhouse effect.

4.2 The average radius of the earth is $R = 6371.22$ km. Let us model the troposphere as an 8-km-thick homogeneous layer at standard conditions of pressure and temperature (http://access.inrp.fr). The average concentrations of CO_2 in 1991 and 2001 were 355.67 and 370.9 ppm, respectively. Between the period 1991 and 2001, it is estimated that 64 Gt of carbon was emitted in human activities.

With this simplified model calculate the amount of carbon contained in the atmosphere in 1991 and 2001.

Calculate the increase in CO_2 concentration during this period using the estimations of the quantity of CO_2 emitted into the atmosphere and discuss the results.

CHAPTER 5

5.1   A water reservoir is located at $h = 300\,m$ above the level of a turbine. The efficiency of the system is supposed to be 85%. In order to get a power of 300 MW, what is the water flow through the turbine?

5.2   Calculate the power for sea waves having a height $h = 4\,m$ and a period $T = 12\,s$. How much power is produced for a costal length of 10 m? Compare that with the case of storm-generated waves of height $h = 15\,m$ and $T = 15\,s$.

5.3   Consider a tidal power station which has a basin of 2 km by 10 km. Assuming an average tidal change of 2 m, what are the theoretical and the available powers? How much energy can be extracted every year? What is the power per square meter?

5.4   Consider a hydro pumping system consisting of an upper reservoir with an area of 1 ha and a depth of 10 m located at 100 m above the base reservoir. What is the potential energy available with this system?

5.5   Electricity consumption in the United States in 2005 was 3720 TWh. What would be the volume of water necessary to produce this amount of electricity if a reservoir has a head of 300 m above the turbine level and the efficiency is 94%. If the reservoir has a depth of 100 m, what would the area of this reservoir have to be?

CHAPTER 6

6.1   The oversimplified photosynthesis reaction

$$\text{Light} + 6H_2O + 6CO_2 \rightarrow C_6H_{12}O_6 + 6O_2$$

requires $4.66 \times 10^{-18}\,J$ per elementary reaction. Calculate the energy per mole in kilowatt-hours and the energy per ton of carbon synthesized.

At least 36 photons are needed to drive the photosynthesis reaction above (in several elementary reactions). Consider a red photon of energy $2.9 \times 10^{-19}\,J$. Calculate the energy of one Einstein of red photons (1 mol). How much energy is contained in the photons?

6.2 We have the following data: 1 toe = 42 GJ = 11.6 MWh. The combustion of 1 toe gives 3.7 tons of CO_2.

One ton of dry wood equals 5.22 MWh. Since there is space between the pieces of wood, the volume corresponding to 1 ton is about $2 \, m^3$.

Assume an oil-fired boiler has an efficiency of 80% and a wood-fired boiler has an efficiency of 75%. Consider a house of $150 \, m^2$ needing $100 \, kWh/m^2$/year for heating. Calculate the quantity of oil needed and the CO_2 emissions per year. Calculate the same quantities for wood.

6.3 Calculate the area necessary to produce the fuels currently necessary for the French fleet of road vehicles (≈ 50 Mt/year) using sugar beets (3.3–3.5 tep/ha), corn (1.5–1.7 tep/ha), or rapeseeds (≈ 3 toe/ha). The total cultivated areas available in France are equal to 18.4 Mha.

6.4 One ton of municipal refuse has a calorific value between 6 and 12 MJ/kg. Collecting and disposing of it require about 5000 MJ/ton. Calculate the number of toe corresponding to 1 ton of refuse. In France, a city of 100,000 inhabitants generates about 35,000 tons of refuse per year. Calculate the amount of energy potentially available and the quantity of oil equivalent.

CHAPTER 7

7.1 Assuming the solar constant to be $1370 \, W/m^2$ and the radius of the earth (supposed to be a sphere) $R = 6370$ km, calculate the power arriving at the top of the atmosphere and the energy received per day and per year.

Electricity production was 1.8×10^4 TWh in 2005. Compare that number to the solar energy arriving each day on top of the atmosphere.

7.2 A 1-m^2 photovoltaic module costs 600 € and produces 100 kWh/year. Assuming a price of electricity from the grid of 0.1 €/kWh (and no additional incentives), calculate the payback time for the module. How much should the price of the module be so that it can pay for itself in a decade?

CHAPTER 8

8.1 A geothermal aquifer supplies hot water at a temperature of 70 °C at a flow rate of 10 L/s. The hot water is used in a district heating system at a temperature of 40 °C. It is assumed that all the thermal energy can be transferred through a heat exchanger. The geothermal aquifer is used for 150 days each year. Calculate the amount of heat used for heating during a year. Calculate the amount of oil necessary to get the same amount of heat assuming an efficiency of 80% for the fuel-fired boiler.

CHAPTER 9

9.1   Consider a 3-MW wind turbine operating at a wind speed of 15 m/s. Its global efficiency (turbine, gearbox, and generator) is supposed to be 30%. What should be the swept area and the size of the rotor?

9.2   Calculate the total energy produced per year by one hundred 3-MW wind turbines of the type in the previous exercise assuming that the wind is available at the nominal velocity during 15% of the time (as is often the case in Germany); during 25% of the time. Assuming that there is a gas-fired plant backup in the case where there is no wind, estimate the amount of gas needed and the CO_2 emission assuming 50% efficiency.

9.3   If it is assumed that the wind is available 25% of the time at a wind speed of 15 m/s, how many wind turbines of 2 MW are necessary to provide the same total energy as a nuclear plant of 1000 MWe producing 8 TWh each year? What is the power of the nuclear plant?

CHAPTER 10

10.1   Calculate, in terajoules per kilogram, the amount of energy released in the fission of 1 kg of ^{235}U. Calculate the amount of electricity produced.

10.2   Each second, 4 million tons of matter disappear in the sun transformed into energy and emitted into space. Calculate the amount of mass converted each year. Do the same for the time period since the sun was born (4.6×10^9 years). How much of the total mass of the sun (2×10^{30} kg) does that represent? Compare this mass with the mass of the earth (6×10^{24} kg).

10.3   Calculate, in MWd/g (megawatt days per gram), the energy released by 1 g of ^{235}U assuming that all the nuclei undergo fission (burn up). Calculate the energy released in the burnup of 1 ton of ^{235}U and of 1 ton of natural uranium assuming that only ^{235}U is burned.

10.4   The Curie is an old unit of radioactivity: $1 Ci = 3.7 \times 10^{10} Bq = 3.7 \times 10^{10}$ disintegrations per second. It was historically supposed to be the activity of 1 g of radium. It is a huge quantity of radioactivity. Calculate the mass per curie of a material made of nuclei of atomic mass A and half-life $t_{1/2}$.

CHAPTER 11

11.1   By the end of 2006 the installed capacity of electricity production in France was 116.2 GW. The total production was 574.5 TWh (478 TWh for domestic consumption). Calculate the average quantity of electricity produced each year for each gigawatt installed. Calculate the quantity of electricity which could be potentially produced if the plants operate 90% of the time. Calculate the overall efficiency of the installed capacity versus the quantity of electricity produced. Calculate the installed power in the case where this efficiency increases to 80% due to the availability of large electricity storage capacities. Calculate the number of nuclear reactors not required if the installed power is correspondingly decreased.

11.2 By the end of 2006 the installed production electricity capacity in France was equal to 574.5 GW. How much power does that correspond to per unit of housing (there are 30.8 million housing units in France). Suppose that each housing unit produces its own electricity and the peak-demand hours are the same for everyone. If on the average 9 kW of power is needed per housing unit, what will be the installed capacity required if each unit produces its own electricity?

11.3 An office is illuminated with 10 incandescent lamps of 100 W for 10 h a day. If these lamps are changed to compact fluorescent lamps of 23 W (equivalent to 130-W incandescent lamps), calculate the energy saving over a year (assuming that the office is open 230 days) and the cost savings in electricity assuming a cost of 0.1 €/kWh.

CHAPTER 12

12.1 A water reservoir is located 500 m above a turbine. Assuming an efficiency (penstock, turbine, and generator) of 80%, calculate the flow rate necessary to generate 500 MW of electrical power. Assuming that the reservoir depth is 20 m, calculate the area necessary to produce 500 MW of electricity during a period of 100 days if there is no rain. If the width of the reservoir is 1 km, what is its length?

12.2 Assuming a pump motor for pumped energy storage with an efficiency of 90% and a turbine generator of efficiency of 95%, calculate the overall efficiency of the pumped hydro system.

12.3 A flywheel which is a homogenous solid cylinder 30 cm in radius and having a mass of 100 kg makes 20,000 rpm. Calculate the amount of kinetic energy stored.

12.4 (a) A LR6 alkaline battery (1.5 V, 2600 mAh) costs 1.5 €. Calculate the energy content in watt-hours and the price of 1 kWh of battery.

   (b) Consider a Li ion rechargeable battery for a digital camera. Its characteristics are 3.6 V and 630 mAh and it costs 41 €. Calculate the energy content in watt-hours and the price per kilowatt-hour of battery electricity if it can be recharged 500 times.

   (c) A silver oxide battery for a small electric appliance has the following characteristics: 1.55 V, 15 mAh, and costs 4 €. Calculate the energy content in watt-hours and the price of 1 kWh of battery electricity.

CHAPTER 13

13.1 Consider an electric car which requires 150 Wh/km. Calculate the energy consumption to cover 100 km and the cost (assume 0.1 €/kWh). Consider a conventional car which needs on the average 7 L/100 km. Assuming that the price of gasoline is 1.5 € per liter, calculate the cost to drive 100 km.

13.2 Consider two cars of the same model driven by two different drivers. The first driver drives nervously and consumes on the average 9 liters of gasoline per 100 km. The second person drives smoothly and the gasoline consumption is on the average 7 liters per 100 km. Both cars are sold at 150,000 km. How much CO_2 has been emitted by the cars during this period of time. How much is the difference in emissions for the two drivers?

13.3 A small two-wheel vehicle requires 4 liters of gasoline per 100 km. It covers an annual distance of 8000 km/year. Calculate the quantity of CO_2 which is emitted.

CHAPTER 14

14.1 Consider a family needing 4000 kWh/year of electricity in their home to be used for needs other than heating, air conditioning, and producing hot water. What area of monocrystalline photovoltaic modules is needed to meet half of that demand?

14.2 Consider an old house of 200 m^2 area needing 330 kWh/m^2/year for heating. If this demand occurs in winter, what is the minimal amount of hot water (assuming 100% yield) stored in summer with a temperature difference of 40 °C which would be needed to meet this demand?

14.3 The heat power P conducted by a material is $P = (A \Delta T)/R$, where A is the area, ΔT the temperature difference between the two sides of the material, and R the R value. Suppose that the external walls of a house have an area of 100 m^2. Suppose that the temperature inside the house is 20 °C and the temperature outside is 0 °C. Calculate the heat power loss assuming an insulation with a material characterized by $R = 2$ m^2/K/W.

CHAPTER 15

15.1   In Germany the standby power consumption per home is about 44 W. For a German home calculate the quantity of electricity used in one year to power these standby modes and the quantity of CO_2 emitted. in Germany 600 g of CO_2 is emitted per kilowatt-hour of electricity.

Consider the same question for France where the standby consumption in a home is on average 27 W. The quantity of CO_2 emitted per kilowatt-hour is 90 g. Do the same for Denmark, with 39 W for the standby mode and 840 g of CO_2 per kilowatt-hour.

15.2   We want to use workers to produce electricity, for example, using a bicycle coupled to a generator. A single worker is assumed to develop a power of 150 W for 8 h. Three teams of people will ensure a 24-h production of electricity. How many people would be needed to produce the same quantity of electricity as a typical nuclear reactor (8 TWh/year). Calculate the cost of 1 kWh assuming that the workers are paid 12 €/h.

15.3   A French refrigerator consumes on the average 380 kWh of electricity per year. The cost of electricity is about 0.1 €. Calculate the cost per year. Now assume we want to produce the required electricity using hydropower to take water from the house level and bring it to a reservoir located at an altitude of 100 m. Usually this job is done using pumps (pumped energy storage), but we shall do it with workers paid 12 €/h. It is assumed that each worker can take 40 kg of water per trip and that it takes half an hour for each round trip. Calculate the number of workers necessary and the cost of this operation. The electricity production system (penstock, turbine, and generator) has an efficiency of 80%.

CHAPTER 16

16.1   Calculate the amount of energy released by the reaction

$$H_2 + \tfrac{1}{2}O_2 \rightarrow H_2O$$

at 1 atm pressure using the following bond enthalpies in normal conditions (298 K, 1 atm): H–H = 432 kJ/mol, O–O = 494 kJ/mol, and H–O = 460 kJ/mol.

16.2   One normal cubic meter of hydrogen weighs about 90 g (show this). To produce 1 Nm^3 of hydrogen by high-temperature electrolysis, one needs about 3 kWh/Nm^3. For conventional electrolysis about 4 kWh/Nm^3 is needed. A car using a fuel cell needs about 1 kg of hydrogen to drive 100 km. Calculate the energy necessary to produce 1 kg of hydrogen.

16.3   Calculate the amount of H_2 contained in $1\,m^3$ of H_2 compressed at 700 bars and in $1\,m^3$ of liquid H_2.

16.4   In France the total distance driven by cars in 2006 was $555 \times 10^{-9}\,km$. Assume that all French cars are to be powered by fuel cells. Further assume that conventional electrolysis will be used to produce the hydrogen needed (energy density $4\,kWh/Nm^3$) for this purpose. If $1\,kg$ of H_2 is needed per $100\,km$ driven, calculate the quantity of electricity needed each year to meet the demand.

▉▉▉▉ SOLUTIONS

CHAPTER 1

1.1 The potential energy is $mgh = 80 \times 9.8 \times 20 = 15{,}680\,J = 4.36\,Wh$. The power is $15{,}680/120 = 130.7\,W$. If the person climbs the stairs four times faster, the energy is the same but the power becomes $15{,}680/30 = 522.7\,W$ [close to 1 HP (736 W)].

1.2 In 1 h: $P = 1.16/1 = 1.16\,kW$. In 5 mn: $P = 1.16/5 \times 60 = 13.9\,kW$. With a 2-kW hotplate and no loss it will take about 35 mn. In fact a longer time is needed because of heat losses.

1.3 The speed when the flowerpot touches the sidewalk is $v = \sqrt{2 \times 9.81 \times 10} = 14\,m/s$ (50.4 km/h). The kinetic energy at the impact point is 490 J. The pedestrian has some chance to survive but could be seriously injured.

1.4 The moment of inertia of a sphere is $J = (2/5)MR^2$. The angular velocity ω is calculated from the rotation of the earth (one day to make one turn). The rotational energy is then $E = \frac{1}{2}J\omega^2$. One finds $E = 2.56 \times 10^{29}\,J$ or $7 \times 10^{22}\,kWh = 7 \times 10^{13}\,TWh$.

CHAPTER 2

2.1 About 6 GJ or 1658 kWh. It is roughly 10 kWh/L. The price was $2/GJ in mid-1990 and $23.3/GJ in mid-2008, a factor of 10 larger.

2.2

$$E = \frac{12n \times 32.8 + 143m}{12n + m}$$

Calculated values for some hydrocarbons are compared with experimental values as follows:

Our Energy Future: Resources, Alternatives, and the Environment
By Christian Ngô and Joseph B. Natowitz
Copyright © 2009 John Wiley & Sons, Inc.

Hydrocarbon	Formula (MJ/kg)	Experiment (MJ/kg)
CH_4	60.4	55.6
C_2H_6	54.8	52
C_3H_8	52.8	50.4

2.3 $T_{hot} = 923\,K$, $T_{cold} = 302\,K$. The maximum yield $\eta = (T_{hot} - T_{cold})/T_{hot} = 0.67$, or 67%.

CHAPTER 3

3.1 On the average approximately 1 kg of CO_2 is produced for each kilowatt-hour of electricity generated. This means that 1.9 Gt of CO_2 corresponds to about 1900 TWh. Since the production is equal to 2200 TWh, this means that U.S. coal fired-plants emit on the average less than 1 kg of CO_2 for each kilowatt-hour produced: They are cleaner than the average.

3.2 For anthracite and bituminous coal, the useful life is 430/3.2 = 134 years. For all types of coal, it becomes 265 years.
   The global energy consumption is equal to 11 Gtoe. Taking the IEA conversion factor 1 toe = 1.43 tce, we get 15.7 Gtce/year. If we use only coal as an energy source, the useful life of the reserves is greatly shortened to 54 years.

3.3 $2.9 \times 10^{10}\,J \approx 8\,MWh$. If the yield is 100%, the plant would produce 8 TWh with 1 ton of coal. The yield is in fact $\frac{3}{8} = 0.375$, or 37.5%. About 3 Mt of CO_2 is emitted per year (1 kg of CO_2 per kilowatt-hour of electricity)

CHAPTER 4

4.1 Applying the Stefan–Boltzmann law to the two temperatures ($-18\,°C = 255.15\,K$ and $15\,°C = 288.15\,K$), we get respectively P values of 245 W/m^2 at $-18\,°C$ and 398 W/m^2 at 15 °C. The difference, 153 W/m^2, is the power absorbed by the atmosphere due to the greenhouse effect.

4.2 The volume of the troposphere is given by $(\frac{4}{3})\pi\left[(R+8)^3 - R^3\right]$. The volume of CO_2 in the troposphere is just that number multiplied by the concentration (number of ppm/10^6). The number of moles is this volume (in liters) divided by 22.4 liters, the volume of 1 mol under standard conditions. Finally the mass of carbon is obtained by multiplying by the mass of 1 mol of carbon (12 g). This gives the following figures:

In 1991: Volume of the troposphere: $4 \times 10^9 \, km^3 = 4 \times 10^{21}$ liters. Volume of CO_2: 1.42×10^{18} liters. Number of moles: 6.35×10^{16}. Mass of carbon: $7.6 \times 10^{17} \, g = 766 \, Gt$ of CO_2.
In 2001: Volume of CO_2: 1.48×10^{18} liters. Number of moles: 6.62×10^{16}. Mass of carbon: $7.95 \times 10^{17} \, g = 795 \, Gt$ of CO_2.

The difference ($795 - 766 = 29 \, Gt$) is smaller than the actual value ($64 \, Gt$) by a little more than a factor of 2, showing the crudeness of the model, although it provides some understanding of the underlying physics.

CHAPTER 5

5.1 Taking into account the efficiency, the power should be equal to $300/0$ $.85 = 353 \, MW = 3.53 \times 10^8 \, W$.

Since the power is equal to mgh, the mass of water m necessary per second is $m = 3.53 \times 10^8/(9.81 \times 300) = 120,000 \, kg/s$. The flow volume is then $120 \, m^3/s$.

5.2 $P \approx 0.5 h^2 T$. In the first case we have $P = 96 \, kW/m$. For 10 m of coastal length the power is close to 1 MW. In the second case $P = 1.7 \, MW/m$. This can damage the installation.

5.3 The theoretical power $P_{th} = 0.22 \times 2 \times 10 \times 2^2 = 17.6 \, MW$ and the available power $P = 0.056 \times 2 \times 10 \times 2^2 = 14.5 \, MW$. The energy produced per year is $E = 365.25 \times 24 \times 14.5 = 0.127 \, TWh$

5.4 The volume of the reservoir is $10^5 \, m^3$. The potential energy is Mgh, or $10^5 \times 10^3 \times 9.81 \times 100 = 9.81 \times 10^{10} \, J = 27,250 \, kWh = 27.25 \, MWh$.

5.5 $1.22 \, m^3$ of water falling from a height of 300 m produces 1 kWh if the efficiency is 100%. This quantity is equal to $1.3 \, m^3$ if the efficiency is 94%.

$3720 \, TWh = 3.72 \times 10^{12} \, kWh$. To produce this quantity of electricity, we would need a volume of $2.86 \times 10^{12} \, m^3$. If the reservoir has a depth of 100 m, its area should be equal to $2.86 \times 10^{10} \, m^2 = 2.86 \times 10^4 \, km^2$. This is a square of 169 km on a side. The large size of the reservoir comes from the fact that renewable energies have small energy densities.

CHAPTER 6

6.1 The energy per mole is 2.8 MJ or 0.78 kWh. This corresponds to 0.13 kWh per atom of carbon. This gives 0.011/g of carbon and 10.8 MWh/t.

The energy of 1 Einstein of photons is 175 kJ. For 36 Einsteins of photons we have 6.3 MJ. That would give a good efficiency. Actually the real situation is more complicated. The photons which are usually absorbed have a higher energy. Furthermore there are other losses, such as reflection or respiration, which lower the efficiency to something like 5.5–7%. Taking into account actual farming conditions over a year leads to an efficiency of about 1% or less.

6.2   To heat the house 15,000 kWh/year is needed. This corresponds to almost 1.3 tons of fuel ($\approx$1.8–1.9 m^3) or 2.9 tons of dry wood (5.8 m^3). The quantity of CO_2 emitted is 4.8 tons. In the case of wood, most of the CO_2 has been previously absorbed from the atmosphere when the trees were growing. For oil there is not this balance.

6.3   The areas necessary are 15.2–14.3 Mha for sugar beets, 33.3–29.4 Mha for corn, and 16.7 Mha for rapeseed. This means that devoting all the available cultivable areas in France to biofuel production would not be enough. We have the choice of eating or driving, but even if we choose driving, we would be limited to much less driving than is currently possible.

6.4   1 toe = 42 GJ. Therefore 1 ton of refuse is equal to 0.14–0.3 toe. 35,000 tons of municipal waste corresponds to values between 210,000 GJ = 58 GWh (5 ktoe) and 420,000 GJ = 117 GWh (10 ktoe). Per inhabitant, the potential energy contained in refuse is on the order of 50–100 koe.

CHAPTER 7

7.1   The earth is seen from the sun as a disk of radius R. The power received is $1370 \times \pi \times R^2 = 1.75 \times 10^{17}$ W. The earth receives 4.2×10^{15} kWh/day or 1.5×10^{18} kWh/year.

   Each day the earth receives 4.2×10^{15} kWh = 4.2×10^6 TWh. This is 233 times the annual electricity consumption.

7.2   Six hundred euros is the cost of 6000 kWh of electricity coming from the grid. Sixty years of module operation is needed to amortize its cost. The lifetime of a module is about 30–60 years. To pay for itself in a decade while generating 1000 kWh of electricity, the module would need to cost 100 €.

CHAPTER 8

8.1   The annual flow rate is 130×10^6 L/year and the heat transferred is that number multiplied by 30 (which is ΔT) $\times$ 1000 (1 cal increases 1 cm^3 of water of 1 °C). This gives 3.9×10^{12} cal = 1.6×10^{13} J = 4.5 10^6 kWh.

   One ton of oil corresponds to 42 GJ, 11,600 kWh, or 10^{10} cal. The amount of oil giving the same amount of heat is equal to 390/0.8 $\approx$ 490 tons of oil.

CHAPTER 9

9.1 The power of the wind needs to be $3 \times 10^6/0.3 = 10^7$ W. The power of the wind is given by $P = \frac{1}{2} \rho A V^3$, where ρ is the mass density of air (1.29 kg/m^3), A the area, and V the nominal speed of the wind (15 m/s). This gives area $A = 4594$ m^2. The rotor diameter is 76.5 m.

9.2 There are 8760 h in a year. The produced energy is 3.9 GWh (15%) or 6.6 GWh (25%). For 100 wind turbines this means 390 and 660 GWh, respectively. In order to get electricity all of the time (2.63 TWh), 2.23 TWh is missing in the first case and 1.97 TWh in the second one. With 50% gas-fired plant efficiency we need 5.3 and 3.9 TWh, respectively. The gross heat from combustion of natural gas is on the order of 10.8 kWh/Nm^3. We would need respectively 2.63×10^9 and 1.97×10^9 m^3 of gas under standard conditions. The CO_2 emissions are (calculated with 0.5 kg/kWh) 1.3×10^9 and 1×10^9 m^3, respectively.

9.3 Because the efficiency for electricity production of a nuclear plant is 33%, a power of 3000 GW wil be needed. A 2-MW wind turbine produces 4.38 GWh per year. To produce the same amount of energy, one would need 1826 wind turbines. However, producing the same amount of energy does not mean that this can meet the demand of the consumer at all times. Indeed, when there is no wind there is no electricity. This means that the wind turbines have to be coupled to an electricity storage system. Pumped hydrostorage can do the job if the wind turbines are located in a region where this facility exists. Since the efficiency of pumped energy storage is not 100%, additional electricity is needed to meet the demand. This means that on the order of 2000 wind turbines of 2 MW are needed to do the same job as a nuclear reactor.

CHAPTER 10

10.1 The fission reaction liberates about 200 MeV. This is equal to 3.2×10^{-11} J. The number of nuclei in 1 kg of ^{235}U is $10^3 \times 6 \times 10^{23}/235$. The energy released in the fission of 1 kg of uranium is equal to ≈ 82 TJ = 22.8 GWh.

10.2 There are 31.5×10^6 s in a year. The amount of mass lost each year equals 1.26×10^{14} tons. Since the sun was born this loss equals 5.8×10^{23} tons. This corresponds to 0.03% of the mass of the sun but around 100 times the mass of the earth.

10.3 The number N of nuclei in 1 g is $N = 6.023 \times 10^{23}/235$. The energy released is about $(N \times 200 \text{ MeV} \cdot 1.60 \times 10^{-13} \text{ J/MeV})/(10^6 \text{ J/MW} \cdot \text{s} \times 86400 \text{ s/q}) \approx 0.95$ MWd/g. The concentration of ^{235}U in natural uranium is 0.7%. This means $\approx 10^6$ MWd/t for pure ^{235}U and 7000 MWd/t for natural uranium.

10.4 If N_0 is the initial number of nuclei at time $t = 0$ and N that number at time t, we have $N = N_0 \exp(-\lambda t)$, where λ is the radioactive constant. For $t = t_{1/2}$, we have $N = N_0/2$. Therefore $t_{1/2} = \log 2/\lambda$. The rate of decay is $\lambda N = \log 2 \times N/t_{1/2}$ becquerels. If m is the mass in grams of the radioactive nucleus, the number of nuclei $N = m \times 6 \times 10^{23}/A$, and we have

$$m = \frac{A t_{1/2}}{1.13 \times 10^{13}} \, g$$

The smaller the period, the smaller the mass necessary to get 1 Ci.

CHAPTER 11

11.1 The average production is $574.5/116.2 = 4.94$ TWh/year. One can potentially produce $8760 \times 0.9 = 7900$ GWh/GW $= 7.9$ TWh/GW. The overall efficiency is $4.94/7.9 = 0.62 = 62\%$.

With an efficiency of 80%, the installed power could be equal to only 90 GW. The difference is equal to 26 GW. This amount of electricity is that produced by about 26 nuclear reactors which are typically 1 GW$_e$.

If the consumption (and production) curve is smoothed, the installed capacity can be reduced.

11.2 The mean installed power per housing unit is $116.2/30.8 = 3.8$ kW.

If every housing unit has 9 kW to satisfy its needs at the peak hours, the total installed capacity is 277 GW, which is much less than for centralized power production.

11.3 The power consumption of incandescent lamps is $100 \times 10 \times 230/1000 = 230$ kWh/year. That of compact lamps is 52.9 kWh. The amount of money saved is 17.7 €/year. Actually a compact lamp costs more than an incandescent lamp, but it lasts much longer.

CHAPTER 12

12.1 Because of the efficiency, the power should be $P = 500/0.8 = 625$ MW $= 626 \times 10^6$ J/s. The potential energy of the water in the reservoir necessary to get this power is $mgh = P$. Therefore, every second a mass $m = 625 \times 10^6/9.81/500 = 1.28 \times 10^5$ kg/s must flow. This corresponds to a flow rate of 128 m^3/s.

For 100 days (=8.64×10^6 s), we need to have 8.64×10^6 s $\times$ 128 m^3/s. This corresponds to 1.1×10^9 m^3. The minimal area of the reservoir should be $1.1 \times 10^9/20 = 55 \times 10^6$ m^2 = 55 km^2. The reservoir should be 55 km long.

12.2 The overall efficiency $\eta = 0.9 \times 0.95 = 0.855$, or 85.5%. For an input of 10 kWh one gets 8.6 kWh back and 1.4 kWh is lost.

12.3 The moment of inertia of a homogeneous solid cylinder of mass M and radius R is $J = \frac{1}{2}MR^2$. The amount of energy stored is $E = \frac{1}{2}J\omega^2$, where ω is the angular velocity. We have $J = \frac{1}{2} \times 100 \times (0.3)^2$ and $\omega = 20{,}000/60 \times 2\pi$. The stored energy $E = 9.8$ MJ, or 2.7 kWh.

12.4 (a) The capacity of the battery is $1.5 \times 2.6 = 3.9$ Wh. The cost of 1 kWh is 385 €.

   (b) The capacity of the Li ion battery is $3.6 \times 630 = 2.27$ Wh and it costs 41 €. The cost per kilowatt-hour if there were only one charge would be 18,222 €. However, if it can be recharged 500 times, the cost drops down to 36 €/kWh.

   (c) The capacity of the battery which is nonrechargeable is 0.023 Wh. The cost of 1 kWh is then 174,000 €. In fact, what is important is not the cost of 1 kWh but the service it provides. For example, the battery can be used to power a watch for a few years for 4 €.

CHAPTER 13

13.1 Electric car consumption is 15 kWh/100 km and it costs 1.5 €. The cost of gasoline to cover the same distance is 10.5 €. Of course there are more taxes on gasoline compared to electricity. Nevertheless, if we just take the price of oil without tax ($0.75 for an oil barrel at $120), we get $5.25/100 km, or approximately 3.8 €. At any rate electrical vehicles are more interesting economically. They are currently more expensive, however.

13.2 The amount of gasoline to drive 150,000 km is 13,500 liters in the first case and 10,500 liters in the second case. The difference is 3000 liters (4500 € if the price of gasoline is 1.5 €/L). The amount of CO_2 emitted is 31.725 tons in the first case and 24.675 tons in the second. The difference is 7 tons of CO_2.

13.3 One liter of gasoline emits 2.35 kg of CO_2. The total amount of CO_2 emitted during the year is 752 kg. The main problem with two-wheel vehicles in Europe is that they also emit other pollutants and in greater quantities than cars because regulation is not as strong as for cars (some of them emit 8–10 times more pollutants). Typical motorcycles consume even more gasoline (6–7 L/100 km).

CHAPTER 14

14.1 One needs to produce 2000 kWh/year. One square meter of photovoltaic cells produces about 100 kWh/year. Therefore 20 m^2 is needed. In Europe this corresponds to a capital cost on the order of 18,000 €.

14.2   A 40 °C temperature difference for water corresponds to 48 kWh/m^3. With a 100% yield one would need about 6.9 m^3/m^2. A house of 200 m^2 would require at least 1380 m^3/year and in fact much more than that because of heat losses. This minimal volume corresponds to the volume of a swimming pool of 40 m × 10 m and 3.5 m depth. This volume can be reduced because some heat production can be done in winter.

14.3 $P = (100 \times 20)/2 = 1000\,\text{W}$.

CHAPTER 15

15.1   During a year, 385 kWh is consumed in standby mode. This corresponds roughly to the average annual energy consumption of a refrigerator. The quantity of CO_2 emitted is equal to 231 kg.

For France one gets 236.5 kWh and 21 kg of CO_2 emitted. For Denmark one gets 342 kWh and 287 kg of CO_2.

Actually it turns out that the large number of wind turbines installed in Germany or Denmark has only a very small influence on CO_2 emissions in the electricity sector. This is because of the great importance of coal-fired plants, which are large CO_2 emitters for electricity production in those countries.

15.2   Each worker produces $150 \times 8 = 1.2\,\text{kWh}$ in 8 h and 438 kWh/year if he or she works all days of the year. To produce 8 TWh during a year one needs 2.28 million workers. If we want to have electricity continuously, this means three teams each day of 760,000 workers. More people are in fact needed if they have to rest or if they get vacations. It should be noted that developing 150 W continuously during 8 h is really a hard job.

The cost would be 80 €/kWh. The cost of nuclear power is 0.03 €/kWh.

15.3   The cost of electricity for the refrigerator is 38 €/year.

One kilowatt-hour corresponds to 3.7 tons of water falling from 100 m. If the efficiency is 80%, one needs 4.6 tons of water to produce 1 kWh. A total of 1748 tons of water is necessary to power the refrigerator for one year. To fill the reservoir 43,700 round trips will be necessary and 21,850 h of work. About 72 round trips per day are necessary. This corresponds to three persons working 8 h a day. The total cost is 262,200 €.

CHAPTER 16

16.1   The energy is estimated from the number of bonds broken and formed. It costs energy to break bonds. Energy is released when bonds are formed:

$$2(460) - 494/2 - 432 = 241 \text{ kJ/mol}$$

16.2 In 1000 liters the number of moles is the volume divided by the volume of 1 mol. One kilogram of H_2 corresponds to 11.1 Nm^3. The energy needed to produce that quantity is equal to 33.3 kWh. This is the energy contained in 3.3 liters of gasoline, for example. In the case of conventional electrolysis the amount of energy is 44.4 kWh (about 4.4 liters of gasoline).

16.3 In 1 Nm^3 of H_2 we have 90 g of H_2. In 1 m^3 of H_2 compressed at 700 bars we have 63 kg of H_2. In 1 m^3 of liquid H_2 we have 70.8 kg of H_2. Compressed hydrogen at 700–800 bars gives an energy density close to that of liquid hydrogen.

16.4 One normal cubic meter contains 90 g of H_2. The amount of electricity to produce 1 ton is $4.4 \times 10^5 \text{ kWh/t}$. To produce 1 Mt of H_2, one needs $4.4 \times 10^{10} \text{ kWh}$ or 44 TWh. Driving a total distance of $5.55 \times 10^{11} \text{ km}$ requires $5.55 \times 10^9 \text{ kg}$ of H_2 or 5.55 Mt of H_2. The quantity of electricity needed for transportation is then 244 TWh, that is, about half the current annual French consumption of electricity. This is the amount of electricity that can be produced by about 25 nuclear reactors or about fifty thousand 2-MW wind turbines.

■■■■ BIBLIOGRAPHY

BOOKS

Acts of the day organized by ECRIN and the OPECST on October 11, 2005, with the French National Assembly, ITER, scientific, technological and socio-economic stakes, available: www.ecrin.asso.fr.

Agator, J. M., Cheron, J., Ngô, C., and Trap, G., *Hydrogène*, Omnisciences, 2007.

Alleau, T., and Haessing, T., *L'hydrogène, energie du future?* EDP Sciences, 2008.

Ballerini, D., *Les biocarburants*, Technip, 2006.

Ballerini, D., *Le plein de biocarburants*? Technip, 2007.

Barré, B., *Tout sur l'énergie nucléaire*, Areva, 2003.

Barré, B., and Bauquis, P. R., *L'énergie nucléaire en 110 questions, DGEMP*, l'énergie nucléaire, Éditions Hire, 2007.

Bastard, P., Fargue, D., Laurier, P., Mathieu, B., Nicolas, M., and Roos, P., *Electricité*, Eyrolles, 2000.

Bataille, C., and Birraux, B., *Les nouvelles technologies de l'énergie*, Rapport OPECST, 2006.

Bauquis, P. R., and Bauquis, E., *Pétrole & gaz naturel*, Éditions Hire, 2004.

Benign energy? *The Environmental Implications of Renewables*, International Energy Agency/Office of Economic Co-operation and Development, Paris, 1998.

Bobin, J. L., Huffer, E., and Nifenecker, H., *L'énergie de demain, techniques, environnement, économie*, EDP Sciences, 2005.

Bobin, J. L., Nifenecker, H., and Stephan, C., *L'énergie dans le monde, bilan et perspectives*, EDP Sciences, 2001.

Bonal, J., and Rossetti, P., *Energies alternatives*, Omnisciences, 2007.

Boucher, S., *La révolution de l'hydrogène*, Le Félin Kiron, 2006.

Boyle, G. (Ed.), *Renewable Energy: Power for a Sustainable Future*, Oxford University Press, 2004.

Chevalier, J. M., *Les grandes batailles de l'énergie*, Éditions Gallimard, 2004.

Christopher, H., and Armstead, H., *Geothermal energy*, E. &, F. N. Spon, 1978.

Our Energy Future: Resources, Alternatives, and the Environment
By Christian Ngô and Joseph B. Natowitz
Copyright © 2009 John Wiley & Sons, Inc.

Da Rosa, A. V., *Fundamentals of Renewable Energy Processes*, Elsevier Academic, 2005.

Davis, S. C., and Diegel, S. W., *Transportation Data Book*, Edition 26, Oak Ridge National Lab, Oak Ridge, TN, 2007.

Delalande, A., *Tout savoir (ou presque) sur l'énergie*, Éditions PYC Livres, 1998.

Fanchi, J., *Energy in the 21st Century*, World Scientific, 2005.

Favennec, J. P., *Géopolitique de l'énergie*, Technip, 2007.

Fay, J. A., and Golomb, D. S., *Energy and the Environment*, Oxford University Press, 2002.

Gladstone, S., *Nuclear Reactor Engineering*, Van Nostrand Company, 1963.

ICAO Environmental Report, available: www.icao.int, 2007.

International Energy Agency (IEA), *Electricity*, IEA, Paris. 2006.

Jouzel, J., and Debroise, A., *Le climat: jeu dangereux*, Dunod, 2007.

L'énergie du vent, sous la direction de P. Rocher, Le cherche midi, 2007.

L'énergie, DGEMP, Collection chiffres clés, 2007.

L'énergie, les entretiens de La Villette, 1994. www.cite-sciences.fr/francais/alacite/evenemen/evll/pages/html.

Lambert, G., Chappellaz, J., Foucher, J. P., and Ramstein, G., *Le méthane et le destin de la terre*, EDP Sciences, 2006.

Le nucléaire expliqué par des physiciens, sous la direction de P.Bonche, EDP Sciences, 2002.

McMullan, J. T., Morgan, R., and Murray, R. B., *Energy Resources*, Edward Arnold, 1983.

Mérenne-Schoumaker, B., *Géographie de l'énergie*, Nathan, 1993.

Multiyear program plan 2007–2012, Office of the Biomass Program, U.S. Department of Energy Report, Washington, DC. 2005.

Nelson, J., *The Physics of the Solar Cell*, Imperial College Press, 2003.

Ngô, C., *Quelles energies pour demain? On se bouge!* Spécifiques Editions, 2007.

Ngô, C., *L'énergie, resources, technologies et environnement*, Dunod, 2002, 2008.

Nouvelles technologies de l'énergie 4, sous la direction de J. C. Sabonnadière, Lavoisier, 2007.

O'Hayre, R., Cha, S. W., Colella, W., and Prinz, F. B., *Fuel Cell Fundamentals*, Wiley, Hoboken, NJ, 2006.

Oliva, J. P., *L'isolation écologique*, La Terre Vivante, 2008.

Patel, M. R., *Wind and Power Systems*, CRC Press, Boca Raton, FL, 1999a.

Patel, M. R., *Wind and Solar Power Systems*, CRC Press, Boca Raton, FL, 1999b.

Petrangeli, G., *Nuclear Safety*, British Library, 2006.

Rebut, P. H., *The Energy of Stars. Controlled Nuclear Fusion*, Odile Jacob, 1999.

Rodrigue, J. P., Comtois, C., and Slack, B., *The Geography of Transport Systems*, Routledge, New York, 2006.

Roux, D., *Comment faire rimer habitable et durable? On se bouge!* Spécifiques Editions, 2008.

Safa, H., *Le nucléaire, quel intérêt pour la planète? On se bouge!* Spécifiques Editions, 2008.

Salomon, T., and Bedel, S., *La maison des négawatts*, La Terre Vivante, 2002.

Sheperd, W., and Sheperd, D. W., *Energy Studies*, Imperial College Press, 2003.

Smil, V, *Energies*, MIT Press, Reading, MA, 1999.

Srinivasan, S., *Fuel Cells*, Springer, 2006.

The Future of Nuclear Power, An Interdisciplinary MIT Study, MIT, Reading, MA, 2003.

The Shanghai 2004 Report, Challenge Bibendum, Michelin.

Transports et émission de CO₂, Conférence Européenne des ministres des transports, Office of Economic Coo-operation and Development, Paris. 2007.

Turner, W. C., *Energy Management Handbook*, Fairmont Press, 2004.

Turner, W. C., *Energy Management Handbook*, Fairmont Press, 2005.

Turpin, L., *Le climat change et nous? On se bouge!* Spécifiques Éditions, 2007.

Weisse, J., *La fusion nucleaire, PUF, collection "que sais je"*, 2003.

Wiesenfeld, B., *L'énergie en 2050*, EDP Sciences, 2005.

Wilson, E. J., and Gerard, D. (Eds.), *Carbon and Sequestration, Integrating Technology, Monitoring and Regulation*, Blackwell, 2007a.

Wilson, E. J., and Gerard, D. (Eds.), *Carbon Capture and Sequestration*, Blackwell, 2007b.

Wingert, J. L., *La vie après le pétrole*, Autrement, 2005.

Wiser, W. W., *Energy Resources*, Springer, 1999.

World Energy Outlook 2006, International Energy Agency, Paris.

INTERNET SITES

A great deal of information can now be accessed using the Internet. Any good search engine will uncover a vast array of information on most energy topics. Since the Net continues to evolve rapidly, some links may quickly become obsolete. New ones appear regularly. Below we present just a few of the many useful sites presently available.

General Sites

http://www.ifremer.fr
www.ademe.fr
www.areva.com
www.brgm.fr
www.ccfa.fr
www.cea.fr
www.eea.europa.eu
www.eia.doe.gov
www.electricitystorage.org
www.en.wikipedia.org
www.energies-renouvelables.org
www.energy.gov
www.eusustel.be

www.iea.org
www.ifp.fr
www.industrie.gouv.fr/energie/sommaire.htm
www.ipcc.ch
www.legrenelle-environnement.gouv.fr
www.nei.org
www.sfen.org
www.wikipedia.com
www.worldenergy.org

More Specialized Sites

(By order of occurrence in the book)
www.holon.se/folke/kurs/Distans/Ekofys/fysbas/LOT/LOT.shtml
http://generationsfutures.chez-alice.fr/
http://www.eia.doe.gov/emeu/international/oilprice.html
www.ecology.com
http://ostseis.anl.gov
www.eia.doe.gov/pub/international/iealf/BPCrudeOilPrices.xls
http://www-energie.arch.ucl.ac.be/CDRom/chauffage/theories/chauthecombustion.
    htm
http://www.itopf.com/
http://waterquality.montana.edu/docs/methane/cbmfaq.shtml#whatiscoalbedmethane
www.wsichina.org/cs5_7.pdf
http://waterquality.montana.edu/docs/methane/roffe_swcs.pdf
Larissa Gammidge www.newcastle.edu.au/
Panorama 2008, www.ifp.fr
www.engineeringtoolbox.com
http://mccoy.lib.siu.edu/projects/crelling2/atlas/macerals/mactut.html
www.iea-coal.org.uk
www.nma.org
http://sequestration.mit.edu/pdf/introduction_to_capture.pdf
http://www.cslforum.org/documents/iea_cslf_Paris_Update_CCS.pdf
http://epa.gov/climatechange/index.html
www.bom.gov.au/info/GreenhouseEffectAndClimateChange.pdf
http://earthsci.org/mineral/energy/wind/wind.html
http://www.british-hydro.org
http://en.wikipedia.org/wiki/Hydropower
http://en.wikipedia.org/wiki/Hydroelectricity
http://www.edf.com/html/panorama/production/renouvelable/hydro/monde.html
http://fr.wikipedia.org/wiki/Barrage_d'Itaipu
http://fr.wikipedia.org/wiki/Barrage_des_Trois_Gorges
http://en.wikipedia.org/wiki/Three_Gorges_Dam

http://www2.ademe.fr/servlet/getDoc?cid=96&m=3&id=30100&p1=00&p2=08& ref=12441

http://www.edf.com/accueil-fr/la-production-d-electricite-edf/hydraulique-120270.html

http://en.structurae.de/structures/data/index.cfm?ID=s0020616

http://www.prim.net/citoyen/definition_risque_majeur/dossier_risque_rupture_ barrage/lesevenementshistoriques.htm

www.hydrocoop.org

http://isitv.univ-tln.fr/~lecalve/oceano/polycop/poly.pdf

http://en.wikipedia.org/wiki/Wave_power

http://www.engin.umich.edu/dept/name/research/projects/wave_device/wave_device. html

http://www.fujitaresearch.com/reports/tidalpower.html

http://exergy.se/goran/cng/alten/proj/98/osmotic/

http://www.earth-policy.org/Updates/2006/Update55_data.htm#fig1

http://fsoso.free.fr/conferences/

http://photovoltaics.sandia.gov/

http://photovoltaics.sandia.gov/docs/PDF/PV_Road_Map.pdf

http://minerals.usgs.gov/minerals/pubs/commodity/

http://www.univ-lehavre.fr/recherche/greah/documents/ecpe/schindler.pdf

http://en.wikipedia.org/wiki/Windmill

http://en.wikipedia.org/wiki/Wind_turbine

http://earthsci.org/mineral/energy/wind/wind.html

http://www.windpower.org/fr/stat/unitsw.htm

www.euronuclear.org/library/public/enews/ebulletinautumn2004/nuclear-reactors.htm

http://www.world-nuclear.org/

http://www.cea.fr/fr/sciences/Iter/

www.itw.uni-stuttgart.de/ITWHomepage/Sun/deutsch/public/pdfDateien/03-09.pdf

http://www1.eere.energy.gov/buildings/residential/

www.eia.doe.gov/emeu/consumption/index.html

http://www.heatpumpcentre.org/

www.tkk.fi/Units/AES/studies/dis/hottinen.pdf

http://www.h2data.de/

http://www.lenntech.com

J. Wang http://cohesion.rice.edu/CentersAndInst/CNST/emplibrary/Wang.ppt.ppt

www.benwiens.com

www.marad.dot.gov/NMREC/